PHYSICAL ORGANIC CHEMISTRY: THEORY AND PRACTICE

PHYSICAL ORGANIC CHEMISTRY: THEORY AND PRACTICE

ALBERTO D'AMORE AND GENNADY E. ZAIKOV

EDITORS

Nova Science Publishers, Inc.

New York

Production Coordinator: Donna Dennis
Editorial Production Manager: Donna Dennis
Senior Production Editor: Susan Boriotti
Office Manager: Annette Hellinger
Graphics: Andrea Charles, Magdalena Nuñez and Annemarie VanDeWater
Editorial Production: Marius Andronie, Maya Columbus, Keti Datunashvili, Melissa Diaz,
    Andrew Kallio, Vladimir Klestov, Lorna Loperfido, Joann Overton and Rusudan Razmadze
Circulation: Ave Maria Gonzalez, Vera Popovic, Luis Aviles, Alexandra Columbus, Raymond
    Davis, Cathy DeGregory, Marlene Nuñez, Jeannie Pappas, Lauren Perl and Katie Sutherland

Library of Congress Cataloging-in-Publication Data
Physical organic chemistry : theory and practice / Alberto D'Amore and Gennady E. Zaikov,
    (editors).
      p. ; cm.
Includes bibliographical references and index.
ISBN 1-59454-275-9 (hardcover)
1. Chemistry, Physical organic.
[DNLM: 1. Organic Chemicals--chemistry. 2. Materials Testing--methods. QD 258 P578 2005] I.
    D'Amore, Alberto. II. Zaikov, Gennadiæi Efremovich.
QD476.P489   2005
547'.13--dc22   2005002166

Copyright © 2005 by Nova Science Publishers, Inc.
         400 Oser Ave, Suite 1600
         Hauppauge, New York 11788-3619
         Tele. 631-231-7269   Fax 631-231-8175
         e-mail: Novascience@earthlink.net
         Web Site: http://www.novapublishers.com

Printed in the United States of America

CONTENTS

Preface — vii

List of Authors — ix

List of Institutions — xi

Chapter 1 — Statistical Copolymers having Aromatic Polycarbonate Structure as Compatibilizers in PS/PC Blends — 1
L. Sartore, M. Penco, S. Della Sciucca and A. D'Amore

Chapter 2 — Isobutylene Copolymerization — 9
Yu. A. Sangalov, K. S. Minsker and G. E. Zaikov

Chapter 3 — Properties and Production of Polyisobutylene — 31
K. S. Minsker, Yu. A. Sangalov and G. E. Zaikov

Chapter 4 — The Condensation of Diethilmethylthienylsilanes with the Polycyclic Bisphenoles — 81
L. M. Khananashvili, E. G. Markarashvili, Z. Sh. Lomtatidze, Ts. N. Vardosanidze, E. I. Khubulava and D. A. Girgvliani

Chapter 5 — Kinetic Study of Polypropylene Nanocomposite Thermal Degradation — 93
Sergei M. Lomakin, Gennadi E. Zaikov, Irina L. Dubnikova and Svetlana M. Berezina

Chapter 6 — Investigation of Physical, Mechanical and Rheological Properties of Compositions Containing Organic and Non-Organic Fillers — 111
O. Legonkova, A. Bokaref and V. Vasiliev

Chapter 7 — Inhibition of Radiation Destruction and Oxidation of Gamma Irradiated Polypropylene — 121
J. N. Aneli, M. A. Donadze, M. S. Kutsia and L. G. Kapanadze

Chapter 8 — Creation of the Conducting Polymer Composites Based on Silicon with Relay Effect — 127
Jimsher N. Aneli, Mamuka S. Kutsia and Medea M. Bolotashvili

Chapter 9   Conductivity of Conducting Rubbers at Flow   **133**
J. N. Aneli, M. M. Bolotashvili and I. G. Ubiria

Chapter 10   Development of Fire and Heat Shield Materials   **141**
M. A. Shashkina, A. A. Donskoi and G. E. Zaikov

Chapter 11   Climatic Natural and Artificial Aging of Materials   **153**
A. A. Donskoi, M. A. Shashkina and G. E. Zaikov

Chapter 12   Properties of Polymeric Waveguides as Information
Transmission Channels   **159**
G. M. Rubinstein, N. G. Lekishvili and G. E. Zaikov

Chapter 13   Degradation of Aliphatic-Aromatic
Polyimides – Polyalkanimides   **167**
E. V. Kalugina, K. Z. Gumargalieva and G. E. Zaikov

Chapter 14   Experimental Proof of the Cluster Model   **211**
G. V. Kozlov and G. E. Zaikov

Chapter 15   Advances in the Field of High-Purity Reagents   **231**
A. M. Yaroshenko and G. E. Zaikov

Chapter 16   Modular Technology of Hydrofluoric Acid   **237**
G. Z. Blyum, A. M. Yaroshenko and G. E. Zaikov

Chapter 17   Bioactive Organic–Inorganic Hybrid Materials
Synthesized by Sol-Gel Method   **247**
M. Catauro, A. D'Amore and A. Marotta

Index   **261**

PREFACE

The real diplomat will think twice
before saying nothing.
Alexander S. Griboedov
The Ambassador of the Russian Empire in Iran in mid 1800's[*]

"…Laudolo sie, mi'signore "Praise be You my Lord
per quelli che perdonno per through those who grant
lo Tuo Amore…" pardon for love of You»
Francesco d'Assisi Saint Francesco from Assisi
«Il Cantico dell Creature»

Contrary to diplomats, scientists must not only think but also inform the scientific society about their ideas (if they are not preposterous).

There is a proverb that very often "severity of the law in many countries is compensated by their dispensable obeying". Contrary to such countries, the laws of physical and organic

The authors would be thankful for any valuable notes and remarks on the current collection contents.

Prof. Alberto D'Amore
The Second University of Naples – SUN
Department of Aerospace and Mechanical Engineering
29 via Roma, 81031, Aversa (CE), Italy

Prof. Gennady E. Zaikov
N.M.Emanuel Institute of Biochemical Physics
Russian Academy of Sciences
4 Kosygin Street, Moscow, 119991 Russia

---

[*] He was beheaded by Iran religious bigots who enraptured the Russian Embassy in Teheran (buried in Tbilisi, Georgia)

LIST OF AUTHORS

Aneli J.N.
Berezina S.M.
Blyum G.Z.
Bokaref A.
Bolotashvili M.M.
Catauro M.
D'Amore A.
Donadze M.A.
Donskoi A.A.
Dubnikova I.L.
Girgvliani D.A.
Gumargalieva K.Z.
Kalugina E.V.
Kapanadze L.G.
Khananashvili L.M.
Khubulava E.I.
Kozlov G.V.
Kutsia M.S.
Legonkova O.A.
Lekishvili N.G.
Lomakin S.M.
Lomtatidze Z.Sh.
Markarashvili E.G.
Marotta A.
Minsker K.S.
Penco M.
Rubinstein G.M.
Sangalov Yu.B.
Sartore L.
Sciucca S. Della
Shashkina M.A.
Ubiria I.G.
Vardosanidze Ts.N.

Vasiliev V.
Yaroshenko A.M.
Zaikov G.E.

List of Institutions

A.N. Nesmeianov Institute of Organoelement Compounds, Russian Academy of Sciences, Moscow, Russia
Bashkirian State University, Ufa, Bashkiria, Russia
Corrosion Center of Georgian Technical University, Tbilisi, Georgia
Department of Aerospace and Mechanical Engineering the Second University of Naples, Italy
Department of Materials and Production Engireering, University of Naples Federico II
Dipartimento di Chimica e Fisica per l'Ingegneria e per i Materiali, Brescia, Italy
I. Javakhishvili Tbilisi State University, Tbilisi, Georgia
Institute of Aviation Materials, Moscow, Russia
Institute of Mechanics of the Georgian Academy of Sciences, Tbilisi, Georgia
Kabardino-Balkarian State University, Nal'chik, Kabardino-Balkaria, Russia
Moscow State University of Applied Biotechnology, Moscow, Russia
N.M. Emanuel Institute of Biochemical Physics, Russian Academy of Sciences, Moscow, Russia
N.N. Semenov Institute of Chemical Physics, Russian Academy of Sciences, Moscow, Russia
Polyplastic Co., Moscow, Russia
State Research Institute of Chemical Reagents and Highly Pure Chemical Substances, Moscow, Russia

In: Physical Organic Chemistry: Theory and Practice
Eds: A. D'Amore and G. E. Zaikov, pp. 1-8

ISBN 1-59454-275-9
© 2005 Nova Science Publishers, Inc.

Chapter 1

STATISTICAL COPOLYMERS HAVING AROMATIC POLYCARBONATE STRUCTURE AS COMPATIBILIZERS IN PS/PC BLENDS

L. Sartore,[*] M. Penco and S. Della Sciucca
Dipartimento di Chimica e Fisica per l'Ingegneria e per i Materiali, Brescia, Italy
A. D'Amore
Department of Aerospace and Mechanical Engineering,
the Second University of Naples-SUN Via Roma 29 81031 Aversa (CE), Italy

ABSTRACT

In this work the use of statistical copolymers having aromatic polycarbonate structure (CPC) as compatibilizers in polystyrene (PS)-polycarbonate of bisphenol A (PCPA) blends has been investigated. Firstly, the compatibility of PS and PCPA with aromatic co-polycarbonates containing bisphenol A (BPA) and tetramethyl bisphenol A (TMBPA) was studied. The simple prevision scheme developed by Sonja Krause was employed to evaluate the effect of the copolymer molecular structure on the miscibility of PS and PCPA. These prevision data were used to select co-polycarbonates of potential interest. Statistical co-polycarbonates (CPC) containing different BPA/TMBPA molar ratios were synthesized by polycondensation reaction between a mixture of the two monomers and phosgene. Binary and ternary blends, prepared by casting from chloroform solution, were studied with differential scanning calorimetry (DSC), dynamical-mechanical thermal analysis (DMTA) and optical microscopy to evaluate the components compatibility, which increases with the TMBPA copolymer content. All PCPA/TMBPA copolymers binary blends showed one Tg due to transesterification reactions between two polycarbonates. A compatibilization effect was highlighted.

Keywords: polycarbonate, polystyrene, blends, compatibilizers.

---

[*] e-mail: luciana.sartore@ing.unibs.it

INTRODUCTION

Polymer blending is a simple and efficient method for designing and controlling the performance of polymeric materials. Unfortunately, a high incompatibility is often shown between the components, and blends having poor properties are obtained. For this reason, considerable efforts in the blends science are aimed to overcome this limitative factor [1]. In general, two main ways are followed for decreasing the interfacial energy of incompatible blends: the addition of a compatibilizer agent or the chemical modification of one of the components. Therefore, the final goal is to increase phase adhesion, which is a necessary condition for obtaining materials with high performance in comparison with starting polymers. This paper studies the effect of the molecular characteristics of statistical aromatic polycarbonates on the compatibility with polystyrene.

Many application fields show a potential interest in binary blends based on polystyrene (PS) and polycarbonate of bisphenol A (PCPA) [2-5]. In particular, polystyrene properties such as thermal stability (HDT), impact strength and optical properties (birefringence-free material) have been increased by blending PS and PCPA [6]. Unfortunately, high incompatibility exists between these polymers [7-9]. On the other hand, it is well-known that PS and polycarbonate of tetramethyl bisphenol A (PCTMP) exhibit good compatibility [10]. PCTMP has a higher glass transition temperature (Tg) than PCPA but is very brittle and thus it is not useful to improve the PS impact strength. The aim of this work is to study the composition effect of statistical co-polycarbonates (CPC) containing BPA and TMBPA units on the compatibility in three components PS/PCPA/CPC blends. The experimental results were compared with data obtained with a simple prevision scheme based on the Flory-Huggins theory [11]. Miscibility was the criterion used to foresee polymer compatibility.

RESULTS AND DISCUSSION

Prevision of Miscibility

The composition and molecular weight of co-polycarbonates were selected using a simple miscibility prediction scheme developed by Sonja Krause [11,12]. According to this scheme, which is based on the Flory-Huggins theory, polymers A and B are miscible if [χ_{AB} - (χ_{AB})$_{cr}$]<0, being χ_{AB} the Flory's interaction parameter. The χ_{AB} and (χ_{AB})$_{cr}$ values were calculated through the following equations:

$$\chi_{AB} = (\delta_A - \delta_B)^2/6 \tag{1}$$

$$(\chi_{AB})_{cr} = \tfrac{1}{2} (1/x_A^{1/2} + 1/x_B^{1/2})^2 \tag{2}$$

where δ is the Hildebrand solubility parameter and x is the number-average degree of polymerization. The solubility parameters of the homopolymers were evaluated using the group molar attraction constants according to Hoy [13], while the copolymer parameters δ_c were calculated from:

$$\delta_c = \sum \delta_i \phi_I \qquad (3)$$

where δ_i and ϕ_i are the solubility parameter and the volume fraction, respectively, of the homopolymer corresponding to monomer i.

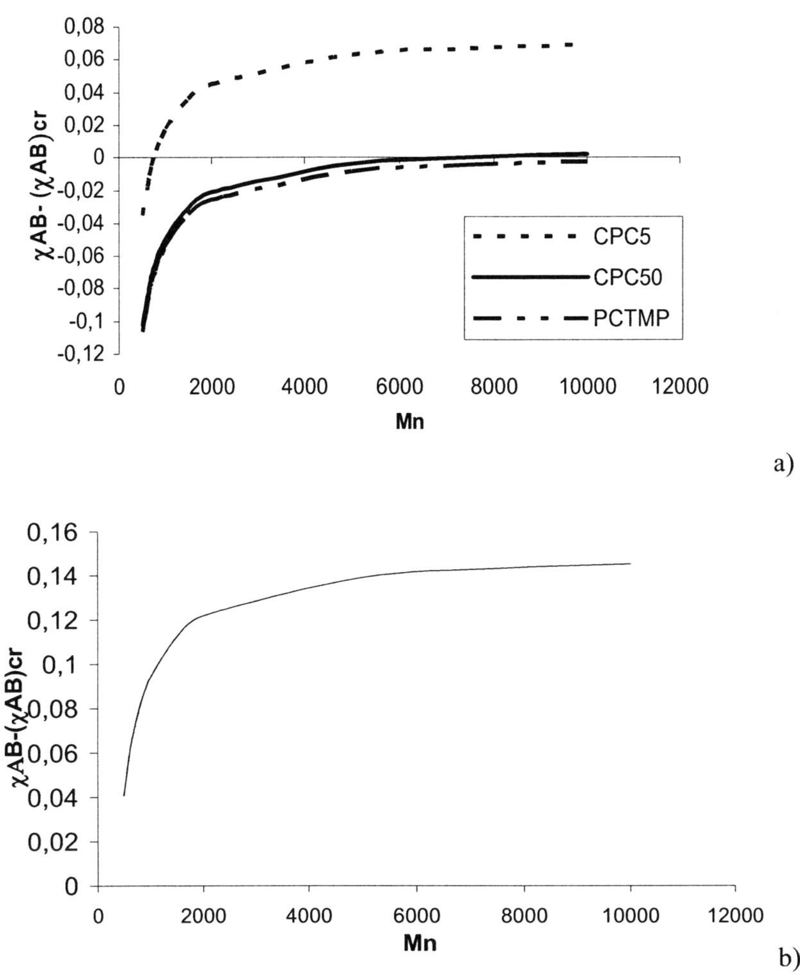

a)

b)

Figure 1. Miscibility prevision curves for PS/PCs and b) PCPA/PCTMP blend.

Values of $[\chi_{AB} - (\chi_{AB})_{cr}]$ versus the $\overline{M}_n$ of copolymer having different composition for blends with PS ($\overline{M}_n$ =132000) are shown in Figure 1a. It should be observed that a negative value of this parameter is always obtained for the PCTMP homopolymer. Our prevision agrees with literature works where some miscibility is reported. Furthermore, it foresees that the miscibility of PS and CPC increases with the TMBPA content. Interestingly, in spite of the large amount of PCPA units present in CPC50, a miscibility similar to that of PCTM is predicted for this copolymer. On the contrary, incompatibility in PCPA/PCTMP blends is foreseen (Figure 1.b).

Synthesis and Thermal Properties of Polycarbonates

Statistical co-polycarbonates containing different BPA/TMBPA molar ratios were synthesized by polycondensation reaction between a mixture of the two bisphenols and phosgene. The composition of all copolymers was evaluated by ^{1}H-NMR and FT-IR spectroscopy while gel permeation chromatography (GPC) and viscosity measurements were employed to determine the molecular masses.

The composition data and molecular characteristics of the polymers are summarized in Table 1. It should be observed that the selected reaction conditions produced polymers having comparable molecular weights ($\approx$8000); this is essential if the composition effect on the miscibility is studied.

The calorimetric properties of the products were investigated by differential scanning calorimetry (DSC) (Table 1). It should be noticed that both CPC5 and CPC50 copolymers are amorphous materials and their T_g increase with the TMBPA content. On the contrary, PCTMP is a high melting point semi-crystalline material with T_m=287.4°C.

Table 1. Molecular and Thermal Properties of Synthesized Polycarbonates

Sample	$TMBPA^a$ mol-%	$\overline{M}_n \cdot 10^3$	$\overline{M}_w \cdot 10^3$	Tg °C	Tm °C
CPC50	50	8.4	23.8	156.2	-
CPC5	5	8.3	20.1	136.4	-
PCTMP	100	9.1	18.8	192.3	287.4
PS	-	132.3	189.6	99.6	-

[a] Tetramethyl bisphenol A.

Binary Blends

PS/CPC Blends

Polystyrene/polycarbonate blends were prepared by casting from chloroform solutions. The solvent was removed first at room temperature, then under vacuum with a multi-step thermal treatment. A high temperature step was carried out to destroy the crystallinity of the polycarbonate phase produced during the evaporation of chloroform.

Blend compatibility was evaluated using DSC and optical microscopy. The calorimetric data of 50:50 weight-% PS/polycarbonate blends are summarized in Table 2. For comparison, data relative to a blend with commercial PCPA and to films of pure polymers prepared under the same conditions are also reported.

In all blends two glass transition temperatures are observed whose values are shifted compared to neat polymers. In particular, it is interesting to observe that some phase separation is present also in PS/PCTMP, even if no appreciable crystallinity is observed. This result agrees with literature data because a considerable shift of the Tg and the absence of crystallinity clearly suggest high compatibility. Compatibility decreases with increasing the bisphenol A content. Even if larger T_g shifts are obtained using the 50:50 copolymer (CPC50), also the PS/CPC5 data are interesting because they evidence that only a slight chemical modification of "typical" polycarbonate produces a fair improvement of compatibility. The morphological analysis confirms the calorimetric results.

Table 2. Thermal Properties of 50/50 wt-% Blends and of Neat Components[a]

Materials	Low T_g (°C)	ΔT^c (°C) low T_g	High T_g (°C)	ΔT^c (°C) high T_g
PS/CPC5	101.7	2.1	131.8	-4.6
PS/CPC50	103.2	3.6	152.7	-3.5
PS/PCTMP	108.3	8.7	168.9	-23.4
PS/PCPA[b]	100.0	0.4	143.7	-0.8
CPC5	-	-	136.4	-
CPC50	-	-	156.2	-
PCTMP	-	-	192.3	-
PS	99.6	-	-	-
PCPA	-	-	144.5	-

a Film.

[b] Commercial PCPA with $\overline{M}_n$ =20100; $\overline{M}_w$ =37400

[c] ΔT= Tg (blend) - Tg (starting material).

PCPA/CPC Binary Blends

The prevision scheme suggests immiscibility in PCPA/PCTMP blends. On the other hand, all blends prepared show only one T_g, which value increases respect to PCPA (Table 3). This fact is the consequence of a transesterification reaction occurring between two components during the thermal treatments. In confirmation of this, after thermal treatment of blend, only one peak is observed in the gel permeation trace (Figure 2).

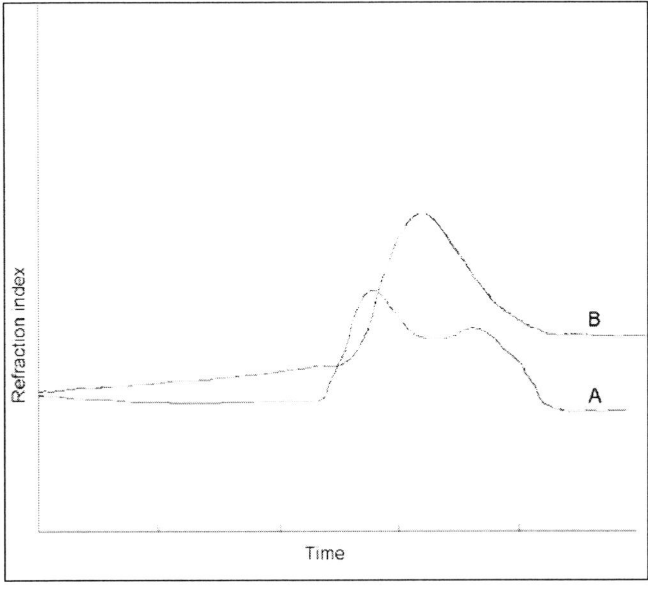

Figure 2: GPC curves of a) untreated and b) thermal treated 50/50 PCPA/PCTMP mixture.

Table 3. 50/50 PCPA/PCs Blends; Thermal Properties

Materials	T_g (°C)
PCPA/CPC50	148.6
PCPA/PCTMP	163.0

Three Components Blends

Three components blends, cast from chloroform solution, are reported in Table 4. For comparison, PS/PCPA binary blends are also reported. All materials were studied with differential scanning calorimetry, optical microscopy and scanning electron microscopy (SEM). Calorimetric data are summarized in the same table. When the compatibilizer is added we observe an increase of both lower and higer T_g, if compared to binary blends. While the effect of CPC on the first T_g is explained with an improved compatibility, the variation of higher T_g can not be easily explained in terms of compatibility but is reasonable due to the transesterification reaction occurring between the polycarbonates. This process converts the PCPA in a copolymer with higher T_g.

Table 4. Thermal Properties of Three Components Blends Cast from Chloroform

Sample	PS %	PCPA %	CPC Type	%	Low T_g (°C)	High T_g (°C)
Blend 1	25	75	-	-	99.8	145.1
Blend 2	50	50	-	-	100.9	141.4
Blend 3	75	25	-	-	98.5	144.7
Blend 4	25	75	PCTMP	5	102.8	150.9
Blend 5	50	50	PCTMP	5	104.2	152.6
Blend 6	75	25	PCTMP	5	101.4	147.0
Blend 7	45[*]	45[*]	PCTMP	10	106.0	152.6
Blend 8	50[*]	25[*]	PCTMP	25	110.7	152.2
Blend 9	25	75	CPC5	5	102.1	145.3
Blend 10	50	50	CPC5	5	101.4	145.2
Blend 11	75	25	CPC5	5	100.0	141.1
Blend 12	25	75	CPC50[a)]	5	102.8	146.7
Blend 13	50	50	CPC50[a)]	5	102.1	145.6
Blend 14	75	25	CPC50[a)]	5	100.7	146.6
Blend 15	25	75	CPC50[b)]	5	103.5	148.8
Blend 16	50	50	CPC50[b)]	5	102.1	145.6
Blend 17	75	25	CPC50[b)]	5	100.0	147.7

[a] Mn = 5980.

[b] Mn = 8470.

Mixtures with PCTMP as compatibilizer were also obtained by melt mixing process using a discontinuous mixer. In particular materials having a weight ratio PS/PCPA 50/50 with different content of PCTMP (Table 5) were prepared. The DSC data confirm the previous conclusions. A typical DMTA curves is showed in Figure 3, where the lower T_g shift is well highlight.

Table 5: Thermal Properties of Three Components Blends Obtained in Brabender Mixer

Sample	PCTMP %	Low Tg °C	High Tg °C
Blend 1M	-	100.0	143.3
Blend 2M	0.5	101.1	141.4
Blend 3M	1	101.8	140.7
Blend 4M	2	102.5	144.6
Blend 5M	5	103.7	144.7

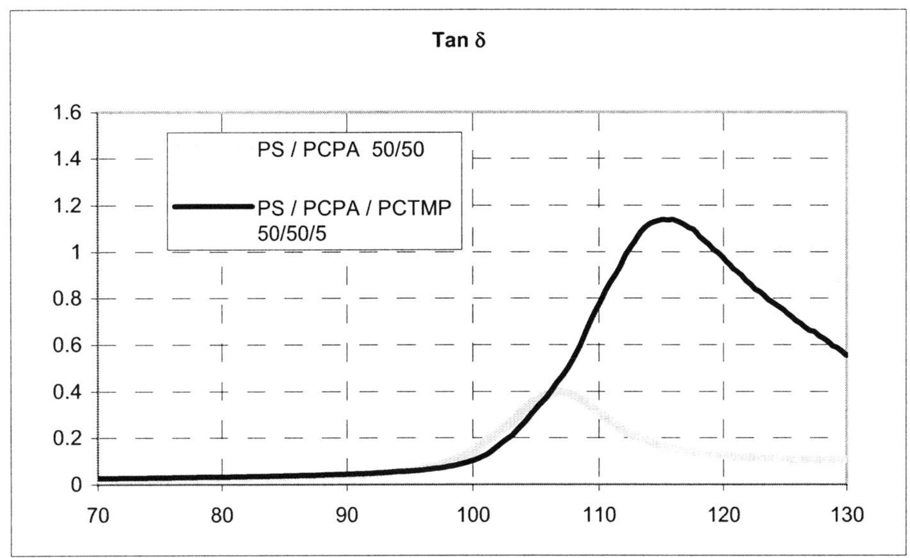

Figure 3: DMTA curves of 50/50 PS/PCPA with and without compatibilizer

CONCLUSIONS

PCTMP and CPC copolymers can be used as compatibilizers in PS/PCPA blends. Thermal treatments involve the raise of transesterification reaction between the polycarbonates, modifying the chemical structure of PCPA. PCTMP is also a useful product for obtaining PS/PCPA blends with higher compatibility by reactive process.

REFERENCES

[1]	L.A. Utraki, *Polymer Alloys and Blends*, C. Hanser: Munich, 1990.
[2]	H. Ohishi, T. U. Ikehara, T. Nishi, *J. Appl. Polym. Sci. 2001*, 80, 2347.
[3]	H. Pu, X. Tang, X. Xu, *Polymer International 1997*, 43, 33.
[4]	H. Ohishi, T. Nishi, *J.Polym.Sci.Polym.Chem.Ed. 2000*, 38, 299.
[5]	C. Belaribi, G. Marin, P. Monge, *Eur. Polym.J. 1986*, 22, 487.

[6]   H. Pu, C. Yanhui, X. Tang, X. Xu, *Polymer International 1997*, 44, 156.

[7]   W. N. Kim, C. M. Burns, *J.Appl.Polym.Sci. 1990*, 41, 1575.

[8]   C. K. Kim, D. R. Paul, *Macromolecules 1992*, 25, 3097.

[9]   G. C. Eastmond, K. Haraguchi, *Polymer 1983*, 24, 1171.

[10]   S. Ziaee, D. R. Paul, *J.Polym.Sci. 1997*, 35, 489.

[11]   D. R. Paul, S. Newman, *Polymer Blends*, vol. 1, Academic Press, New York 1978, p.45.

[12]   J. C. Huang, D. C. Chang, R. D. Deanin, *Advances in Polymer Technology, 1993*, 12, 81.

[13]   K. L. Hoy, *J. Paint Technol. 1970*, 42, 76.

In: Physical Organic Chemistry: Theory and Practice
Eds: A. D'Amore and G. E. Zaikov, pp. 9-29

ISBN 1-59454-275-9
© 2005 Nova Science Publishers, Inc.

Chapter 2

ISOBUTYLENE COPOLYMERIZATION

Yu. A. Sangalov and K. S. Minsker
Bashkirian State University, Ufa, Bashkiria, Russia
G. E. Zaikov
N.M.Emanuel Institute of Biochemical Physics, Moscow, Russia

SPECIFIC FEATURES OF CATIONIC COPOLYMERIZATION

To determine relative rates of copolymerization reactions between M1 and M2, usual correlations, deduced for free-radical copolymerization are appropriate.

In the cationic copolymerization in binary systems, the growing carbocation is able to react with any of the copolymers M1 and M2. In this case, two types of carbonium ions occur, able to start four differing chain propagation reactions:

$$\cdots\!{>}C_1^+A^- + M_1 \xrightarrow{\ k_{11}\ } \cdots\!{>}C_1\!{-}\!{>}C_1^+A^-$$

$$\cdots\!{>}C_1^+A^- + M_2 \xrightarrow{\ k_{12}\ } \cdots\!{>}C_1\!{-}\!{>}C_2^+A^-$$

$$\cdots\!{>}C_2^+A^- + M_2 \xrightarrow{\ k_{22}\ } \cdots\!{>}C_2\!{-}\!{>}C_2^+A^-$$

$$\cdots\!{>}C_2^+A^- + M_1 \xrightarrow{\ k_{21}\ } \cdots\!{>}C_2\!{-}\!{>}C_1^+A^-$$

Composition of a copolymer molecule is determined by copolymerization constants r_1 and r_2 ($r_1 = k_{11}/k_{12}$, $r_2 = k_{22}/k_{21}$), which bind rates of carbonium ion interaction between one monomer and another one, existing in the mixture. Composition of the copolymer also depends on appropriate initial concentrations of monomers in the reaction mixture. The composition of resulting copolymer, based on copolymerization constants r_1 and r_2 and concentrations of monomers is described by the Mayo–Lewis equation:

$$\frac{d[m_1]}{d[m_2]} = \frac{[M_1]}{[M_2]} \cdot \frac{r_1[M_1]+[M_2]}{[M_1]+r_2[M_2]},$$

(1)

where m_1 and m_2 are monomer units 1 and 2 in the copolymer; $[M_1]$ and $[M_2]$ are their initial concentrations in the initial mixture. In more suitable form, equation (4.1) is written down via molar parts of monomers in the initial mixture f_1 and f_2 and in copolymer F_1 and F_2 (the Feinman–Ross equation):

$$F_1 = 1 - F_2 = \frac{d[m_1]}{d[m_1]+d[m_2]},$$

$$f_1 = 1 - f_2 = \frac{[M_1]}{[M_1]+[M_2]}.$$

(3)

Substituting these expressions into equation (1), we obtain:

$$F_1 = \frac{f_1(r_1 f_1 + f_2)}{r_1 f_1^2 + rf_1 f_2 + r_2 f_2^2},$$

(4)

$$\frac{f_1(1-2F_1)}{(1-f_1)F_1} = r_2 + \left[\frac{f_1^2(F_1-1)}{(1-f_1)^2 F_1}\right] \cdot r_1.$$

(5)

Graphically (the Feinman–Ross method), equation (1) is approximated by a straight line with tangent equal to r_1, which cuts r_2 on the ordinates axis.[1]

Two important consequences follow from equations (1) and (4): a) individual rates of homopolymerization are of no importance for predicting the composition of copolymer; b) quantitative data (r_1 and r_2) are tolerable, if they are obtained in experiments during which initial concentrations of monomers can be assumed constant. Usually, this requirement is fulfilled with suitable approximation only at low transformation degrees (3–5%) or with continuous (by a special program) loading of a monomer.

To obtain dependence of the copolymer composition on conversion in the periodic regime of synthesis in accordance with (1), it is necessary to fulfill the requirement:

$$\frac{[M_1]}{[M_2]} = X, f_1 = \text{const.}$$

---

[1] At the present time computers are widely used for determination of r_1 and r_2 values. There are special mathematical methods, which enable the difference between experimental and theoretical (calculated) values to be minimized at all experimental points.

The amount $\Delta[M_1]$ to be added to the system up to the current moment of time τ per specific volume of the reaction mixture, under the condition that copolymer is absorbed by the monomer M_1, and under selected conditions it must be added while the reaction proceeds, is estimated from the expression [1, p. 197]:

$$\Delta[M_1] = \left|M_1^0\right| \left(\frac{r_1 X + 1}{r_2 + X} - 1\right)^{\left(1 - e^{-k_{eff} \cdot \tau}\right)} , \tag{6}$$

where $\left|M_1^0\right|$ is the initial concentration of the monomer 1; k_{eff} is the effective constant of the expense rate $[M_2]$ (k_{eff} includes concentrations of both active centers and appropriate propagation constants).

To apply equation (6), it is necessary to determine the k_{eff} value. For this purpose, the initial rate of decline of both monomers must be measured by the kinetic curve tangent, and the value obtained must be divided by the initial concentration of the monomers.

The following integral equation is obtained by integrating equation (1):

$$\lg\frac{[M_2]_P}{[M_1]_P} = \frac{r_2}{1 - r_2}\lg\frac{\left[M_2^0\right]\cdot[M_1]_t}{\left[M_1^0\right]\cdot[M_2]_t} - \frac{1 - r_1\cdot r_2}{(1 - r_1)(1 - r_2)}\cdot\lg\frac{(r_1 - 1)\cdot\dfrac{[M_1]_t}{[M_2]_t} - r_2 + 1}{(r_1 - 1)\cdot\dfrac{\left[M_1^0\right]}{\left[M_2^0\right]} - r_2 + 1} .$$

Composition for composition, copolymers may differ by the sequence of unit distribution in the chain, that is to say, by the microstructure of the chain [1, p. 198].

The limiting case is a strictly alternating copolymer ($r_1 = r_2 = 0$). If $r_1 = 0$, the monomer M_1 participates in the chain as single units. In all other cases, length of the block of each comonomer changes during the process. Average lengths of chain blocks of M_1 ($\overline{v}_1$) and M_2 ($\overline{v}_2$) are equal to

$$\overline{v}_1 = r_1\cdot\frac{[M_1]}{[M_2]} + 1, \tag{8}$$

$$\overline{v}_2 = r_2\cdot\frac{[M_2]}{[M_1]} + 1. \tag{9}$$

The distribution function for blocks is approximated by the equation:

$$\Psi_{M_1} = \frac{\left(r_1 \cdot \dfrac{[M_1]}{[M_2]} \right)^{n-1}}{\left(1 + r_1 \cdot \dfrac{[M_1]}{[M_2]} \right)^n} . \qquad (10)$$

In some cases, formation of composite-inhomogeneous macromolecules or heterogeneity in the composition during copolymerization of two monomers is observed. This may be indicated by a change of "immediate" composition of the copolymer as the reaction proceeds. However, even the copolymer obtained under conditions of constant composition, for example, in an azeotropic mixture, is inhomogeneous in composition due to accidental chemical reaction. There always exists some probability that the present macromolecule accidentally contains more chains of one monomer than expected from the average composition. Heterogeneity of the composition of this type depends on the mean molecular mass of the polymer and on the size of blocks of comonomers, i.e. on values of r_1 and r_2 and concentrations of monomers. The lower is the mean molecular mass and the higher are values of r_1 and r_2 (the longer are blocks of copolymers), the higher is the composite heterogeneity of the copolymer with respect to composition.

Much higher composite heterogeneity is possible with ionic polymerization due to probable presence of several types of AC and different values of r_1 and r_2.

A specific feature of cationic copolymerization compared with the radical one is the complex character of dependence of the monomers' activity on their structure, nature of solvent, catalyst and temperature. As in the case of any joint polymerization, the process is described by equation (1), if the ratio of constants of the cross-chain propagation k_{12}/k_{21} differs from unity by not more than a decimal degree, i.e. activity of the monomers or stability of appropriate growing ions differ little from each other. To put it differently, $k_{12} \cdot C_1^+ \cdot M_2 = k_{21} \cdot C_2^+ \cdot M_1$ which follows from the principle of stability of the chain propagation. However, modeling of reactions of the cross-chain propagation has shown that even if the stability principle is fulfilled for every comonomer separately, their mutual influence in the cross-chain propagation leads to non-stationary kinetics [2]. Reactivity of growing ions on account of influence of substituents in the α-position changes more than in the case of growing radicals. This supposes noticeable differences in activity of cation-active monomers. If, for example, the reactivity M_1 is much higher than M_2, the reaction rate $k_{12} \cdot C_1^+ \cdot M_2$ will be low, and the condition of the cross-chain propagation is not fulfilled. In this case, full differential equations must take into account partial rates of initiation W_n' and W_n'' of the monomers M_1 and M_2 and partial rates of termination W_0' and W_0'' of cations C_1^+ and C_2^+ [3]:

$$\frac{dC_1^+}{dt} = W_n' + k_{21} C_2^+ M_1 - k_{12} C_1^+ M_2 - W_0' = 0,$$

$$\frac{dC_2^+}{dt} = W_n'' + k_{12}C_1^+ M_2 - k_{21}C_2^+ M_1 - W_0'' = 0.$$

At greater differences in activities of monomers, the rates of initiation and termination will affect the ratio of concentrations of propagating chains C_1^+/C_2^+ and, in its turn, composition of the copolymer. If the usual equation of the copolymer composition is used, wrong values of constants of relative activity of monomers are obviously obtained.

In cationic polymerization, the effect of the medium must be taken into account. Dielectric constant ε (polarity) of the medium, which determines the state of an ionic pair together with other factors, must change during the process, excluding cases of azeotropic copolymerization, equality of ε values of both monomers or sufficient prevalence of ε of the solvent over that of the monomers. That is why the rate constants of four reactions of propagation can change in accordance with non-constancy of the mixture composition during copolymerization. In the general case, more than two monomers and more than two types of AC are contained in the systems, which limits application of the equation of copolymerization and the importance of numerical values of copolymerization constants r_1 and r_2.

It is thought that in non-polar solvents an ionic pair is selectively solvated by a more polar monomer, which results in growth of the rate of its interaction with a cation. Polar solvents supersede monomers from the solvate cover; that is why relative activity of monomers is mainly determined by features of their chemical composition. The following sequence of activities of monomers in attachment to carbonium ions was obtained: vinyl esters >> isobutylene > styrene > vinyl acetate > isoprene > butadiene. As is obvious, isobutylene is one of the most active comonomers.

In concordance with the acid–base nature of cationic polymerization, monomers react with an electrophilic center in accord with their nucleophilicity. Addition of a strong nucleophilic monomer to a cation with increased electrophilic characteristics is more probable than attachment to their own cations. That is why the difference between relative reactivities of monomers must be greater than that between appropriate rate constants of homopolymerization of applied monomers k_{11} and k_{22}.

Data on the counterion influence on the copolymer composition are ambiguous. Counterions of different chemical nature can differently "permeate" comonomers to a propagating macrocation on account of steric or electrostatic obstacles. This influence is more typical for non-polar solvents, whereas in some cases in polar solvents it is absent, because ionic pairs are separated in polar solvents, and the solvate cover contains no monomer.

A most important role is usually played by the cocatalyst forming structures of the counterion, which may be determined by comparatively large radii of counterions $[MeX_{n+1}]^\delta$ and changes of their values at the transition from one counterion to another, for example [4, p. 138]:

Counterion	BF_3OH^-	$AlCl_3OH^-$	$TiCl_4OH^-$
Ionic radius, nm	0.28	0.38	0.42

The composition of copolymer is significantly affected by the temperature of the process. Usually, the product $r_1 \cdot r_2$ grows with temperature and often significantly exceeds unity.

The influence of isobutyl-vinyl alcohol with isobutylene on the composition of copolymer, produced under the effect of $AlBr_3$ (hexane, 195 K), with small amounts of an electron-donor (concentration of $AlBr_3$ is $4.8 \cdot 10^{-3}$, isobutylene — 11.0, vinyl ester — 0.8 mol/l, donor agent — 0.67 vol.%), is noted in the work [5]. In particular, content of isobutylene chains in the copolymer in the presence of diethyl ester is 60, dimethylsulfoxide is 52, dimethylformamide is 89 mol%, whereas in the absence of an electron-donor it is only 10 mol%.

Hence it can be stated that the cationic copolymerization is highly sensitive to changing of components of appropriate systems and conditions under which the process occurs. At the present time, there is no general theory that unites, explains and predicts experimentally observed changes of r_1 and r_2, i.e. changes of the reactivity of monomers in cationic polymerization.

REACTIVITY OF MONOMERS
IN COPOLYMERIZATION WITH ISOBUTYLENE

For isobutylene $Q = 0.033$, $e = -0.96$. Data on constants of copolymerization — the relative reactivity of isobutylene with other monomers — are generalized in a number of monographs and reviews, for example, [6, pp. 466–494]. Published data on constants of isobutylene copolymerization with some comonomers, correlated with the most important conditions of the experiment, are summarized in Table 1. As mentioned above, the constants r_1 and r_2 significantly depend on experimental conditions and, with consideration for features of the cationic polymerization discussed in the previous Section, are determined with greater errors than for free-radical copolymerization of two monomers. The effect of chain alternation in copolymers is practically absent for isobutylene. In most cases, the product $r_1 \cdot r_2 > 1$, and structure of the copolymer is characterized by existence of blocks of homopolymers. In the case of some pairs of monomers, for example, isobutylene–isobutylvinyl ester, no formation of copolymers occurs [3]. Critical consideration of data on cationic copolymerization and, in particular, isobutylene copolymerization [3], was performed with the help of the Kelen–Tudos linear graphical method — a modification of the well-known Feinman–Ross one [26]. It is based on the approximate type of equation describing cationic copolymerization in the presence of more than two AC in the system ($\sim> C_1^+$ and $\sim> C_2^+$), which leads to a dependence of determining copolymerization constants on the mixture composition and concentration of monomers. If the data on the copolymer composition can be presented by a linear dependence in η—ξ coordinates of the equation

$$\eta = \left(r_1 + \frac{r_2}{a} \right)\xi - \frac{r_2}{a},$$

where a is a constant $(a > 0)$, $\eta = \dfrac{G}{a+F}$, $\xi = \dfrac{F}{a+F}$, $G = \dfrac{x(y-1)}{y}$, $F = \dfrac{x^2}{y}$, $x = \dfrac{M_1}{M_2}$, and $y = \dfrac{d[m_1]}{d[m_2]}$, the process of copolymerization is characterized by values of r_1 and r_2 (otherwise they are meaningless).

Table 1. Constants r_1 and r_2 in the Cationic Copolymerization of Isobutylene (M_1)

Monomer M_2	r_1	r_2	$r_1 \cdot r_2$	Catalyst	Solvent	Temperature, K	Reference	Note
Styrene	1.10	0.99	1.09	AlBr$_3$	Benzene	273	[4]	
	14.90	0.53	7.90	AlCl$_3$	Nitrobenzene	273	[4]	
	12.20	2.80	34.16	SnCl$_4$·H$_2$O	Benzene	273	[4]	
	8.60	1.25	10.75	SnCl$_4$·H$_2$O	Nitrobenzene	273	[4]	
	1.60	0.17	0.25	SnCl$_4$	Ethyl chloride	273	[7]	a
	3.1	1.1	3.41	SnCl$_4$	Sulfur dioxide	195	[7]	
	2.2	1.1	2.42	SnCl$_4$	Sulfur dioxide–nitroethane	195	[7]	
	1.5	1.0	1.5	SnCl$_4$	Sulfur dioxide–dichloroethane	195	[7]	
	0.8	0.5	0.4	SnCl$_4$	Sulfur dioxide	195	[7]	
	1.9	0.6	1.14	SnCl$_4$	Sulfur dioxide–benzene	195	[7]	
	3.5	0.33	1.15	γ-radiation	Ethyl chloride	195	[6, 7]	
	9.02±0.77	1.99±0.24	1.79	AlCl$_3$ (0.12%)	Methyl chloride	181	[6, 7]	
	2.51±0.55	1.21±0.06	3.04	AlCl$_3$ (0.5%)	Methyl chloride	243	[6, 7]	
	2.36±0.06	0.76±0.22	1.79	AlCl$_3$ (0.14%)	Methyl chloride	243	[6, 7]	
	1.66±0.02	0.42±0.22	0.70	AlCl$_3$ (0.5%)	Methyl chloride	183	[6, 7]	
	1.79±0.22	0.24±0.22	0.43	AlCl$_3$ (0.14%)	Methyl chloride	183	[6, 7]	
	3±1	0.6±0.3	1.8	AlCl$_3$	Methyl chloride	170	[6]	
	2.8±0.2	1.4±0.2	3.92	C$_2$H$_5$AlCl$_2$	Methyl chloride	173	[7]	
	1.78±0.10	1.20±0.10	2.13	TiCl$_4$	Toluene	195	[7]	
	0.37±0.07	2.41±0.12	0.89	TiCl$_4$	Hexane	195	[7, 8]	
	0.54±0.24	1.20±0.11	0.64	TiCl$_4$	Hexane	253	[7]	
	2.63±0.52	5.50±0.55	14.46	TiCl$_4$	Hexane–methylene (75/25)	253	[7, 8]	
	3.25±0.25	2.75±0.25	8.93	TiCl$_4$	Hexane–methylene chloride (50/50)	253	[7, 8]	
	4.11±0.19	1.70±0.07	6.98	TiCl$_4$	Hexane–methylene chloride (25/75)	253	[7, 8]	
	4.48±0.28	1.08±0.07	4.83	TiCl$_4$	Methylene chloride	253	[7, 8]	
	1.51±0.03	2.44±0.15	3.68	TiCl$_4$	Hydrogen sulfide	253	[7]	
	1.53±0.03	2.61±0.4	3.99	TiCl$_4$	Decalin	253	[7]	
	0.75±0.15	1.30±0.15	0.97	TiCl$_4$	Methylcyclohexane	253	[7]	
n-Chloro-styrene	1.01	1.02	1.03	AlBr$_3$	Hexane	273	[6, 7]	
	14.7	0.15	2.20	AlBr$_3$	Nitrobenzene	273	[6, 7]	
	22.5	0.7	15.70	AlBr$_3$	Nitromethane	273	[6, 7]	
	8.6	1.2	10.3	SnCl$_4$	Nitrobenzene	273	[6, 7]	
Butadiene	115±15	0.01±0.01	—	AlCl$_3$	Methyl chloride	170	[6, 7]	b
	43	0	—	C$_2$H$_5$AlCl$_2$	Methyl chloride	173	[7, 9]	c
1-Chloro-butadiene	5.5	0.1	0.55	C$_2$H$_5$AlCl$_2$–tert-C$_4$H$_9$OH	Methylene chloride	193	[10]	d
Isoprene	2.5±0.5	0.4±0.1	1	AlCl$_3$	Methyl chloride	176	—	
	0.80	1.28	1.02	C$_2$H$_5$AlCl$_2$	Hexane	193	[7]	

**Table 1. Constants r_1 and r_2
in the Cationic Copolymerization of Isobutylene (M_1) (Continued)**

Monomer M_2	r_1	r_2	$r_1 \cdot r_2$	Catalyst	Solvent	Temperature, K	Reference	Note
	1.17	1.08	1.26	$C_2H_5AlCl_2$	Hexane–methyl chloride (88/12)	193	[7]	
	1.90	1.05	1.99	$C_2H_5AlCl_2$	Hexane–methyl chloride (50/50)	193	[7]	
	2.15	1.03	2.21	$C_2H_5AlCl_2$	Hexane–methyl chloride (12/88)	193	[7]	
	2.5±0.5	—	—	$C_2H_5AlCl_2$-Cl_2	Methyl chloride	238	[7]	
	2.25	0.4	0.9	$C_2H_5AlCl_2 \cdot 0.3 H_2O$	Methyl chloride	188	—	
	0.98±0.03	0	—	$C_2H_5AlCl_2 \cdot H_2O$	Isopentane	188	—	
	0.99±0.03	0	—	$(C_2H_5)_{1.5}AlCl_{1.5} \cdot H_2O$	Isopentane	188	—	
	2.4±0.35	4.5±3.0	—	$BCl_3 \cdot C_6H_5C(CH_3)_2OCOCH_3$	Methyl chloride, methylene chloride	243	[11]	e
α-Methyl-styrene	0.27±0.08	1.41±0.25	0.38	$TiCl_4$	Methylene chloride	195	[7]	
	1.2	5.5	6.6	$TiCl_4$	Toluene	195	[7]	
	1.1±0.2	2.2±0.2	2.42	$TiCl_4$	Methylene chloride	201	[7]	
Cis-1,3-pentadiene	5	0	—	$(C_2H_5)_2AlCl$-Cl_2 ($tert$-C_4H_9Cl)	Methyl chloride	218	[5, 12]	f
Trans-1,3-pentadiene	2.3	0	—	$(C_2H_5)_2AlCl$-Cl_2 ($tert$-)C_4H_9Cl	Methyl chloride	218	[5, 12]	f
Cyclo-pentadiene	0.60±0.15	4.5±0.5	2.7	$BF_3 \cdot O(C_2H_5)_5$	Toluene	195	[7]	
	0.73±0.17	1.86±0.20	1.35	$BF_3 \cdot O(C_2H_5)_5$	Methyl chloride	195	[7]	
	0.21±0.02	6.3±0.6	1.32	$SnCl_4$-tetracyanoethylene	Toluene	195	[7]	
	0.80±0.15	1.55±0.20	1.24	$SnCl_4$–tetracyano-ethylene	Methylene chloride	195	[7]	
Cyclo-hexadiene	0.77±0.20	1.05±0.25	0.80	$SnCl_4$–tetracyano-ethylene	Toluene	195	[7, 13]	
	0.98±0.20	1.07±0.45	1.04	$SnCl_4$–tetracyano-ethylene	Methylene chloride	195	[7, 13]	
α-Cyclopen-tadiene-ethyl-ferrocene	1.50	0.07	0.10	$AlCl_3$	Methylene chloride	183	[14]	
Trans,trans-2,4-hexadiene	6.0	0.88	5.28	$C_2H_5AlCl_2$	Methylene chloride–pentane (50/50)	343	[15]	g
Trans,cis-2,4-hexadiene	0.74	1.8	1.3	$C_2H_5AlCl_2$	Methylene chloride–pentane (50/50)	343	[15]	g
Cis,cis-2,4-hexadiene	2.3	1.5	3.45	$C_2H_5AlCl_2$	Methylene chloride–heptane	343	[15]	g
Trans-hexa-triene-1,3,5	0.58	1.57	0.92	$C_2H_5AlCl_2$	Methyl chloride–heptane (50/50)	203	[16]	h
Vinyl acetylene	8	0.13	1.04	BF_3	Methyl chloride	173	[6]	i

Table 1. Constants r_1 and r_2
in the Cationic Copolymerization of Isobutylene (M_1) (Continued)

Monomer M_2	r_1	r_2	$r_1 \cdot r_2$	Catalyst	Solvent	Temperature, K	Reference	Note
Vinylbenzyl chloride	4.5±1	0.7±0.1	3.15	BF_3	Ethyl chloride	170	[6]	
Isoprenylben-zyl chloride	0.4	1	0.4	BF_3	Ethyl chloride	173	[6]	
α-Pinene	11.0	0.09	0.99	$C_2H_5AlCl_2$	Ethyl chloride	168	[17]	
β-Pinene	0.27	3.0	0.81	$C_2H_5AlCl_2$	Ethyl chloride	223	[7, 18]	
	0.52	1.9	0.99	$C_2H_5AlCl_2$	Ethyl chloride	195	[18]	
	0.77	1.5	1.15	$C_2H_5AlCl_2$	Ethyl chloride	173	[18]	
	1.0	0.95	0.95	$C_2H_5AlCl_2$	Ethyl chloride	163	[18]	
	1.0	1.0	1.0	$C_2H_5AlCl_2$	Ethyl chloride	153	[18]	
	1.0	1.0	1.0	$C_2H_5AlCl_2$	Ethyl chloride	143	[18]	
	5.4	0.2	1.08	BF_3	Methyl chloride	173	[19]	
5-Chlorobi-cyclo-[2,2,1]-heptene-2	7.3±1.5	0.09±0.02	0.65	BF_3	Heptene	195	[20]	
Indene	1.1±0.02	2.2±0.02	2.42	$TiCl_4$	Methylene chloride	195	[21]	
Isobutyl-vinyl ester	0.033	28.0	0.92	$AlBr_3 \cdot 0.8(C_2H_5)_2O$	Toluene	195	[22]	
	0.030	12.6	0.37	$AlBr_3 \cdot 1.2(C_2H_5)_2O$	Toluene	195	[22]	
	0.37	22.4	8.28	$AlBr_3 \cdot 1.2(C_2H_5)_2O$	Toluene	195	[22]	
	0.12	12.6	1.51	$AlBr_3 \cdot 2.5(C_2H_5)_2O$	Toluene	195	[22]	
	0.52	13.25	6.89	$AlBr_3 \cdot 1.1HC(O)N(CH_3)_2$	Toluene	195	[22]	
	0.46	10.5	4.83	$AlBr_3 \cdot 2.3HC(O)N(CH_3)_2$	Toluene	195	[22]	
	0.19	55.0	10.45	$AlBr_3 \cdot 0.8(C_2H_5)_2O$	Hexane	195	[22]	
	0.18	50.0	9.0	$AlBr_3 \cdot 1.2(C_2H_5)_2O$	Hexane	195	[22]	
	0.13	12.5	1.62	$AlBr_3 \cdot 1.2(CH_3)_2SO$	Hexane	195	[22]	
	0.033	4.6	0.14	$AlBr_3 \cdot 2.5(CH_3)_2SO$	Hexane	195	[22]	
	0.27	52.0	14.0	$AlBr_3 \cdot 0.6HC(O)N(CH_3)_2$	Hexane	195	[22]	
	0.51	47.0	23.97	$AlBr_3 \cdot 1.2HC(O)N(CH_3)_2$	Hexane	195	[22]	
Trimethyl-sylylvinyl ester	0.37±0.2	5.82±0.86	2.15	Sulfocationite KU-2-8–$C_2H_{55}AlCl_2$ ($SO_3H/Al = 6$)	Toluene	253	[23, 24, 25]	
Dimethyl-(butoxy)-sylylvinyl ester	1.09±0.10	6.1±3.5	6.64	Sulfocationite KU-2-8–$C_2H_{55}AlCl_2$ ($SO_3H/Al=6$)	Toluene	253	[23, 24, 25]	

a — estimation by the Kelen–Tudos method testifies an approximate value of r_1 and r_2; b — it is supposed that $r_1 \cdot r_2 = 1$; c — polymerization of diene into trans-1,4- and 1,2-positions; d — polymerization of diene into 3,4-position;e — statistic copolymer with 1,4-structure of the diene part under low content of M_2 and a block-copolymer with partially cycled structure under high content of M_2; f — diene reacts into the 1,2-position; g — diene reacts into the 2,5-position; h — diene reacts into the 2,6-position; i — determination by the first experiment; j — approximate values due to heterogeneity of the catalyst.

Hence the Kelen–Tudos method enables the suitability of the usual two-parameter model of copolymerization and correctness of determination of constants to be estimated, and also serves to refine their values. Consideration of a number of binary systems, based on the example of isobutylene–styrene, isobutylene–alkylvinyl ester ones in the presence of various catalysts, has shown that in most cases the values of copolymerization constants published are evidently displayed with great errors or are meaningless. Taking into consideration difficulties in application of the copolymerization equation to cationic systems, it would be more correct to compare relative activities of monomers or parameters of copolymerization of monomers as "indexes of relative reactivity" (for a sequence of typical reactions) and to use non-linear mathematical methods to estimate them [27]. These indexes display no connection with the composition of the monomeric mixture and resulting copolymer, but can serve to conduct a high-quality analysis of reactions of the chain cross-propagation, in particular, of selectivity of this reaction (the ratio r_1/r_2 or $\lg(r_1/r_2)$).

Applicability of this approach was demonstrated using the example of isobutylene copolymerization with isobutylvinyl ester (see Table 1). When electron-donor compounds are introduced into a hardly polymerizing mixture of these monomers, besides total reduction of the reactivity, convergence of relative reactivities of macroions is observed, because the more electrophilic carrier of the chain must be solvated better than the less electrophilic one. In this case, an increase in the proportion of the less active monomer (isobutylene) in the copolymer (fractionated samples) was observed. In its turn, proportion of the true resulting copolymer reduces with growth of an ester monomer content in separate fractions. Increase of concentration of the donor agent enables selectivity of the reaction of the chain cross propagation to be reduced. An analogous effect is also caused by growth of copolymerization temperature. The connection between selectivity and nature of the catalyst and the solvent is complex.

The processes observed are qualitatively coincident with the behavior of carbocationic centers of both hard and soft acids. The above-mentioned growth of selectivity in selection of monomers, closely related to their activity, on coordination carbonium ions with a soft aromatic base promotes increase in the role of the softness factor. On the contrary, coordination of hard bases (O-, N-, and S-containing compounds) cause the opposite effect: growth of hardness of ionic centers and, as a consequence, reduction of selectivity of the chain cross-propagation during copolymerization take place. The temperature influence is analogous: reduction of dissociation of ionic pairs with the temperature increase correlates with growth of hardiness of polymeric carbonium ions. Table 2 shows values of reactivity indexes for the isobutylene–isobutylvinyl ester pair, calculated with the help of linear graphical and non-linear mathematical methods [27].

Even if a refined calculation using a computer is performed, significantly different values of indexes of the relative reactivity of monomers are obtained in relation to the chosen method. Consequently, various structures can be ascribed to isobutylene copolymers (a wide range of changes of r_1/r_2 and $r_1 \cdot r_2$).

Table 2. Comparison of Indexes of the Relative Reactivity for Isobutylene–isobutylvinyl Ester System[*] (n-hexane, $AlCl_3$ ($4.82 \cdot 10^{-3}$ mol/l), Dimethylsulfoxide ($2.1 \cdot 10^{-3}$ mol/l))

Method	r_1	R_2	r_1/r_2	$r_1 \cdot r_2$
Feinman–Ross	0.000±0.007	0.811±0.009	0	0
Kelen–Tudos	0.251±0.025	7.956±5.210	0.031	1.996
Non-linear (numerical integration by the trapezium method)	0.134±0.047	1.726±0.350	0.072	0.231
Non-linear (integral)	0.880±0.296	9.635±1.783	0.092	8.478

[*] Constants r_1 and r_2 were calculated using a computer.

The approximate understanding of the reactivity of isobutylene in copolymerization with some monomers (styrene [28], butadiene [29], etc.) is confirmed by non-coincidence of experimentally measured content of dyads and triads of chains in polymers with the ones calculated by copolymerization constants, existing in the literature. All this testifies to the necessity to conduct further deeper studies in the field of cationic copolymerization of isobutylene, for which formation of true copolymers can be thought proved for single cases only. Among them may be systems with low-active cyclic and diene monomers. In this aspect, of interest is the example of isobutylene copolymerization with vinylsilyl esters (see Table 1). Although the heterogeneous type of catalyst somehow limits consideration of r_1 and r_2 as the real constants of copolymerization, its application gives the possibility of conducting regulated synthesis of products [30]. Simplicity of separation of homopolymers of isobutylene and trimethylsilylvinyl (dimethylbutoxisilylvinyl) ester by the method of fractional dissolution (the heptane–water system) and the negative result obtained in analysis of products of isobutylene polymerization in the presence of vinylsilyl esters confirm formation of the real copolymers. The solid catalytic surface, irreversibly sorbing more active silyl ester and preventing its homopolymerization, plays the role of regulator of activity of comonomers. The presence of isobutylene simplifies desorption of polymeric products and, hence, is the condition of participation of vinylsilyl monomer in the copolymerization.

Data on isobutylene copolymerization can be presented as relative reactivities of monomers in relation to substituted *tert*-butyl cation, expressed by values opposite to r_1 (k_{12}/k_{11}). The relative reactivity of the carbocation with its own monomer equals 1 (Table 3).

Approximate data on the relative activity of isobutylene and other monomers in cationic copolymerization are also obtained by application of appropriate values of the parameter e from the Alfrei–Price equation, obtained for free-radical polymerization [6, p. 483]. If the effect of end chains is neglected at the stage of polymeric chain propagation, it is assumed that $r_1 = k_{11}/k_{12} = \exp(e_2 - e_1)$ (where e_1 and e_2 are parameters characterizing polar properties of comonomers). Published values of parameters e are: -1.77 (isobutylvinyl ester), -1.27 (α-methylstyrene), -1.03 (indene), -0.96 (isobutylene), -0.80 (styrene), -0.33 (*n*-chlorostyrene). They reflect correctly the behavior of isobutylene in copolymerization with the mentioned monomers.

Table 3. Relative Reactivities of Some Monomers in Respect to the Carbonium ion of Isobutylene

Monomer (M_2)	Catalyst	Solvent	Temperature, K	Relative reactivity
Isobutylene	$SnCl_4$	Carbon tetrachloride	273	1.0
Styrene	$SnCl_4$	Ethyl chloride	273	0.62
Isoprene	$AlCl_3$	Methyl chloride	173	0.44
Butadiene	$AlCl_3$	Methyl chloride	173	0.009
β-Pinene	$C_2H_5AlCl_2$	Ethyl chloride	173	1.3
Isopropenyl-benzyl chloride	BF_3	Ethyl chloride	173	2.5

SPECIFIC METHODS OF DERIVING COPOLYMERS OF ISOBUTYLENE

Progress towards initiation of isobutylene polymerization, in particular, application of complex catalysts, high-energy radiation, or combined methods of influence on monomeric systems, broadens the possibilities of synthesis of copolymeric products. Recently, data have appeared on free-radical and other methods of the synthesis of isobutylene copolymers of various structure, which enable the number of monomers copolymerizing with it to be increased. Contrary to the traditional initiation by cationic catalysts, they lead to isobutylene copolymers of strictly alternate structure or with increased affinity to alternation of various monomeric chains (copolymerization constants smaller than 1).

1:1-Copolymers of isobutylene with chlorotrifluoro- and tetrafluoroethylene were synthesized using free-radical initiation in the mass, solution or emulsion, radiation by γ-rays [31–34]. For isobutylene–vinylidene chloride and isobutylene–dichlorodifluoroethylene, copolymerization was activated by γ-radiation from ^{60}Co of preliminarily prepared channel complexes of thiocarbamide with appropriate monomers [34]. In the case of vinyl chloride, the copolymer is enriched by a chlorine-containing monomer ($r_{iso-C_4H_8}$ = 0.34; $r_{C_2H_3Cl}$ = 2.11; 333 K) [35]. 1:1-Copolymers of isobutylene with maleic anhydride, a diethylfumarate, were obtained, which are interesting as emulsifying agents, sorbents of metals from solutions, and modifiers of rubbers [36]. The structure of the polymeric products does not depend on the composition of the initial mixture of monomers and the type of initiator. Copolymers of isobutylene–cyclopentene–SO_2, polyisobutylene–norbornene–SO_2 block-copolymer and isobutylene with butadiene were synthesized by applying the principle of alternation applied [37, 38].

Alternating copolymerization proceeds in isobutylene–acrylic monomer double systems [39, 40] or isobutylene–methacrylate–maleic anhydride triple systems [41], isobutylene–methylacrylate–vinylbenzyl chloride [42] under simultaneous effect of the source of free radicals and the complexing additive HAOS. The generalized mechanism of formation of polymeric products by radical, cationic and other methods has been suggested [43, 44]. A complex of accepting monomer and HAOS is considered as a complex monomer with reduced electronic density on the double bond. Radical initiator activates the main state of the complex and generates stable radicals, which enter the donor–acceptor interaction with isobutylene. Thus, a hexagonal cyclic structure is formed. The donor–acceptor interaction between the monomer existing in the complex with a radical and copolymer (1:1-complex of acrylate with HAOS) enables a double complex to be formed at the propagating end of the chain. As a conjugated system is formed in this case, the chain propagation can proceed according to the adjusted reaction as a simultaneous addition of two monomers at once. An important role is played by aluminum, linked to the propagating end of a macromolecule because, at its detachment, the localized radical becomes quite labile in order to provide inclusion of two monomeric units at once into the chain. Usage of complex formers of the optimum acidity provides the required degree of stabilization of the propagating radical.

Hence the factor of AC complexing is important for both classic cationic copolymerization and alternating copolymerization with participation of isobutylene, which proceeds not necessarily as a cationic process. The mentioned analogy seems pertinent, because it has been shown that the principles of possible realization of continuous cationic polymerization under material initiation and "living" free-radical polymerization are the same.

In the case of alternating cationic polymerization [43], propagation AC are formed at proton detachment from a molecule of a monomer, solvent or complex by a donor–acceptor complex, which includes a molecule of a donor monomer (isobutylene) and a complex of an acceptor monomer with the Lewis acid. In the stage of chain propagation, the monomer complex is introduced into the contact ionic pair of AC. Alternating copolymers of isobutylene with polar and diene monomers are capable of forming various polymer-analogous transformations and vulcanization.

Another group of isobutylene copolymers is represented by products, produced with application of modified methods of cationic initiation. In this manner, properties of known copolymers were improved, and in some cases, new polymeric products were synthesized. For example, application of HAOS combined with various additives, and mixed (polar–non-polar) solvents enables properties (content of gel fractions, molecular mass, degree of unsaturation, ability to be vulcanized, oxidized, etc.) of isobutylene copolymers with diene monomers (butadiene, isoprene, pyperilene, cyclohexadiene, cyclopentadiene, etc.) to be widely regulated [45, 46].

Radiation copolymerization of isobutylene with isoprene in mass gives products with much higher molecular mass, than at the material initiation. In this characteristic they approach the molecular mass of polyisobutylene, synthesized under analogous conditions [47].

Of interest for modification and thermal (chemical) vulcanization of isobutylene polymers is copolymerization with a cyclic diene (1,2-dimethylene cyclobutane, 1-methyl-3-methylene cyclobutane-1) and triene (*trans*-1,3,5-hexatriene, 5-methylheptatriene-1,3,6, 1,6-diphenylhexatriene-1,3,4, octatriene-2,4,6) monomers [48, 49].

Introduction of 2-allyloxytetrahydropyranic units into a polyisobutylene chain during polymerization gives products that can be used directly as additives to oils.

Development of principles of catalysis by Friedel–Crafts halides, in particular, by MeX_n–RHal systems, formed the basis of a method to synthesize block- and grafted copolymers of isobutylene. Similar to reactions of arene alkylation [50] or ionic telomerization of olefins [51], activation of the R–Hal bond with the help of Lewis acids is the effective way to generate cationic particles $R^{\delta+}$, which initiate polymerization and other electrophilic transformations of isobutylene:

$$R\text{–}Hal + MeX_n \longrightarrow R^{\delta+} MeXnHal^{\delta-}$$

$$\downarrow C_6H_6 \qquad \downarrow CH_2{=}C(CH_3)_2 \qquad \downarrow nCH_2{=}C(CH_3)_2$$

$$C_6H_5R \qquad RCH_2\text{–}C(CH_3)_2Hal \qquad R\text{–}(CH_2\text{–}C(CH_3)_2)_{n-1}\text{–}C_4H_8$$

If a polymer activated by MeX_n serves as a carrier of a mobile halogen atom, this leads to formation of block- or graft copolymers. In particular, methods were developed, based on introduction of active atoms of halogens into the polymer at the stage of initiation, transmission and termination of the chain with further excitation of polymerization by the following schemes (A – isobutylene, B and B′ - other monomers, Hal and Hal′ – halogens) [52–54]:

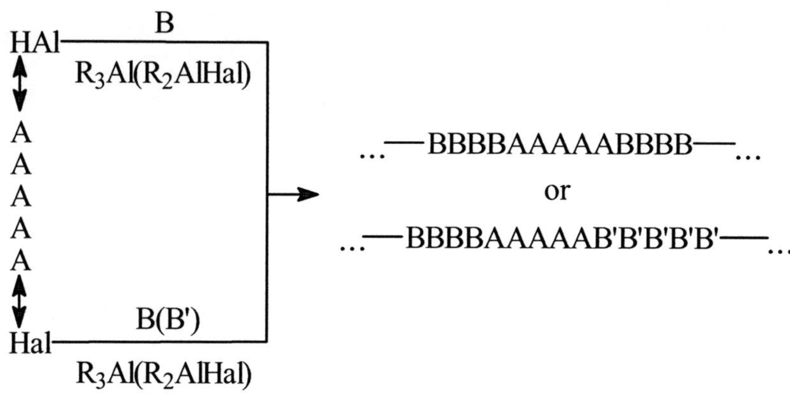

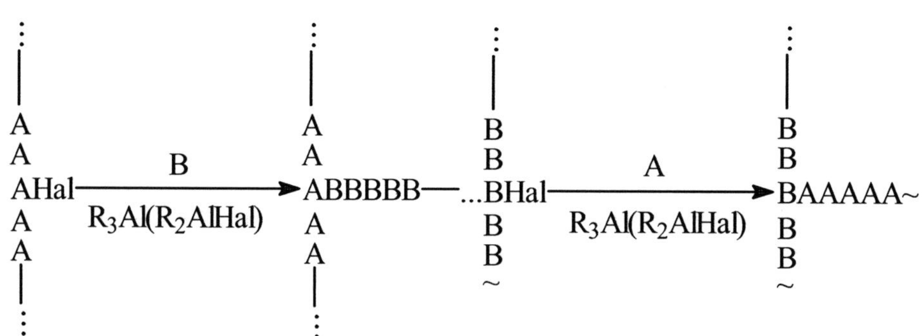

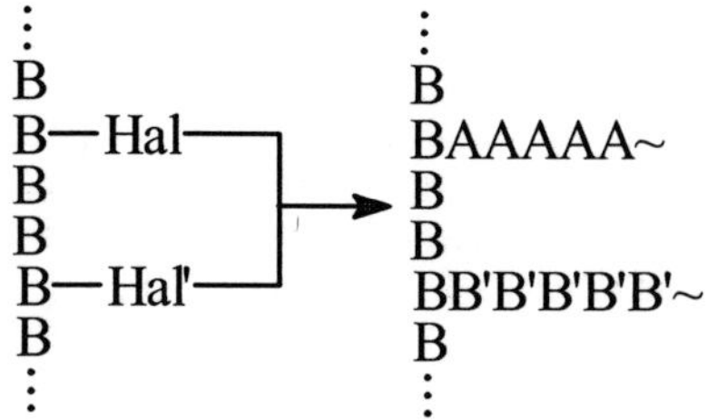

According to the above-mentioned scheme, as variants of synthesis of block- and graft copolymers of isobutylene, the methods are suggested, based on application of a wider range of Lewis acids (BCl$_3$, TiCl$_4$ and others), other, beside halogens, active functional groups (OH,

COOH, acetoxyalkyl, unsaturated, etc.), macromolecular compounds with polyisobutylene chain, polycondensation methods, which include reactions of α,ω-bifunctional oligomers and isobutylene polymers with other functional ones, non-polymerizational methods, via interaction of functional groups of partners. These methods use achievements in the branch of cationic polymerization of isobutylene and other monomers, conduction of polymerization by the principle of "living chains", in particular, in synthesis of telehelic polymers, the effect of inifers, proton traps, etc. [7, 54], as well as polymer-analogous transformations of polymers, in particular, at the synthesis of polyisobutylene with various functional groups [55]. On the other hand, a broad selection of methods of synthetic organic chemistry (polyaddition, condensation, alkylation, etc.) was used for synthesizing block- and grafted copolymers.

The most typical examples of block- and graft copolymers of isobutylene, synthesized according to various schemes, are shown in Table 4.

As follows from the data displayed in Table 4, methods II and III of production of block- and graft isobutylene copolymers dominate among others. They are directly connected with the achievements in the controlled synthesis of polyisobutylene, mentioned above. Let us note some general and particular positions concerning the theme under consideration.

Partners of polyisobutylene in formation of copolymers represent a wide range of polymeric products: polymers and copolymers of olefins [16, 17, 26–29], polydienes [4–6, 20–23, 25, 36–39], polymers of vinyl monomers [1, 2, 7–12, 14, 17–19, 25, 30–32, 36, 37, 40, 41, 50], heteroatom-containing polymers [13, 42–48], inorganic [15, 33–35] and natural [49] polymers. Practically all possible types of block- and graft copolymers were obtained: double, triple, polyblock, star-like with blocks, approximately equal by in size, and fragmental (with a single shortened block) block-copolymers with blocks of the same and different nature (rigid, elastic), etc. Control of size of blocks, grafting degree and other parameters enables products with various physicomechanical and other properties to be produced, which are of interest from the point of view of both modification of polyisobutylene properties (for example, elimination of the cold flow, stability increase) and its application for improving properties of other polymers (increase of chemical stability of polar products of the polyurethane type, etc.). Additional information about properties of copolymers and their application are discussed in [1 - 10].

Among separate representatives, block-copolymers of isobutylene with isoprene should be noted, which give films with unusually high adhesion to metals at chemical and photochemical structuring [60]; fragmental block-copolymer of isobutylene with hydroxyalumoxane [68], as examples of copolymerization of polyisobutylene with a monomer, generated *in situ*; and three-ray copolymer of isobutylene with styrene (α-methylstyrene) [63–65] as demonstration of the possibilities of the fine organic synthesis applied to polymeric objects.

Notice also structural investigations of the three-block copolymer poly-*para*-chlorostyrene-polyisobutylene-poly-*para*-chlorostyrene, in which cylindrical domains of styrene fragments were observed in the polyisobutylene matrix [100].

In conclusion, let us note a number of macromolecular compounds of polyisobutylene, synthesis of which demonstrates progress and possibilities of the "living" cationic polymerization [101] (Table 5).

Table 4. Block- and Graft Polymers of Isobutylene

No.	Type of copolymer	Production method	Reference
1	Poly(styrene-block-isobutylene)	I	[7, 56-58]
2	Poly(styrene-block-isobutylene-block-styrene)	I	[7]
3	Poly(isobutylene-block-propylene)	II	[7]
4	Poly(isobutylene-block-butadiene)	II	[59]
5	Poly(isobutylene-block-isoprene)	II	[60]
6	Poly(isoprene-block-isobutylene-block-isoprene)	II	[60]
7	Poly(isobutylene-block-styrene)	II	[61-63]
8	Poly(isobutylene-block-α-methylstyrene)	II	[64]
9	Poly(styrene-block-isobutylene-block-styrene)	II	[62, 64]
10	Poly(α-methylstyrene-block-isobutylene-block-styrene)	II	[64, 65]
11	Poly(isobutylene-block-styrene) – three-ray	II	[63]
12	Poly(isobutylene-block-α-methylstyrene) – three-ray	II	[64, 65]
13	Poly(dimethylsiloxane-block-isobutylene-block-dimethylsiloxane)	II	[66]
14	Poly(isobutylene-block-methylmethacrylate)	II	[67]
15	Fragmental poly(isobutylene-block-almoxane-block-isobutylene)	II	[68]
16	Poly(ethylene-graft-isobutylene)	III	[69]
17	Poly(tetrafluoroethylene-graft-isobutylene)	III	[69]
18	Poly(vinyl chloride-graft-isobutylene)	III	[69-73]
19	Poly(vinyl chloride-graft-isobutylene-isoprene)	III	[74]
20	Poly(chloroprene-graft-isobutylene)	III	[75]
21	Poly(chloroprene-graft-isobutylene-graft-α-methylstyrene)	III	[76, 77]
22	Poly(butadiene-graft-isobutylene)	III	[7]
23	Poly(cyclopentadiene-graft-isobutylene)	III	[78]
24	Poly(styrene-graft-isobutylene)	III	[79]
25	Poly(styrene-butadiene-graft-isobutylene)	III	[80]
26	Poly(ethylene-propylene-1,4-hexadiene-graft-isobutylene)	III	[81, 82]
27	Poly(ethylene-propylene-1,4-hexadiene-graft-styrene-graft-isobutylene)	III	[81, 82]
28	Poly(ethylene-propylene-1,4-hexadiene-graft-α-methylstyrene-graft-isobutylene)	III	[83]
29	Poly(ethylene-propylene-dicyclopentadiene-graft-isobutylene)	III	[84]
30	Poly(methylmethacrylate-graft-isobutylene)	III	[85]
31	Poly(methylmethacrylate-2(dimethylamino)-ethylmethacrylate-graft-isobutylene)	III	[86]
32	Poly(butylacrylate-graft-isobutylene)	III	[87]
33	Silica-graft-polyisobutylene	III	[88]
34	Silicagel-graft-polyisobutylene	III	[89]
35	Silicagel-graft-isobutylene-isoprene copolymer	III	[89]
36	Poly(isobutylene-isoprene-graft-styrene)	IV	[61]
37	Poly(isobutylene-graft-α-methylstyrene)	IV	[61]
38	Poly(isobutylene-isoprene-graft-chloroprene)	IV	[61]
39	Poly(isobutylene-isoprene-graft-indene)	IV	[61]
40	Poly(isobutylene-graft-vinyl fluoride)	IV	[89]
41	Poly(isobutylene-block-styrene-block-isobutylene)	V	[74]
42	Poly(isobutylene-block-dimethylsiloxane)	V	[91, 92]
43	Poly(isobutylene-block-dimethylsiloxane-block-isobutylene)	V	[91, 92]
44	Fragmental poly(isobutylene-block-dimethyl-block-siloxane-block-isobutylene)	V	[92]
45	Poly(isobutylene-block-ethylene glycol)	V	[93]
46	Poly(ethylene glycol-block-isobutylene-block-ethylene glycol)	V	[94]

Table 4. Block- and Graft Polymers of Isobutylene (Continued)

No.	Type of copolymer	Production method	Reference
47	Poly(isobutylene-block-ethylene glycol)	V	[95]
48	Fragmental poly(isobutylene-block-urethane)	V	[96, 97]
49	Starch-graft-polyisobutylene	V	[98]
50	Poly(styrene-graft-isobutylene)	V	[99]

I – block-copolymerization of isobutylene with other polymers; II – block-copolymerization of polyisobutylene with other monomers; III – grafting of polyisobutylene to a polymer of different nature; IV – grafting of other polymers to polyisobutylene; V – interaction of functionalized polyisobutylene with a polymer of different nature.

Existence of styrene, olefin and other active groups in polyisobutylene macromolecules gives a possibility of synthesis via copolymerization with other monomers of various comb-like copolymers.

Combined with progress in the field of cationic polymerization of isobutylene, the controlled synthesis of block- and graft copolymers of isobutylene represents the visual confirmation of progress in molecular engineering in the field of polyisobutylene, which beside pure scientific advantages, is no doubt interesting from the point of view of obtaining thermal elastoplastics, based on isobutylene, modification of polymers by the method of surface grafting, creation of new composite materials, and solving of other applied tasks.

Table 5. Characteristics of Macromonomers Based on Polyisobutylene

End group	Method*	$\overline{M}\cdot 10^{-3}$	$\overline{f}_n$	Reference
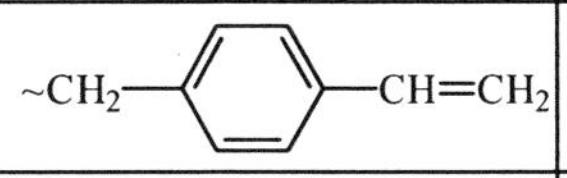 $\sim CH_2$–(C$_6$H$_4$)–$CH{=}CH_2$	A	10-30	—	102
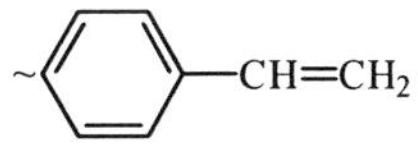 $\sim$(C$_6$H$_5$)–$CH{=}CH_2$	A	4.2-34.2	0.9-1.1	103
$\sim CH_2{-}C({=}CH_2)Me$	B	1.6-3.2	0.95-1.04	104
$\sim CH_2OCOC({=}CH_2)Me$	B	1.6-13.9	0.98-1.05	104
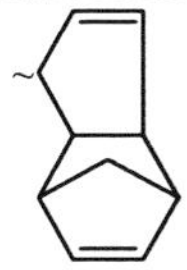 $\sim$ (indenyl/bicyclic structure)	C	27.0	1.0	84
$\sim CH_2OCOCH{=}CH_2$	B	3.6	1.95**	105

* Methods of introduction of end groups: A – initiation, B – reactions by end groups, C – the method with application of combined agent-initiator and chain transmitter ("inifer").

** Bifunctional derivative.

REFERENCES

[1] Berlin, Al.Al., Wolfson, S.A., and Enikolopyan, N.S., *Kinetics of Polymerization Processes*, 1978, 320 p., Moscow, Khimia. (Rus)

[2] Heublein, G., Faserforsch. Textitech Z. *Polymer Forsch.*, 1976, Bd. 27(2), S. 57.

[3] Kennedy, J.P., Kelen, T., and Tudos, F., *J. Polymer Sci., Polymer Chem. Ed.*, 1975, v. 13, pp. 2277–2289.

[4] Erusalimskii, B.L. and Lubetskii, S.G., *Processes of Ionic Polymerization*, 1974, 256 p., Leningrad, Khimia. (Rus)

[5] Heublein, G. and Romer, W.J., *J. Plast. Chem.*, 1973, v. 315(5), pp. 801–809.

[6] *Cationic Polymerization*, Ed. P. Plesha, 1966, 584 p., Moscow, Mir. (Rus)

[7] Kennedy, J.P. and Merechal, E., *Carbocationic Polymerization*, 1982, 510 p., N.Y.

[8] Imanishi, Y., Higashimura, T., and Okamura, S., *J. Polymer Sci.*, 1965, Pt. A, v. 3(7), pp. 2455–2474.

[9] Corno, C., Priola, A., and Cesca, S., *Macromolecules*, 1979, v. 12(3), pp. 411–418.

[10] Harzallah, F., Tardi, M., Polton, A., and Sigwalt, P., *Eur. Polym. J.*, 1984, v. 20(2), pp. 163–170.

[11] Faust, R., Fehervari, A., and Kennedy, J.P., *Rubber Chem. and Technol.*, 1977, v. 60(4), p. 794.

[12] Priola, A., *Makromol. Chem.*, 1975, Bd. 176(7), S. 1969–1981.

[13] Faust, R., Fehervari, A., and Kennedy, J.P., *Brit. Polym. J.*, 1987, v. 19(3–4), pp. 379–386.

[14] Gulimov, V.I., Vishiakov, T.P., and Vlasova, U.D., Izvestiya VUZov, 1988, v. 31(6), pp. 136–138, *Khimiai Khim.* Tekhnol. (Rus)

[15] Priola, A., Corno, C., and Cesca, S., *Polymer Bull.*, 1982, Bd. 7(11–12), S. 599–606.

[16] Priola, A., Corno, C., Bruzzone, M., and Cesca, S., *Polymer Bull.*, 1981, Bd. 4(12), S. 735–742.

[17] Kennedy, J.P. and Nakao, M., *J. Macromol. Sci.*, 1977, Pt. A, v. 11(9), pp. 1621–1636.

[18] Kennedy, J.P., *Makromol. Chem.*, 1974, Bd. 175(4), S. 1101–1115.

[19] Kennedy, J.P. and Chou, T., *Adv. Polymer Sci.*, 1976, v. 21(1), pp. 1–30.

[20] Sangalov, Yu.A., Nelkenbaum, Yu.Ya., Khusnutdinov, R.I., Jemilev, U.M., and Minsker, K.S., *Highmolec. Comp.*, 1984, Ser. A, v. 26(3), pp. 600–606. (Rus)

[21] Marechal, E.-C., R. *Acad. Sci.*, 1968, Ser. C, v. 266(19), pp. 1427–1430.

[22] Heublein, G. and Romer, W., *Makromol. Chem.*, 1973, Bd. 163(1), S. 143–153.

[23] Sangalov, Yu.A. and Nelkenbaum, Yu.Ya., *Physical and Chemical Grounds of Synthesis and Processing of Polymers*, 1984, pp. 23–26, Gorky. (Rus)

[24] Gladkikh, I.F., Sangalov, Yu.A., and Komarov, N.V., Izvestiya VUZov, 1984, v. 27(6), pp. 707–710, *Khimia and Khim.* Tekhnol. (Rus)

[25] Sangalov, Yu.A., Gladkikh, I.F., Komarov, N.V., and Zisovin, E.G., *Intern. Polym. Sci. and Technol.*, 1984, No. 10, pp. 41–43.

[26] Tudos, F., Kelen, T., and Berezhnikh, T.F., *J. Polym. Sci., Polymer Symp.*, 1975, No. 50, pp. 109–132.

[27] Heublein, G. and Wondraczek, R., *J. Polymer. Sci., Polymer Symp.*, 1976, No. 56, pp. 359–372.

[28] Kennedy, J.P. and Chou, T., *J. Macromol. Sci.*, 1976, Pt. A, v. 10(7), pp. 1357–1369.

[29] Corno, C., In: *Proc. Europ. Conf. MMR Macromol.*, Sassari, 1978, S. 1, s.a. 185–191.

[30] Sangalov, Yu.A., Gladkikh, I.F., and Minsker, K.S., *Highmolec. Comp.*, 1982, Ser. B, v. 24(5), pp. 356–358. (Rus)

[31] Sebenic, A., Guhaniyogi, S., Santee, E.R., and Harwood, H.J., *Amer. Chem. Soc. Polym. Prepr.*, 1986, v. 27(2), pp. 138–139.

[32] Kostov, G.K., *J. Polymer Sci., Polymer Chem. Ed.*, 1979, v. 17(12), pp. 3991–4001.

[33] Ishigyre, K., *Macromolecules*, 1976, v. 9(2), pp. 290–293.

[34] Schneider, Ch. and Bohlmann, H.P., In: IUPAC Macro, Florence, 1980, *Intern. Symp. Macromol*, 1980, Prepr., v. 2, Pisa, s.a. 134–137.

[35] Takemoto, T., Kunststoffe, 1967, Bd. 57(9), S. 697–701.

[36] Hummel, D.O. and Bestgen, J., Angew. Chem., *Intern. Ed. Engl.*, 1974, v. 13(6), p. 414.

[37] Raetsch, M. and Bormann, G., *Plaste und Kautschuk*, 1976, Bd. 23(3), S. 173–176.

[38] Priola, A., Ferraris, G., and Cesca, S., *Makromol. Chem.*, 1979, Bd. 180(12), S. 2859–2874.

[39] Kuntz, I., Chamberlain, N.F., and Stehling, F.J., *Amer. Chem. Soc., Polymer Prepr.*, 1978, v. 19(1), pp. 216–221.

[40] Florjanczyk, Z., *Makromol. Chem.*, 1982, Bd. 183(5), S. 1081–1091.

[41] Hirooka, M., Appl. 74-103983, Japan; C.A., 1975, v. 82, 98781 w.

[42] Kuntz, I., Pat. 2515381 (FRG), 1975; C.A., 1976, v. 84, 60866 f.

[43] Gaylord, N.G., *J. Polymer Sci., Polymer Chem. Ed.*, 1975, v. 13(2), pp. 467–482.

[44] Hirooka, M., In: 23rd Intern. Congr. Pure and Appl. Chem., Boston, 1971, *Macromol. Prepr.*, v. 1, S. 1, s.a. 311–318.

[45] Thaler, W.A. and Buckley, D.J., *Rubb. Chem. Technol.*, 1976, v. 49(4), pp. 960–966.

[46] Pat. 4171414 (USA), 1979.

[47] Williams, F.,, Shinkava, A., and Kennedy, J.P., *J. Polymer Sci., Polymer Symp.*, 1976, No. 56, pp. 421–430.

[48] Tsao, J.H. and Lenz, R.W., *J. Polymer Sci., Polymer Chem. Ed.*, 1979, v. 17(2), pp. 331–349.

[49] Priola, A., Como, C., Bruzzone, M., and Cesca, S., *Polymer Bull.*, 1981, Bd. 4(12), S. 743–750.

[50] Olah, G.A., *Friedel-Crafts Chemistry*, 1973, 581 p., N.Y.

[51] Petrov, A.A. and Genusov, M.L., *Ionic Telomerization*, 1968, 296 p., Leningrad, Khimia. (Rus)

[52] Kennedy, J.P., *J. Macromol. Sci.*, 1980, Pt. A, v. 14(1), pp. 1–10.

[53] Kennedy, J.P., *J. Macromol. Sci.*, 1979, Pt. A, v. 13(5), pp. 695–714.

[54] Kennedy, J.P., *Cationic Graft Polymerization (Appl. Polymer Symp., No. 30)*, 1979, 193 p., N.Y., Wiley.

[55] Sangalov, Yu.A. and Yasman, Yu.B., *Uspekhi Khimii*, 1985, v. 44(7), pp. 1208–1229. (Rus)

[56] Kennedy, J.P. and Melby, E.G., *Polymer Prepr.*, 1974, v. 15, p. 180.

[57] Kennedy, J.P. and Melby, E.G., *J. Polym. Sci., Chem. Ed.*, 1975, v. 13, p. 29.

[58] Kennedy, J.P., Feinberg, S.C., and Huang, S.Y., *J. Polym. Sci., Chem. Ed.*, 1978, v. 16(1), pp. 243–259.

[59] Nemes, S. and Kennedy, J.P., *J. Macromol. Sci.-Chem.*, 1971, v. A28(3, 4), pp. 311–328.

[60] Puskas, J.E., Kaszas, G., and Kennedy, J.P., *J. Macromol. Sci.-Chem.*, 1991, v. A28(1), pp. 65–80.

[61] Kennedy, J.P. and Charles, J.J., *J. Appl. Polym. Sci., Appl. Polym. Symp.*, 1977, v. 30, pp. 119–139.

[62] Fodor, Z.S., Kennedy, J.P., Kelen, T., and Tudos, F., *J. Macromol. Sci.*, 1987, v. A24(7), pp. 735–747.

[63] Kaszas, G., Puskas, J.E., Kennedy, J.P., and Hager, W.G., J. Polym. Sci., Pt. A, Polym. Chem., 1991, v. 29, pp. 427–435.

[64] Kennedy, J.P., Guhaniyogi, S.G., and Ross, L.R., *J. Macromol. Sci.-Chem.*, 1982, v. A18(1), pp. 119–128.

[65] Kennedy, J.P. and Smith, R.A., *J. Polym. Sci., Polym. Chem. Ed.*, 1980, v. 18(5), pp. 1539–1546.

[66] Wilczek, L., Michka, M.K., and Kennedy, J.P., *J. Macromol. Sci.-Chem.*, 1987, v. A24(9), pp. 1033–1049.

[67] Kennedy, J.P. and Chen, F., *J. Amer. Chem. Soc., Polymer Prepr.*, 1979, v. 20(2), pp. 310–315.

[68] Tolstikov, G.A., Sangalov, Yu.A., Nelkenbaum, Yu.Ya., and Prokofiev, I.K., *Metal-organic Chemistry*, 1988, v. 1(3), pp. 676–679. (Rus)

[69] Kabanov, V.Ya., Sidorova, L.P., and Spitsyn, V.J., *Eur. Polym. J.*, 1974, No. 10, p. 1153.

[70] Chapiro, A., J. Polym. Sci., *Polym. Symp.*, 1976, No. 56, pp. 431–436.

[71] Kabanov, V.Ya. and Sidorova, L.P., *Highmolec. Comp.*, 1972, v. 14(10), pp. 2091–2095. (Rus)

[72] Kennedy, J.P. and Davidson, D.L., *J. Appl. Polym. Sci., Appl. Polym. Symp.*, 1977, v. 30, pp. 51–72.

[73] Gupta, S.N. and Kennedy, J.P., *Polym. Bull.*, 1979, No. 1, pp. 253–259.

[74] Minsker, K.S. and Sangalov, Yu.A., *Isobutylene and Its Polymers*, 1986, 224 p., Moscow, Khimia. (Rus)

[75] Kennedy, J.P., Plamthottam, S.S., and Ivan, B., *J. Macromol. Sci.*, 1982, v. A17(4), pp. 637–651.

[76] Kennedy, J.P. and Plamthottam, S.S., *J. Macromol. Sci.-Chem.*, 1979, v. A14, p. 729.

[77] Kennedy, J.P. and Plamthottam, S.S., *Polymer Bull.*, 1982, v. 7(7), pp. 337–344.

[78] Heublein, G. and Freitag, W., *J. Prakt. Chem.*, 1979, Bd. 321(4), S. 544–548.

[79] Jiang, Y. And Frechet, J.M.J., *197th ACS Nat. Mect.*, Dallas, Tex, April 9–14, 1989, Abstr. Pap., 1989, p. 833.

[80] Oziomek, J. and Kennedy, J.P., J. *Appl. Polym. Sci., Appl. Polym. Symp.*, 1977, v. 30, pp. 91–104.

[81] Kennedy, J.P. and Vidal, A., J. *Polym. Sci., Chem. Ed.*, 1975, v. 13(7), pp. 1765–1773.

[82] Kennedy, J.P. and Vidal, A., *J. Polym. Sci., Chem. Ed.*, 1975, v. 13(10), pp. 2269–2276.

[83] Vidal, A. and Kennedy, J.P., *Polym. Lett.*, 1976, v. 14, p. 489.

[84] Farona, M.F. and Kennedy, J.P., *Polym. Bull.*, 1984, v. 11(4), pp. 359-363.

[85] Kennedy, J.P. and Hiza, M., *J. Polym. Sci., Polym. Chem. Ed.*, 1983, v. 21(4), pp. 1033–1044.

[86] Chen, D. and Kennedy, J.P., *Macromol. Sci.*, 1988, v. A25(4), pp. 389–401.

[87] Kennedy, J.P., Feisch K.C., Ir. Preprints, Macro Florence, *IUPAC*, 1980, v. 2, p 162.

[88] Vidal, A., Guyot, A., and Kennedy, J.P., *Polym. Bull.*, 1980, v. 2(5), pp. 315–320.

[89] Vidal, A., Guyot, A., and Kennedy, J.P., *Polym. Bull.*, 1982, v. 6(7), pp. 401–407.

[90] Hrublkin, G., Rottmayer, H.H., Stadermann, D., and Vogel, J., *Acta Polym.*, 1987, v. 38(10), S. 559–561.

[91] Prokofiev, I.K., Nelkenbaum, Yu.Ya., and Sangalov, Yu.A., Zh. *Prikl. Khim.*, 1989, No. 2, pp. 436–438. (Rus)

[92] Nelkenbaum, Yu. and Sangalov, Yu., Macro 87, *31st IUPAC Macromol. Symp.*, Merseburg, 1987, Abstr. Pap. Microsymp. 2, Microsymp. 3, S. 1, s.a. 45.

[93] Macosko, C.W. and Saam, J.C., *Polym. Bull.*, 1987, v. 18(5), pp. 463–471.

[94] Kennedy, J.P. and Hongu, Yu., *Polym. Bull.*, 1985, v. 13(2), pp. 115–121.

[95] Percec, V., Guhaniyog, S., Kennedy, J.P., and Ivan, B., *Polym. Bull.*, 1982, v. 8(1), pp. 25–32.

[96] Kennedy, J.P., *Chemtech.*, 1986, v. 16(1), pp. 694–697.

[97] Miyabayashi, T. and Kennedy, J.P., *J. Appl. Polym. Sci.*, 1986, v. 31(8), pp. 2523–2532.

[98] Takemoto, K., Kondo, M., and Imoto, M., *Memoirs Fac. Eng.*, 1964, v. 6(2), pp. 211–224, Osaka City Univ.

[99] Kennedy, J.P. and Chang, Y.L., *J. Polym. Sci., Polym. Chem. Ed.*, 1971, v. 19(11), pp. 2737–2744.

[100] Kennedy, J.P. and Kurian, J., *J. Polym. Sci.*, 1990, Pt. A, v. 28(13), pp. 3725–3738.

[101] Kazanskii, K.S., Kubisa, P., and Penchek, S., *Uspekhi Khimii*, 1987, v. 46(8), pp. 1360–1386. (Rus)

[102] Kennedy, J.P. and Frisch, K.S., *IUPAC Int. Symp. On Macromolecules*, Preprints, Florence, 1980, v. 2, p. 162.

[103] Kennedy, J.P. and Lo, C.Y., *Amer. Chem. Soc. Polym.* Preprints, 1982, v. 23, p. 99.

[104] Kennedy, J.P. and Hiza, M., *J. Polym. Sci., Polym. Chem. Ed.*, 1983, v. 21(4), pp. 1033–1044.

[105] Liao, T.P. and Kennedy, J.P., *Polym. Bull.*, 1981, v. 6, p. 135.

In: Physical Organic Chemistry: Theory and Practice   ISBN 1-59454-275-9
Eds: A. D'Amore and G. E. Zaikov, pp. 31-79   © 2005 Nova Science Publishers, Inc.

Chapter 3

PROPERTIES AND PRODUCTION OF POLYISOBUTYLENE

K. S. Minsker and Yu. A. Sangalov
Bashkirian State University, 32 Frunze st., Ufa, 450074, Bashkiria
G. E. Zaikov
N.M.Emanuel Institute of Biochemical Physics, 4 Kosygin st., Moscow, Russia

PHYSICAL PROPERTIES

A wide spectrum of oligo- and polyisobutylenes with the polymerization degree $\overline{P}$ from 3 to 4000 and higher is obtained during polymerization of isobutylene.

Oligoisobutylenes with molecular mass up to 2,000 are quite mobile liquids; low-molecular products with molecular mass from 5,000 to 50,000 are viscous liquids; high-molecular polyisobutylenes with molecular mass over 70,000 are elastomers, possessing cold flow and ability to crystallize at stretching. Packing density of crystalline and amorphous polyisobutylene is 0.362–0.342; Densities of crystalline and amorphous phases are 937 kg/m^3 and 912 kg/m^3, respectively. T_{melt} of the crystalline phase is 401 K. If crystallized [1, p. 444], polyisobutylene forms a rhomboidal three-dimensional lattice with sizes (nm): $a = 0.694$, $b = 1.196$, $c = 1.863$ (the identity period is 1.85 nm); C(CH$_3$)$_2$- and CH$_2$-groups are dislocated helicoidally (the spiral 8$_5$), which corresponds to the minimum of the potential energy [2] (Figure 1); the valence angle is 114°; the spiral step is 0.233 nm. An elementary cell contains two molecular chains with the "head-to-tail" type of attachment. Chains are pseudo-hexagonal packed $\left(b = c\sqrt{3}\right)$. Works [3, 4] note formation of the spiral 8$_5$ with the valence angle of 122°. A statistic segment of polyisobutylene is composed of ~4 chains. Rotation of the probe molecule of 4-hydroxy-2,2,6,6-tetramethylpyridine-1-oxyl is determined by conformational mobility of small segments of macromolecules (correlation time $\tau = 1.9 \cdot 10^{-16}$ s at the activation energy $E = 43$ kJ/mol [5]).

As shown with the help of model compounds (2,2,4,4-tetramethylpentane, 2,2,4,4,6,6-hexamethylpentane and 2,2,4,4,6,6,8,8-octamethylnonane), elasticity of polyisobutylene at low temperatures is due to the fact that, despite steric hindrance of macromolecules, a change of angles is possible at the sacrifice of rotation of C–C bonds around each other by up to 30° [4] (Figure 1).

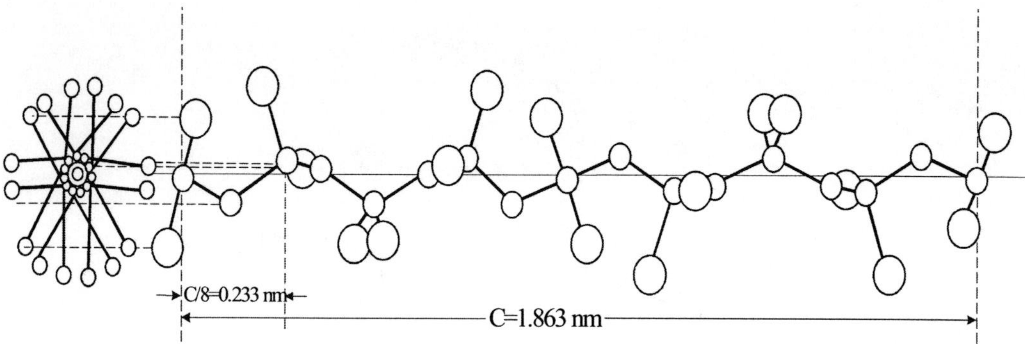

Figure 1. Schematic depiction of chains of crystalline polyisobutylene.

Thermodynamic flexibility of polyisobutylene macromolecules (the ratio of unperturbed sizes of the molecule to those of the chain of free rotation) or the factor of deceleration of the internal rotation is 2.2. The inertness z-average radius $\left\langle S_z^2 \right\rangle^{1/2}$ of polyisobutylene is (7.5 ± 0.5) nm in block and (7.7 ± 0.5) nm in a Θ-solvent [6], which indicates the identity of configurations of macromolecules in block and in the Θ-solvent [7].

Beside other characteristics of thermodynamic flexibility smaller than for other non-polar polymers (polyethylene, polypropylene, etc.), for example, length of the segment equal to 1.83 nm, they indicate quite high flexibility of polyisobutylene in solution. On the other hand, concentrated polyisobutylene possesses a relatively rigid chain, the index of which is a low amplitude of vibration implying fast motion of internuclear vector C–H of the methylene group [8]. This marks great steric hindrances of this group, with which, beside classic segmental motion, appearance of a number of secondary relaxation processes (β-process, rotation of CH_3-group, transition of the liquid-liquid type, etc.) in the local dynamics of polyisobutylene is connected [9].

The main physical properties of oligo- and polyisobutylenes are displayed in Table 1.

The Flory–Huggins parameter (at 293–298 K) of thermodynamic interaction χ_1, which determines changing of the free energy, necessary for transfer of solvent molecule into the polymeric phase, for polyisobutylene–solvent systems equals: polyisobutylene–benzene 0.741; polyisobutylene–toluene — 0.488; polyisobutylene–cyclohexane — 0.390. The upper and the lower critical temperature of mixing for polyisobutylene–benzene systems is 296 K (T_{ucts}) and 433 K (T_{lcts}). Critical temperature of mixing at M → ∞, Θ-temperature, at which the most of thermodynamic properties do not depend on the solvent concentration for the polyisobutylene–benzene system is equal to 297 K (the upper Θ-temperature) [10].

Table 1. Physical Properties of Oligo- and Polyisobutylenes

Index	Oligoisobutylenes, mol. mass < 50,000	Polyisobutylenes, mol. mass = 70,000–225,000
Density, kg/m^3	830–910	910–930
Refraction index, n_d^{298}	1.506	1.5089
Temperature, K, of		
Softening	218–295	373
Glass transition	185–205	199
Heat conductivity coefficient, W/(cm·K)		0.12–0.14
Specific heat capacity, kJ/(kg·K)		1.94
Temperature coefficient of bulk expansion at 298 K, K^{-1}		$6.2 \cdot 10^{-6}$
Strength at elongation, mPa		1.5–6
Relative elongation at break, %		500–1000
Hardiness by Shore		25–35
Elasticity by recoiling, %		10–12
Dielectric permeability at 1kHz	2.0–2.2	2.4–2.9
Specific volumetric electric resistance, TOhm·m	10^{-1}–10^4	10^3–10^4
Dielectric tangent		
at 50 Hz	10^{-3}–10^{-4}	$2 \cdot 10^{-4}$
at 1 kHz	10^{-3}	$5 \cdot 10^{-4}$
Breakdown strength at 353 K, kV/mm	15–20	16–25
Viscosity at 373 K, mm^2/s	75–550	
Dipole moment, D	0	0
Density of cohesion energy, J/cm^3		266
Adhesion, N/cm:		
To copper foil		1.5–3.3
To iron foil		2.45–5.55

For polyisobutylene with molecular mass of 13,000–1,600,000, the value of the second virial coefficient (A_2) decreases with growth of the molecular mass according to correlations: $A_2 = 6.88 \cdot 10^{-3} \cdot M_w^{-0.203}$ (cyclohexane, 296 K) and $A_2 = 3.08 \cdot 10^{-3} \cdot M_w^{-0.188}$ (heptane, 296 K). Diffusion coefficients D depend on the molecular mass of polyisobutylene (298 K): $D = 1.78 \cdot 10^{-4} \cdot M_w^{-0.572}$ (cyclohexane), $D = 4.1 \cdot 10^{-4} \cdot M_w^{-0.560}$ (heptane) and $D = 5.94 \cdot 10^{-5} \cdot M_w^{-0.493}$ (isoamyl ester of isovalerianic acid). The inertness radius R of polyisobutylene changes depending on the molecular mass in accordance with equations (in nm): $R = 1.66 \cdot 10^{-2} \cdot M_w^{-0.58}$ (cyclohexanone), $R = 1.34 \cdot 10^{-2} \cdot M_w^{-0.584}$ (heptane) and $R = 3.08 \cdot 10^{-2} \cdot M_w^{-0.5}$ (isoamyl ester of isovaleric acid).

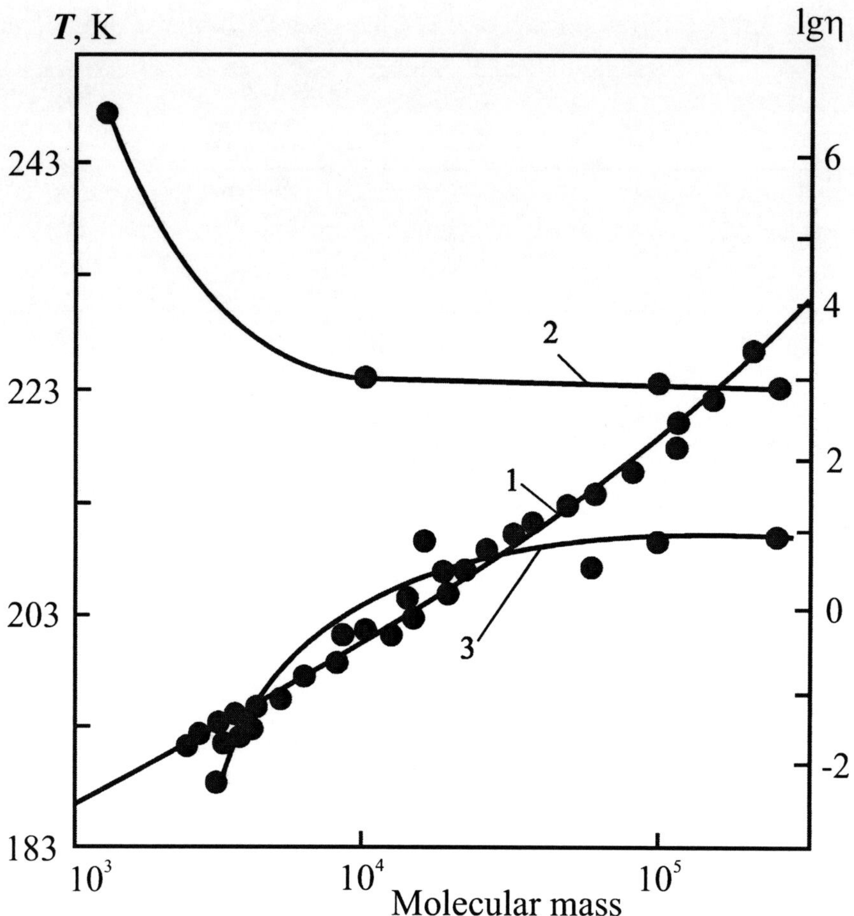

Figure 2. Dependence of glass transition temperature T_g (1), and friability T_{fr} (2), and viscosity η (3) on the molecular mass of polyisobutylene.

Changing of many main properties of polyisobutylene is bound to the molecular mass and width of the MMD of polymers based on isobutylene. The glass transition temperature T_g for oligoisobutylenes is greatly increased with the molecular mass, approaching to a limiting value, typical for polyisobutylene in the $\overline{M}_n$ range of 1,000-2,000 (Figure 2). Consequently, temperature of friability, determined by the strength material, decreases [1, p. 595]. In these cases, the size of a segment is determined as 30 ± 10 monomeric units. Flow temperature T_f also depends on the molecular mass of polyisobutylene, the following correlation being true in the range of large MM values [1, p. 600]:

$$\lg T_f = 3.3 \lg M + \text{const.} \tag{1}$$

Figure 2 (curve 3) displays the dependence of viscosity of polyisobutylene melt on the molecular mass at constant temperature [1, p. 601]. The curve possesses clear bending at molecular masses exceeding the limiting values:

$$\lg \eta = 3.4 \lg \overline{M}_\eta - \lg K \ . \tag{2}$$

The ratio of viscosities $\eta(T)/\eta(T_0)$ does not depend on the molecular mass of polyisobutylene for values greater than 2,000 [1, p. 602]. Some physical properties of the polymer are much more significantly affected by the structure of the main chain. Contrary to the usual polyisobutylene (attachment of monomeric units by the "head-to-tail" structure), a polymer with different disposition of monomeric units ("head-to-head or "tail-to-tail"), obtained under special conditions, is a partially crystalline product (with crystallinity degree of 50%) with $T_{melt} = 460$ K and $T_g = 360$ K [11]. The maximum rate of decomposition of this polymer (about 590 K) is some 70° lower than that for the usual standard polyisobutylene.

Changing of mechanical properties of polyisobutylene in the range of its softening is shown in Figure 3 [1, p. 586]. Dynamic modulus of shear G_d abruptly (by more than by 3.5 degrees) decreases with temperature growth at constant frequencies. On moving to higher frequencies, "modulus-temperature" curves shift to the side of lower temperatures.

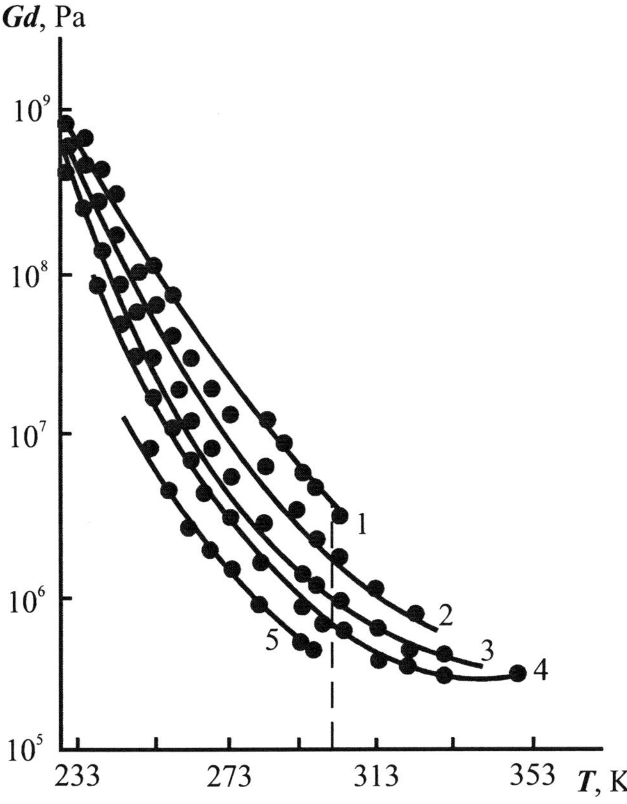

Figure 3. Dependence of dynamic modulus G_d of polyisobutylene in the range of softening with temperature rise for constant frequencies: 1 – 3,000 Hz, 2 – 1,000 Hz, 4 – 100 Hz, 5 – 30 Hz.

Modulus–temperature curves, as well as those for modulus–frequency, may be superimposed on each other as a result of their parallel transposition along the abscissa axis, which defines polyisobutylene as a thermodynamically simple polymer. Under these conditions, the general dependence of the modulus on time and temperature is true. This

                K. S. Minsker, Yu. A. Sangalov and G. E. Zaikov

dependence is set with the help of two functions, which themselves depend on a single variable only: a) total "modulus–frequency" curve at the reference temperature T_0 (298 K); and b) the function of reduced variables ("time–temperature").

The shear modulus at frequency v and temperature T is expressed by a simple dependence relationship [1, p. 589]:

$$G_d(v, T) = G_d = (\alpha v, T_0), \tag{3}$$

where $G_d(v, T)$ is the shear modulus at the reference temperature T_0 (298 K). The dependency $\alpha = \alpha(T, T_0)$ represents a function of reduced time–temperature variables and depends specially on T and T_0. To put it differently, all values of relaxation times increase by one and the same value, aliquot to α, at temperature change from T_0 to T:

$$\tau(T)/\tau(T_0) = \alpha(T, T_0). \tag{4}$$

In all cases, the function of reduced variables can be represented by the expression [1, p. 590]:

$$\lg\alpha(T, T_0) = -C_1(T - T_0)/(C_2 + (T - T_0)). \tag{5}$$

Values of characteristic constants C_1 and C_2 depend on the initial temperature chosen T_0 (Table 2).

Table 2. Parameters of Equation (4) for Polyisobutylene

T_0, K	C_1	C_2, K	T_g, K	$f_g \cdot 10^{-2}$	$\alpha_0 \cdot 10^{-2}$	f_g^2/α
298	8.3	193	202	2.6	2.7	2.5
298	7.2	184	202	2.9	3.3	2.5

Here $f_g = a(C_2 - T_g - T_0)$ is the relative free volume at T_0.

Polyisobutylene is able to dissolve in hydrocarbons, chlorinated hydrocarbons, and esters, and partially in n-butyl alcohol. It swells in animal and vegetable oils and fats. It is insoluble in ethyl and isopropyl alcohols, acetone, methylethylketone, glacial acetic acid.

Solvents form the following sequence according to entropy of mixing: toluene > carbon tetrachloride > iso-octane > cyclohexane. Up to 375 K, polyisobutylene is stable in water and displays low gas permeability $[M^2/(s \cdot N/m^2)]$: $2.0 \cdot 10^{-18}$ for water; $2.2 \cdot 10^{-18}$ for nitrogen; $4.9 \cdot 10^{-18}$ for hydrogen; $9 \cdot 10^{-18}$ for oxygen.

At 298 K polyisobutylene is non-toxic. The polymer, P-200 trademark, in particular, is permitted for application in the food industry. It is fire hazardous. At air the temperature of 268 K and wind velocity of 5 m/s, the rate of burnout of polyisobutylene (P-118 trademark) is 1.25 kg/(m$^2 \cdot$min); the rate of flame spreading on the surface is 0.25 m/min. When it combusts, the material spreads forming a smoke with fly ash.

CHEMICAL PROPERTIES

The structure of polyisobutylene macromolecules (linear polymer usually containing a single end C=C bond per macromolecule) determines its chemical properties. Polyisobutylene is one of the most chemically inert of the known polymers. Polyisobutylene is resistant to dilute and concentrated acids: H_2SO_4 (98.8%), HNO_3 (50%), HCOOH, HCl (37%), CH_3COOH; ammonia, alkali (NaOH – 40%), solutions of salts, hydrogen peroxide. On heating, it is resistant to the effect of HNO_3 and other acids and is carbonized interacting with H_2SO_4. Already at 290 K, polyisobutylene is unstable to liquid and gaseous Cl_2 and Br_2, their water solutions, ozone, and some energetic influences as well.

Heating above 620 K leads to degradation and decomposition of the polymer. The monomer yield in volatile reaction products is about 20–30 mass%, and yield of C_5 hydrocarbons is higher — 65–70 mass% at depolymerization rate of 2.5–3%/min (623 K) [12]. Among the most important products obtained, besides isobutylene, are: di-, tri- and tetramers of isobutylene, formed at intramolecular chain transmission according to the free-radical mechanism.

Under the effect of ionizing radiation, polyisobutylene decays with random scission of macromolecules by the law of accident and release of volatile products [13–15]. In solution, the process proceeds much faster than in the absence of solvents and is accelerated with diminution of the polymer content. The energy of fission of a single C–C bond in the main chain by electrons or γ-rays is (eV): 77 K — 45, 193 K — 27, 293 K — 20, 343 K — 12, and 363 K — 10 [13, p. 108]. Besides isobutylene, great amounts of H_2 and especially CH_4 are formed (obviously, at the sacrifice of detachment of H and framing CH_3-groups). The rate of formation of CH_4 and H_2 is proportional to the depth of polyisobutylene degradation, and the total yield does not depend on temperature; therewith, formation of isobutylene is increased with the radiation dose (and temperature, as well), which is the consequence of its formation as a result of detachment of end groups of a molecule of the initial polymer, as well as the end groups appearing under the radiation effect. Each single act of chain fission matches formation of 1.87 vinylidene groups $\sim CH_2C(CH_3)=CH_2$.

The ESR-spectrum of polyisobutylene, radiated by γ-rays at 77–173 K, represents a doublet appropriate to formation of the free radical $\sim C(CH_3)_2–CH–C(CH_3)_2\sim$, which disappears completely at polymer heating up to room temperature [14, p. 225]. Kinetics of termination of free radicals is described by a reaction equation of the second order with effective values E_a = 77.5 $\pm$ 5.4 kJ/mol and pre-exponential multiplicand $3.98 \cdot 10^{-5}$ cm^3/(spin·s) [16]. Detachment of a H atom from methyl group does not occur, but even if it does take place, no long living radicals are formed. At 123 K, owing to bipolar widening of signals, the doublet can transform into a singlet in the spectrum.

Generally, transformation of polyisobutylene under the effect of γ-radiation is a complex process. Depending on the conditions, either deep decomposition of the polymer is realized, or processes of degradation and cross-linking of macrochains proceed simultaneously. Conjugated diene structures are formed simultaneously at the sacrifice of transformations of macroradicals [15].

Polyisobutylene is relatively stable in UV-light. However, on long-term photo-radiation (by Hg-lamp for several days at 77 K with further heating up to 195 K) five types of free radicals were observed, which indicated fission of both C–C and C–H bonds [15]. Primary products include formation of isolated double and conjugated double bonds, H_2 and CH_4. Subsequent transformations of polyisobutylene give isobutylene, 2,4,4-trimethyl-1-pentene and 2,4,4-trimethylpentane. Compared with radiolysis, lower yields of products suggest a limited role for triplet excited ionic states.

The main products of oxidative degradation of polyisobutylene in the presence of O-plasma are H_2, H_2O, CO and CO_2 [15]. The presence of even small amounts of ozone (3 mass%) in oxygen promotes formation of CO_2 and H_2O; as a result, C=O and COOH-groups appear in the composition of macromolecules of the polymer, and >C=C< double bonds disappear.

Under various mechanochemical interactions, such as flaring, grinding in a ball mill, vibrogrinding, or ultrasonic radiation, polyisobutylene degrades [17, 18]. The process is accompanied by reduction of the molecular mass, but only down to a definite limiting value, which depends on the initial molecular mass and conditions of the polymer degradation. In this case, the chemical nature of the end groups affects stability of the polymer. During the reaction, not only MM, but also MMD of polyisobutylene changes. Mechanochemical transformations of polyisobutylene are of interest from the point of view of a possibility to obtain block-copolymers or modified (from the end of the chain) isobutylene polymers. The stability index of the polymer to mechanical degradation (by decrease of viscosity) is regulated for all trademarks of polymers, used as thickening additives.

For isobutylene polymers, reactions at both unsaturated C=C bonds and saturated C–C bonds in the chain are typical, including specific processes of degradation by the law of accident and the law of end groups. There is also the possibility of cross-linking of polymeric chains, but this occurs to only a limited extent.

Polyisobutylene Reactions at Unsaturated Bonds

Existence of end C=C bonds in polyisobutylene raise the possibility of chemical reactions that are typical for olefin hydrocarbons. However, reactivity of end groups decreases with growth of molecular mass of the polymer, which, in turn, usually limits the possibility to make macromolecules functional [19, 20].

The known reactions of chlorination, phosphorylation, sulfonation, sulfochlorination, ozonation, epoxidation of C=C bonds (see Section 1) are accompanied by changing of not only the chemical structure of the end

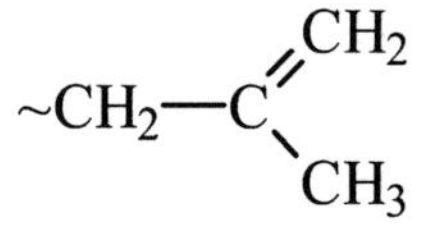

$$\sim CH_2-C\overset{\displaystyle CH_2}{\underset{\displaystyle CH_3}{\Big\langle}} \qquad \sim CH=C(CH_3)_2,$$

groups, but also often of the chemical nature of the main chain, as well as changing of its length.

Various sulfur-containing derivatives may be obtained by adding sulfur dichloride, 5-sulfenylchloride-4-chloro-3-methyltetrahydrophthalic anhydride or 10-sulfenylchloride-9-chlorostearic acid to polyisobutylene [21, 22]. Reactions with SH_2, S_2Cl_2 and other compounds of this kind are known.

Of interest are hydrosilylation reactions [23, 24], in particular, by trichlorosilane or alkylchlorosilanes in the presence of H_2PtCl_6 as the catalyst:

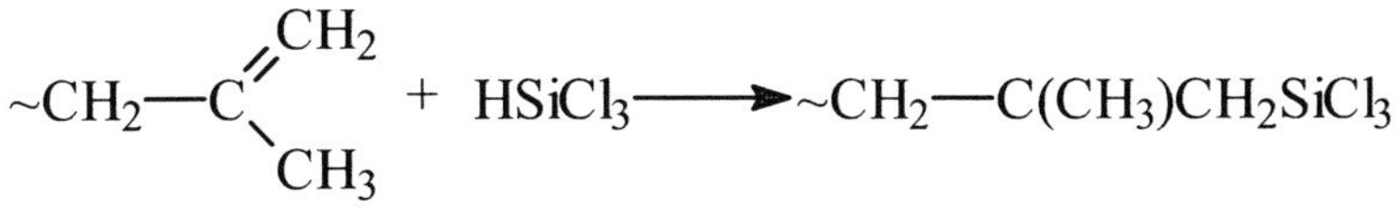

$$\sim CH_2-C\underset{CH_3}{\overset{CH_2}{\diagup}} + HSiCl_3 \longrightarrow \sim CH_2-C(CH_3)CH_2SiCl_3$$

In the presence of electrophilic catalysts, polyisobutylene alkylates arenes up to the highest alkyl-aromatic compounds; as a result, partial depolymerization of polyisobutylene with subsequent alkylation or depolyalkylation of macromolecules can take place [25] with regard to acidity of the catalyst applied.

The main conditions that determine the ratio of reactions of direct (non-degradation) alkylation and partial depolymerization of polymeric chains with subsequent alkylation (depolyalkylation) are temperature, acidity of the catalyst, and ratio of components [26, 27], i.e. the same factors that are important for arene alkylation, conjugated with the chain propagation at isobutylene polymerization.

Processes of non-degradation alkylation and depolyalkylation of polyisobutylene proceed according to Scheme 1 [25, 28]:

Scheme 1

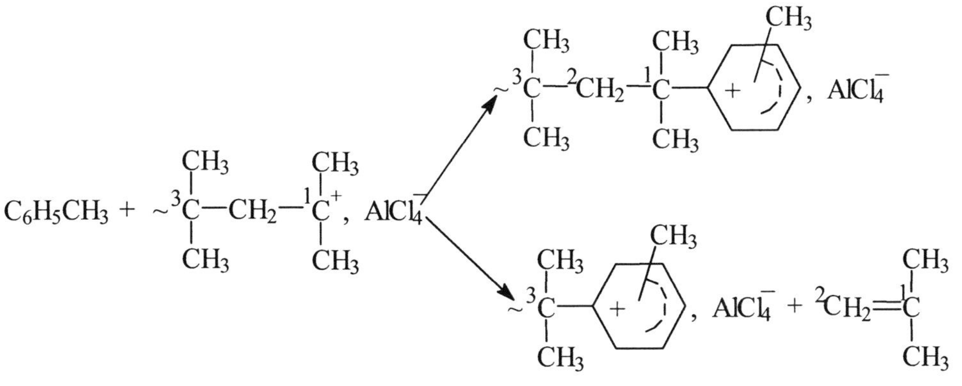

i.e. by the type of reaction:

$$C_4H_8 + C_6H_5CH_3 \xrightarrow{\text{the Gustavsson complex}} C_4H_9C_6H_4CH_3.$$

The role of reaction I (Scheme 1) reduces with growth of temperature and catalyst concentration due to strengthening of interaction between carbocation and counterion. The process of polyisobutylene depolymerization prevails because of interaction of a carbonium center with electrons in the β-position to the C–C bond, which leads to fragmentation of polyisobutylene, i.e. to β-decay of macromolecules by reaction II (Scheme 1). The condition for carbonium ion fragmentation in solution is high stability of resulting end carbocations, for example, at the sacrifice of processes of internal stabilization of conjugation or induction. In the case of catalytic degradation of polyisobutylene, arene, probably, plays the role of external stabilizer of carbonium ions, making fragmentation of the polymer by reaction II (Scheme 1) easier. In this case, a macromolecular fragment of the initial carbocation, formed during degradation of polyisobutylene, enters the reaction of conjugated alkylation with formation of arenonium structures of polyisobutylene–aromatic compounds possessing molecular mass $M < M_0$. In moving from benzene and toluene to more basic arenes, the depth of polyisobutylene degradation decreases, which is connected to growth of steric obstacles to polymer fragmentation. Simultaneously, yielding isobutylene alkylates a new molecule of arene forming *tert*-butyltoluene by reaction III (Scheme 1). Chemical linking of isobutylene by toluene (similar to removal of the monomer from the reaction zone in a different way) decreases the equilibrium concentration of the monomer and causes a decrease of T_{border} at polyisobutylene degradation.

Consequently, arene activates thermal degradation of polyisobutylene, catalyzed by acidic catalysts, assisting alkylation in conjugation with depolymerization or depolyalkylation of polyisobutylene.

Of practical significance are reactions of polyisobutylene with maleic anhydride [29]: hydrogenation (catalyst is mixed Ni–W sulfide, 645 K, 2.4 MPa) and alkylation of phenols in the presence of electrophilic catalysts (300–360 K) [30–36]. Modification of polyisobutylene by phenol in the presence of benzosulfoacid (375–395 K) gives the maximum yield of polyisobutylenyl-phenol of 31 mass% [37]. Significantly better results are obtained at alkylation of phenols by polyisobutylene in the presence of $Me[AlCl_4] \cdot H_2O$ (Me — Li, Na, K), differed by lower (than for $AlCl_3$) relative acidity (mg equiv. OH/g of catalyst): $AlCl_3$ — 5.0; $Li[AlCl_4]$ — 2.2; $Na[AlCl_4]$ — 1.6; $K[AlCl_4]$ — 0.9 [38] (Table 3).

Table 3. Alkylation of phenols by polyisobutylene 3.5 hours, heptane, at phenol: polyisobutylene: catalyst ($Na[AlCl_4]$) molar ratio of 2:1:0.5

Phenol (385 K)	Yield	Aminophenol (355 K)	Yield
Phenol	98	2-Aminophenol	90
2-Methylphenol	95	4-Aminophenol	95
4-*iso*-Propylphenol	96	2-Benzylidenaminophenol	98
3-Methylphenol	91	4-Benzylidenaminophenol	96
2,6-di-*tert*-Butylphenol	85		
Pyrocatechol	94		
Resorcinol	98		
Hydroquinone	91		

The yield of polyisobutylene-phenols decreases with growth of the molecular mass of polyisobutylene.

Functionalization of polyisobutylene by phenols and aminophenols proceeds easily with high yield (2–4 hours; 373–393 K) in the presence of complex salts of $Me[AlCl_4]$ type, where Me is Li, Na, K, the yield of the product being practically independent on nature of alkaline metal in the catalyst. If formation of products of substitution into the aromatic ring is typical for phenols in alkylation by polyisobutylene by the Friedel–Crafts reaction, obtaining N-substituted compounds is typical for amiophenols [31]:

High-molecular phenols and aminophenols are capable of subsequent inter- and intramolecular polymer-analogous transformations, in particular, with 3,4-dibromine-4-methyltetrahydropyrane (dibromide). Synthesis of phenyl esters is performed by interaction of phenols with dibromide in hexane medium in the presence of K_2CO_3 (24 hours).

Allyl esters at 423 K (diethylaniline, 5 hours) are regrouped by the Claisen method into appropriate substituted phenols with quantitative yield [32]:

In the case of application of high-molecular Schiff bases, the process proceeds analogously:

Basing on two-atomic phenols, production of their mono-ethers is possible under the condition of preliminary protection of one of hydroxy groups by the reaction of benzoilation of appropriate phenols [32]:

Existence of two hydroxy groups implies occurrence of two isomers in products of benzoylation: A and B. Condensation of two-atomic phenols, substituted by one HO-group, with dibromide proceeds easily (12 hours). After hydrolysis in a water-alcohol solution of NaOH, protection was removed (373 K, 6 hours), and then isomerization of esters by the Claisen reaction took place (443 K, 5 hours):

Obtaining full esters from high-molecular two-atomic phenols is also possible, if the ratio of reagents is changed:

PIB
OH
OH
H3C
Br
Br
2
O
, 2K2CO3
-4KBr
PIB
CH3
O
O
CH3
O
O
Δ
CH3
PIB
OH
CH3
O
O
OH

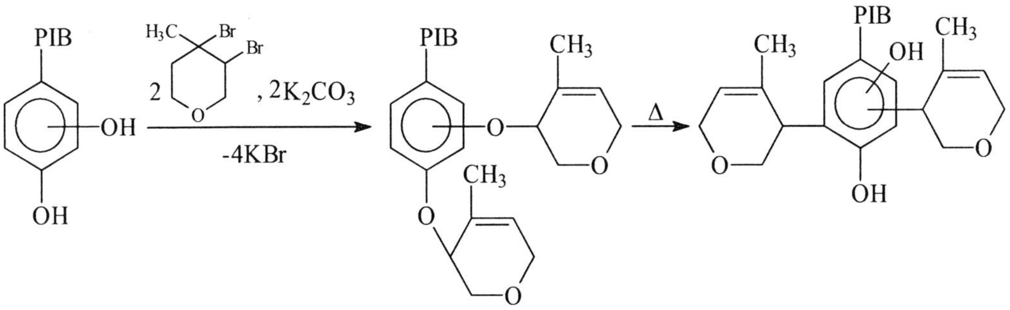

Regrouping by the Claisen reaction in the sequence of allyl–aryl esters based on high-molecular two-atomic phenols (full esters) (443 K, 5 hours) proceeds with lower yields, compared with isomerization of non-full esters of two-atomic phenols (mono-esters), which is stipulated apparently by steric hindrances with regard to introduction of the second pyranyl substituent of non-planar structure into the aromatic ring of the phenol [33].

Intermolecular polymer-analogous transformations of polyisobutylene-phenols with triphenylphosphite enabled the class of high-molecular esters of phosphoric acid to be obtained. Re-etherification of one-atomic polyolephenylphenols under soft conditions (363 K, 1.5–3.0 hours) and with quantitative yield leads to formation of mono-phosphites:

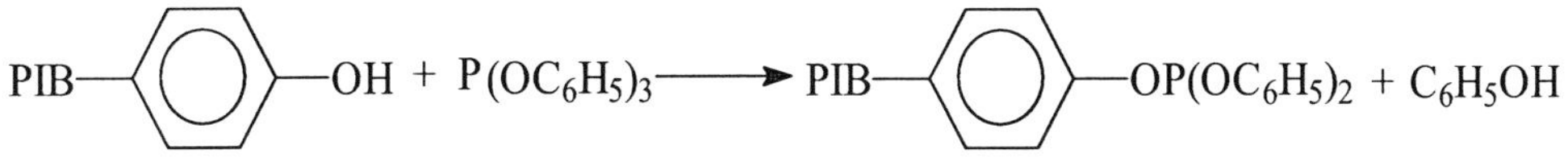

PIB
OH + P(OC6H5)3
PIB
OP(OC6H5)2 + C6H5OH

For two-atomic high-molecular phenols, depending on the ratio of initial substances, a sequence of mono- and di-substituted phosphites can be obtained; Condensed cyclic phosphites are formed that are derivatives of polyisobutylene-pyrocatehols [34]:

P(OC6H5)3
−C6H5OH
OH
1/2 R
OH
OP(OC6H5)2
R
OP(OC6H5)2
R = PIB; 4-OH;
R = PIB; 3-OH
OH
R
OH
R
O
O
POC6H5
R = PIB

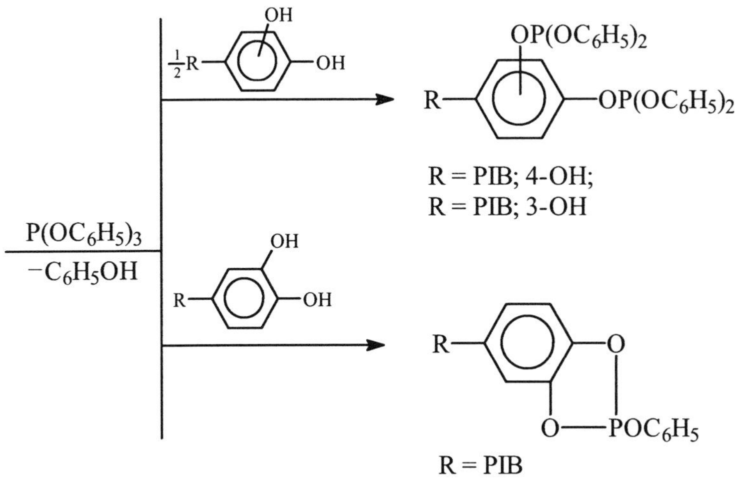

Functionalization of high-molecular phenols by silicon-containing compounds was conducted by re-etherification of framing OC_2H_5-groupings in polyethoxysiloxane molecule (ethylsilicate-40) by polyisobutylene-phenols and polyisobutylenyl-aminophenols, which enables a new class of high-molecular compounds to be formed — polyolephenyl(aminophenoxy)siloxanes [35]:

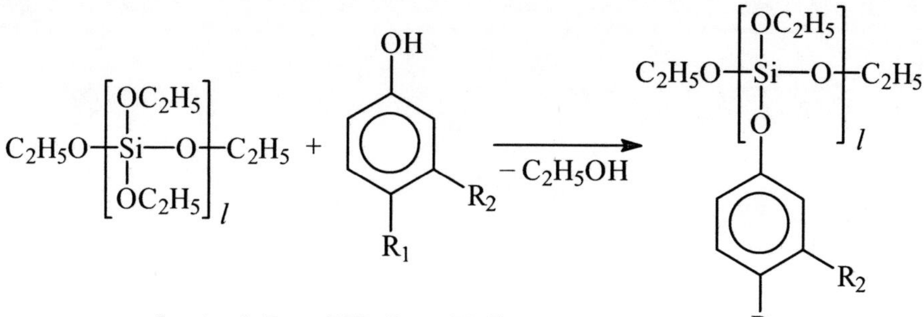

$$l = 4 - 5;\ R_1 = PIB,\ R_2 = H\ (I);$$
$$R_1 = PIB,\ R_2 = OH\ (II);$$
$$R_1 = -NH\text{-}PIB,\ R_2 = H\ (III);$$

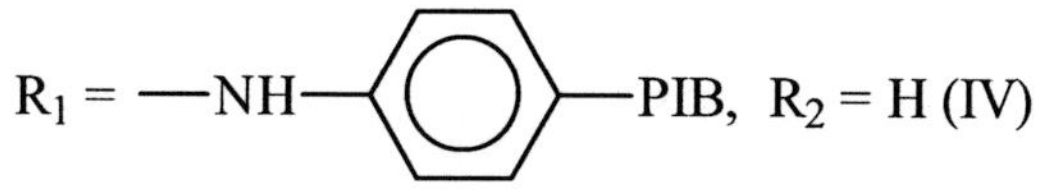

$$R_1 = -NH-\!\!\left\langle\bigcirc\right\rangle\!\!-PIB,\ R_2 = H\ (IV)$$

Existence of two reactive centers (HO- and H_2N-groups) in aminophenols gives the possibility of interaction between aminophenols and 3,4-dibromine-4-methyl tetrahydropyrane in two directions, prescribing the mechanism of the subsequent regrouping by the Claisen reaction and chemical structure of resulting compounds (Scheme 2):

Scheme 2

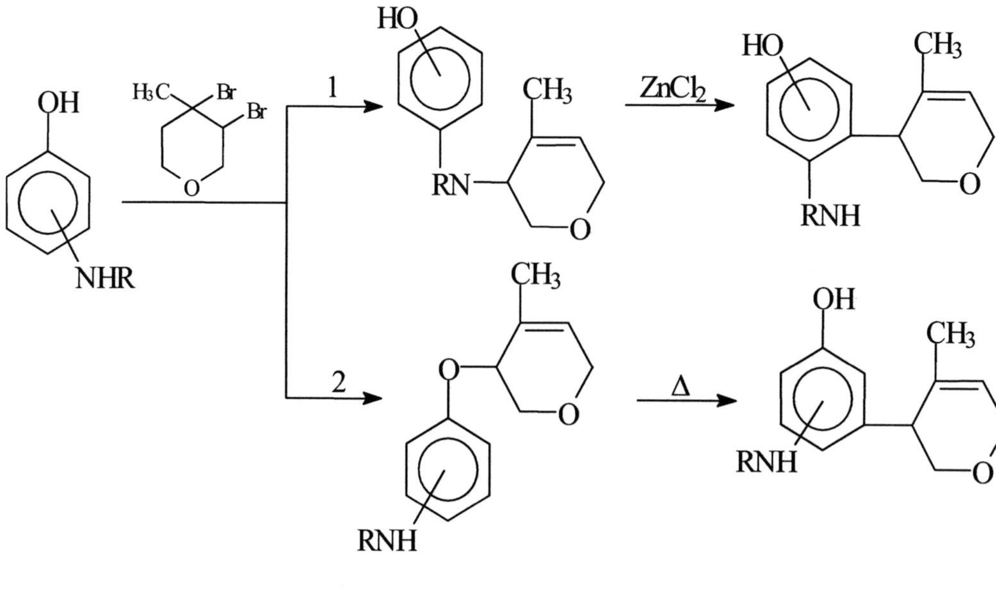

Condensation of aminophenols with dibromide (direction 1 in Scheme 2) proceeds in triethylamine at 360 K, which participates first in formation of a bromine-allyl bond. Subsequently, 5-bromine-4-methyl-5,6-dihydro-2H-pyrane enters the reaction of nucleophilic substitution of mobile allylic bromine with formation of products of the interaction by the amino group [36]:

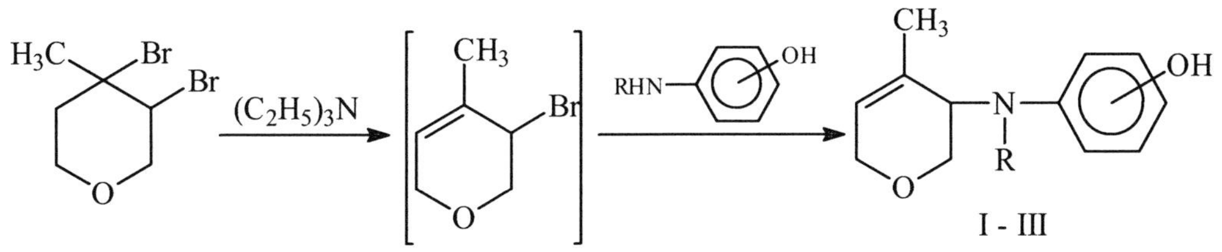

R = H, 2-OH (I); R = H, 4-OH (II); R = C$_6$H$_5$, 4-OH (III).

Alkylation of N-allyl-substituted aminophenols by polyisobutylene (348 K, 3 hours) enables high-molecular analogues to be synthesized, isomerization of which proceeds in nitrobenzene (390 K) only in the presence of catalytic amounts of Lewis acids, ZnCl$_2$, in particular:

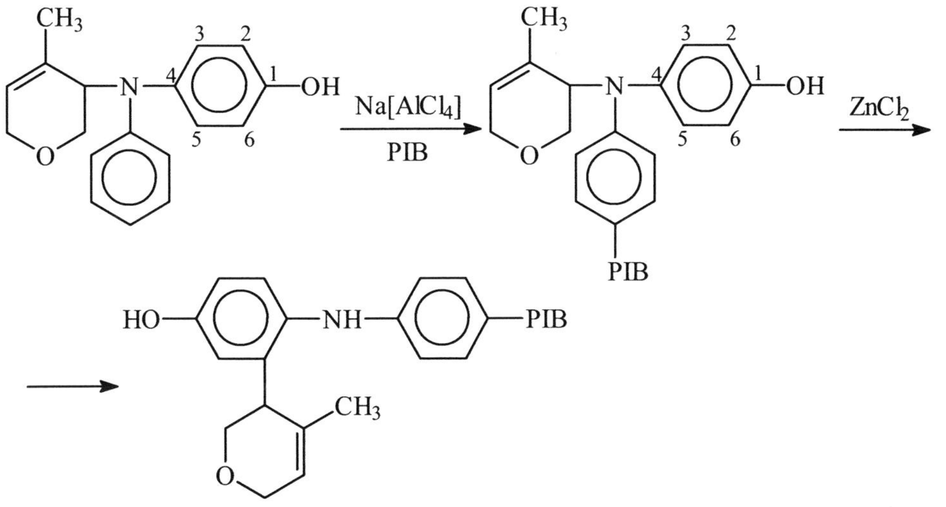

Contrary to esters from the sequence of dihydropyranes, for which thermal regrouping (443 K) into dihydropyranylphenols with quantitative yield in the base medium (N, N-diethylaniline) is typical, the Claisen regrouping of N-allylaminophenols proceeds with a yield of 81-85%.

To realize direction 2 (Scheme 2), i.e. condensation of dibromide with aminophenol by hydroxy group

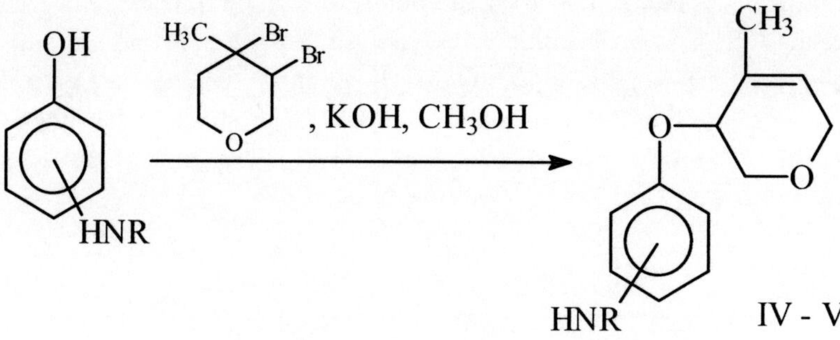

R = H, 2-OH (IV); R = H (V); R = C$_6$H$_5$, 4-OH (VI).

it is necessary to use harder bases (for example, KOH in methanol), which perform the same functions as triethylamine at proceeding of condensation by direction 2. Contrary to the reaction of dibromotetrahydropyrane with aminophenols by H$_2$N-groups, the reaction by HO-group in aminophenol proceeds at lower temperature (353 K, 1.0–1.5 hours). If in the case of dibromide condensation with aminophenols by H$_2$N-group, which leads to formation of practically a single product – [N-(4-methyl-5,6-dihydro-2H-pyranyl-5)amino]phenols, no N-substituted compounds – [N-(4-methyl-5,6-dihydro-2H-pyranyl-5)amino]phenols in the mol/mol ratio of 2:1 of interaction products by HO- and H$_2$N-groups are formed during proceeding of the process by the HO-group together with (4-methyl-5,6-dihydro-2H-pyranyl-5-oxy)- aminobenzenes. Separation of isomeric substances is performed with the help of the column chromatography.

High-molecular analogues were obtained by alkylation of compounds from the class of (dihydropyranyloxy)aminobenzenes of polyisobutylenes (383 K, 3 hours), and their regrouping was conducted subsequently:

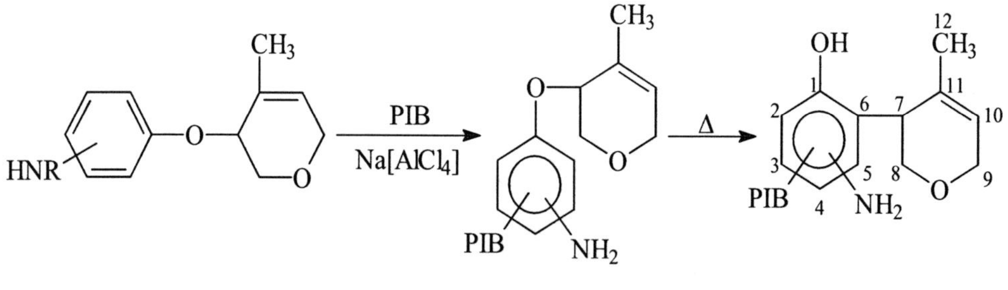

Isomerization of (4-methyl-5,6-dihydro-2H-pyranyl-5-oxy)-polyisobutylenyl-aminobenzenes proceeds by the classic scheme [3,3] of sigmatropic transformations in N,N-diethylalanine at 443 K over 4 hours with practically quantitative yield. By virtue of the fact the that allyl fragment in the molecule of (4-methyl-5,6-dihydro-2H-pyranyl-5-oxy)-aminobenzenes, used in the Claisen isomerization, is allyl-symmetric, the product of regrouping is identical to the initial structure, and inversion cannot be performed.

End functional groups of new high-molecular compounds, synthesized during the past 20 years by modification of polyisobutylene by >C=C< bonds by various low-molecular substances are displayed in Table 4.

Table 4. Modification of Polyisobutylene by End Groups [19, 31-38]

Initial groups in PIB[*]	Reagents	Groups in modified PIB
$tert\text{-}C_4H_9{\sim}CH_2{-}C(CH_3){=}CH_2$	HCl (cat.) C_6H_5R (cat.) H_2 (cat.) (maleic anhydride) $HSiCl_3$ RSCl	$tert\text{-}C_4H_9{\sim}C(CH_3)_2Cl$ $tert\text{-}C_4H_9{\sim}C_6H_4R$ $tert\text{-}C_4H_9{\sim}CH_2CH(CH_3)_2$ $tert\text{-}C_4H_9{\sim}$ ${\sim}CH{=}C(CH_3)CH_2{-}$ (succinic anhydride) $tert\text{-}C_4H_9{\sim}SiCl_3$ $tert\text{-}C_4H_9{\sim}CH_2C(CH_3)(Cl)CH_2SR$
$ClC(CH_3)_2{\sim}CH_2C(CH_3)_2Cl$	$tert\text{-}C_4H_9OK$	$CH_2{=}C(CH_3)CH_2{\sim}CH_2C(CH_3){=}CH_2$
$ClCH_2C(CH_3)_2{\sim}CH_2{-}{-}C(CH_3)_2Cl$ $CH_2{=}C(CH_3)CH_2{\sim}CH_2C(CH_3){=}CH_2$	MeOH $m\text{-}Cl{-}C_6H_4CO_3H$ BH_3; O_2(NaOH)	$ClCH_2C(CH_3)_2{\sim}CH_2C(CH_3){=}CH_2$ (diepoxide) $CH_2{-}CH(CH_3)CH_2{\sim}CH_2CH(CH_3){-}CH_2$ $HOCH_2CH(CH_3)CH_2{\sim}CH_2CH(CH_3)CH_2OH$
$tert\text{-}C_4H_9{\sim}C_6H_4R$	RuO_4	$tert\text{-}C_4H_9{\sim}COOH$
$RC_6H_4{\sim}C_6H_4R$	RuO_4	$HOOC{\sim}COOH$
$HOOC{\sim}COOH$	$SOCl_2$ $LiAlH_4$ HOROH H_2NRNH_2 $CH_2{-}CH{-}R$ (epoxide, O)	$ClOC{\sim}COCl$ $HOCH_2{\sim}CH_2OH$ $HOROC(O){\sim}C(O)OROH$ $H_2NRNHC(O){\sim}C(O)NHRNH_2$ $HOCH_2CH(R)OC(O){\sim}C(O)OCHCH_2OH$ with R ${\sim}C(O)OCH(R)CH_2OH$
$ClC(O){\sim}C(O)Cl$	$HS{-}P({=}S)(OR)_2$	$(RO)_2P({=}S){-}C(O){\sim}C(O){-}P({=}S)(RO)_2$ $ROC(O){\sim}C(O)OR$

Table 4. Modification of Polyisobutylene by End Groups [19, 31-38] (Continued)

Initial groups in PIB	Reagents	Groups in modified PIB
$HOCH_2\sim CH_2OH$	HOROH	
	$ClSi(CH_3)_3$ C_6H_5NCO $ClC(O)NCO$	$(CH_3)_3SiOCH_2\sim CH_2OSi(CH_3)_3$ $C_6H_5NHC(O)O\sim OC(O)NHC_6H_5$ $OCNCOO\sim OCONCO$
[structure: diepoxide]	$ZnBr_2$	[structure: dialdehyde]
[structure: isopropenyl end group]	$Me[AlCl_4]$ (cat.), Me = Li, Na, K, [phenol]	[structure: $\sim C(CH_3)_2$-phenyl-OH]
[structure: isopropenyl end group]	[structure: 2-aminophenol]	[structure: $\sim C(CH_3)_2$-NH-phenol]
[structure: isopropenyl end group]	[structure: salicylaldimine]	[structure: $\sim C(CH_3)_2$-phenol-N=CHC_6H_5]
[structure: $\sim C(CH_3)_2$-phenyl-OH]	K_2CO_3 (cat.), [dibromo pyran]	[structure: aryl ether product]
[structure: $\sim C(CH_3)_2$-catechol]	C_6H_5COCl	[structures: $\sim C(CH_3)_2$-phenol-OCOC_6H_5 isomers]

Table 4. Modification of Polyisobutylene by End Groups [19, 31-38] (Continued)

Initial groups in PIB	Reagents	Groups in modified PIB
$\sim C(CH_3)_2-$ phenol with OH and $N=CHC_6H_5$	H_3C, Br, Br (pyran); K_2CO_3,	$\sim C(CH_3)_2-$ aryl-O-pyranyl (CH_3), $N=CHC_6H_5$
$\sim C(CH_3)_2-$ with OH and $OCOC_6H_5$; $\sim C(CH_3)_2-$ with $OCOC_6H_5$ and OH	K_2CO_3 (cat.), H_3C, Br, Br (pyran)	$\sim C(CH_3)_2-$ with O-pyranyl (CH_3) and $OCOC_6H_5$; $\sim C(CH_3)_2-$ with O-pyranyl (CH_3) and $OCOC_6H_5$
$\sim C(CH_3)_2-$ with O-pyranyl (CH_3) and $OCOC_6H_5$; $\sim C(CH_3)_2-$ with O-pyranyl (CH_3) and $OCOC_6H_5$	$NaOH, C_2H_5OH$	$\sim C(CH_3)_2-$ with O-pyranyl (CH_3) and OH; $\sim C(CH_3)_2-$ with O-pyranyl (CH_3) and OH
$\sim C(CH_3)_2-$ with O-pyranyl (CH_3) and OH; $\sim C(CH_3)_2-$ with O-pyranyl (CH_3) and OH	$t°$	$\sim C(CH_3)_2-$ with H_3C pyranyl, OH, OH; $\sim C(CH_3)_2-$ with OH, OH, pyranyl CH_3

Table 4. Modification of Polyisobutylene by End Groups [19, 31-38] (Continued)

Initial groups in PIB	Reagents	Groups in modified PIB
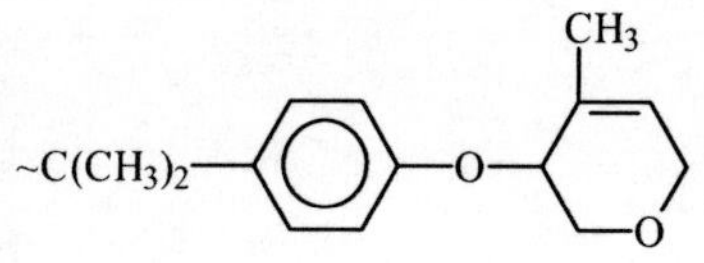 ~C(CH₃)₂—(phenyl)—O—(4-methyl-3,6-dihydro-2H-pyranyl)	t°	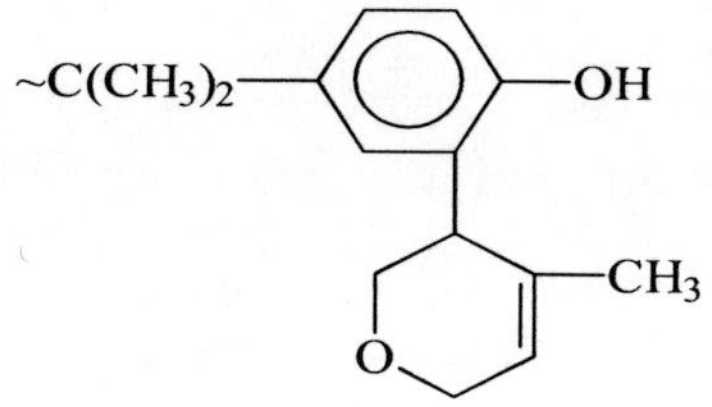 ~C(CH₃)₂—(phenyl with OH)—(4-methyl-3,6-dihydro-2H-pyranyl)
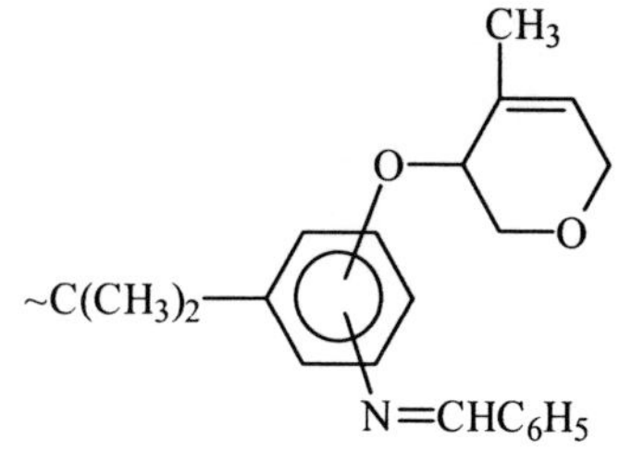 ~C(CH₃)₂—(phenyl, N=CHC₆H₅)—O—(4-methyl-3,6-dihydro-2H-pyranyl)	t°	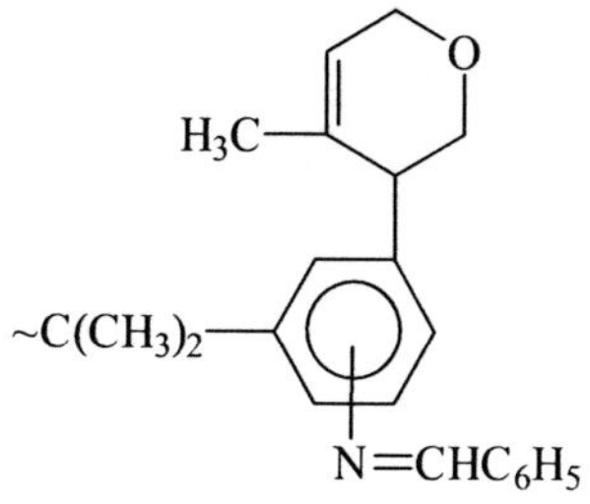 ~C(CH₃)₂—(phenyl, N=CHC₆H₅)—(4-methyl-3,6-dihydro-2H-pyranyl, H₃C)
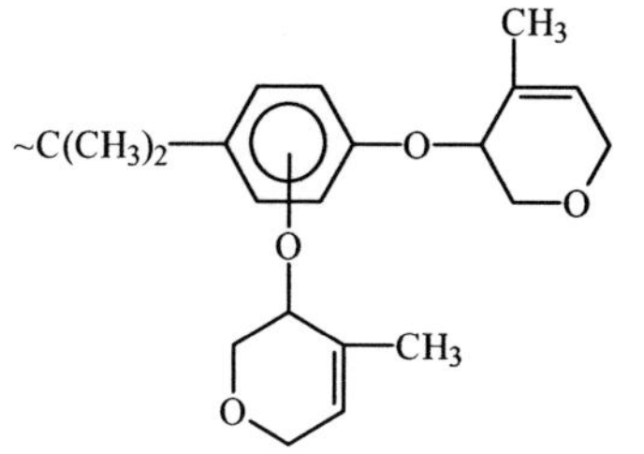 ~C(CH₃)₂—(phenyl, O—4-methyl-3,6-dihydro-2H-pyranyl)—O—(4-methyl-3,6-dihydro-2H-pyranyl)	t°	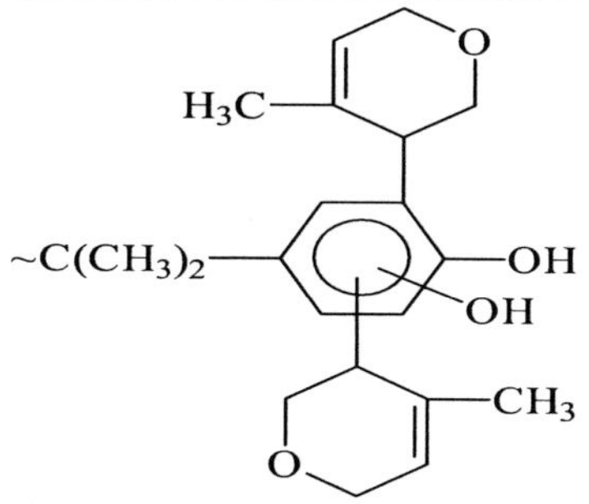 ~C(CH₃)₂—(phenyl with two OH and two 4-methyl-3,6-dihydro-2H-pyranyl groups, H₃C)
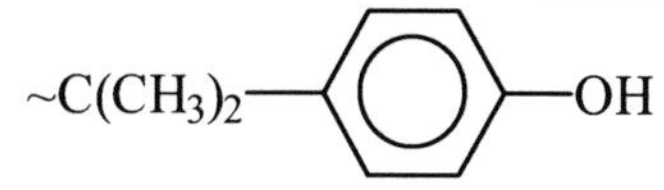 ~C(CH₃)₂—(phenyl)—OH	P(OC₆H₅)₃	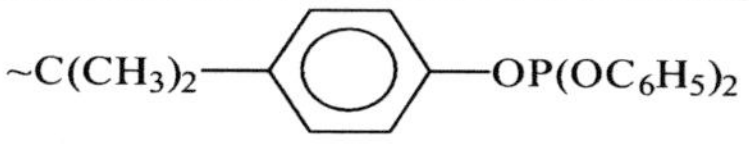 ~C(CH₃)₂—(phenyl)—OP(OC₆H₅)₂
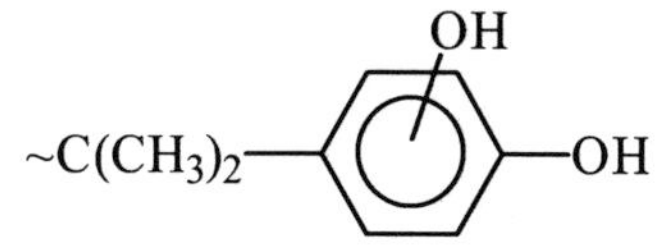 ~C(CH₃)₂—(phenyl, OH)—OH	P(OC₆H₅)₃	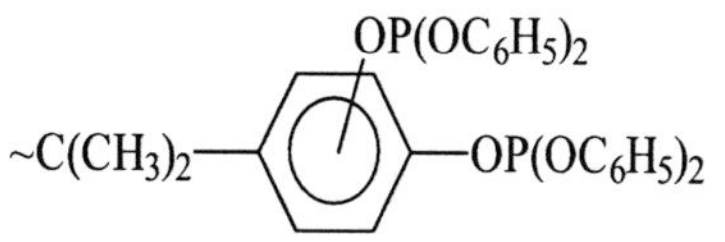 ~C(CH₃)₂—(phenyl, OP(OC₆H₅)₂)—OP(OC₆H₅)₂
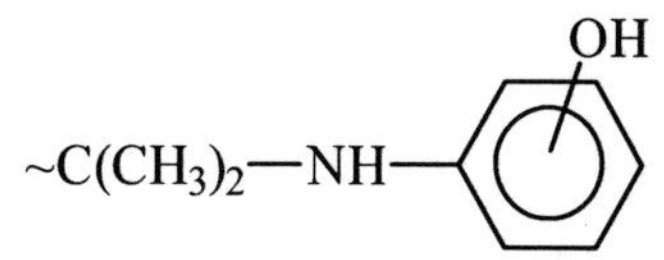 ~C(CH₃)₂—NH—(phenyl)—OH	P(OC₆H₅)₃	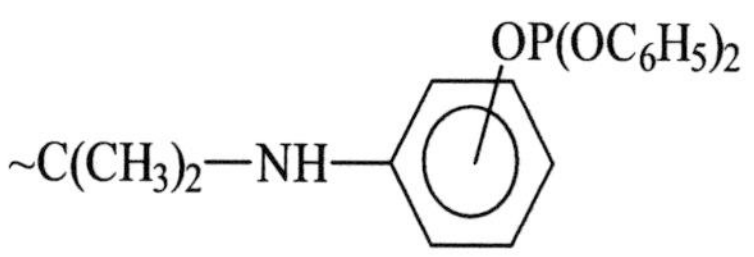 ~C(CH₃)₂—NH—(phenyl)—OP(OC₆H₅)₂
~C(CH₃)₂—(phenyl)—OH	HNO₃	~C(CH₃)₂—(phenyl)—OP(=O)(OC₆H₅)₂

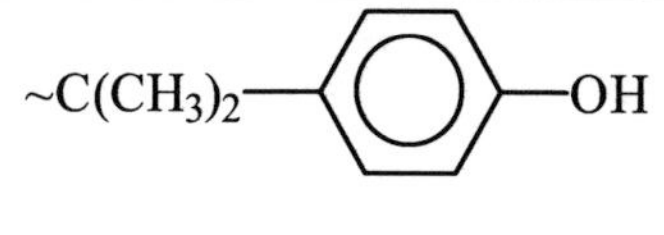

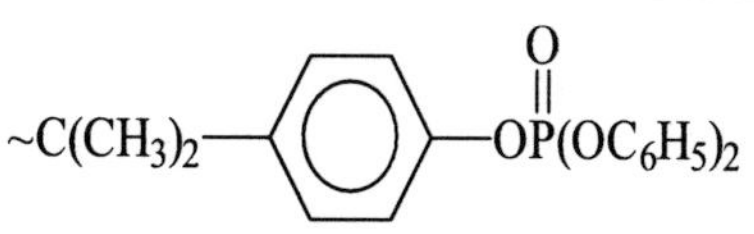

Table 4. Modification of Polyisobutylene by End Groups [19, 31-38] (Continued)

Initial groups in PIB	Reagents	Groups in modified PIB
~C(CH₃)₂—C₆H₃[OP(OC₆H₅)₂][OP(OC₆H₅)₂]	HNO_3	~C(CH₃)₂—C₆H₂[OP(O)(OC₆H₅)₂][OP(O)(OC₆H₅)₂]
~C(CH₃)₂—C₆H₄—OH	C_2H_5—O—[Si(OC₂H₅)₂—O]—C₂, $L = 4\text{-}5$	~C(CH₃)₂—C₆H₄—O—[Si(OC₂H₅)₂—O—C₂H₅]
~C(CH₃)₂—NH—C₆H₄—OH	C_2H_5—O—[Si(OC₂H₅)₂—O]—C₂H₅	~C(CH₃)₂—NH—C₆H₃[O—Si(OC₂H₅)₂—O—C₂H₅]
~C(CH₃)₂—C₆H₄—N(H)—(4-methyl-3,6-dihydro-2H-pyran-3-yl)	$t°$, $ZnCl_2$ (cat.)	~C(CH₃)₂—C₆H₂[OH][NH₂][(4-methyl-3,6-dihydro-2H-pyran-3-yl)]
~C(CH₃)₂—C₆H₄—N[C₆H₄—OH]—(4-methyl-3,6-dihydro-2H-pyran-3-yl)	$t°$, $ZnCl_2$ (cat.)	~C(CH₃)₂—C₆H₄—N(H)—C₆H₃[OH][(4-methyl-3,6-dihydro-2H-pyran-3-yl)]
~C(CH₃)₂—C₆H₃[NH₂]—O—(4-methyl-3,6-dihydro-2H-pyran-3-yl)	$t°$	~C(CH₃)₂—C₆H₂[OH][NH₂][(4-methyl-3,6-dihydro-2H-pyran-3-yl)]
~C(CH₃)₂—C₆H₄—N(H)—C₆H₄—O—(4-methyl-3,6-dihydro-2H-pyran-3-yl)	$t°$	~C(CH₃)₂—C₆H₄—N(H)—C₆H₃[OH][(4-methyl-3,6-dihydro-2H-pyran-3-yl)]

* Symbol ~ marks –(CH₂-C(CH₃)₂-)ₙ group.

Polyisobutylenylphenols display high activity as stabilizers/antioxidants for many polymers and oils.

Aromatic hydrocarbons with mobile atoms of hydrogen or CH-acid, for example, *iso*-propylenebenzenes, under appropriate conditions cause ionic hydrogenation of polyisobutylenes.

The reaction of ionic (electrophilic) hydrogenation of polyisobutylene, representing consecutive electrophilic (H^+) and nucleophilic (H^-) of the olefin attack [39–43],

$$H^+, A^- + R^= + HR' \rightarrow R^{\delta+}, A^{\delta-} + HR' \rightarrow RH + R'^{\delta+}, A^{\delta-}$$

proceeds, if the cation R'^+ is more stable than R^+, and the acidic anion differs by low nucleophility. Selection of appropriate donors of hydride-ions, catalyst and conditions of the reaction determines the direction and efficiency of carbocation transmission compared with competing conjugated processes – depolymerization or depolyalkylation of polymer, alkylation and other reactions. In particular, dimers and trimers of isobutylene in the presence of 1,4-methylisopropylbenzene and acidic catalyst transform with high conversions into appropriate alkanes (I) and alkyl-derivatives of indane (II) by the reaction

Main products of the reaction are compounds I and II independent of the ratio of vinylidene and three-substituted end C=C bonds in macromolecules. These products are formed in approximately equimolar amounts. The presence of polyisobutylene with end indane groups is indicated by the data of UV-spectra of the products (absorption with the maximum at 270 nm, $\lg\varepsilon = 3.01$), identical UV-spectra of 1,1,3,5-tetramethyl-3-neopentylindane.

Table 5 displays experimental data on the reaction of ionic hydrogenation of polyisobutylene in the presence of various hydrogenating agents.

Table 5. Ionic Hydrogenation of Isobutylene Dimer under the Effect of $Al_2Cl_6 \cdot HCl \cdot 3C_6H_3(CH_3)_3$ and Various Isopropylbenzenes (the Ratio Polyisobutylene:Hydrogenating Agent = 1:5, 298 K, 1 Hour)

Hydrogenating agent	Yield of products of hydrogenation, mass%, for catalyst:substrate ratios		
	0.01	0.03	0.1
Isopropylbenzene	0	—	33
1,4-Methylisopropylbenzene	94	82	62
1,3,5-Triisopropylbenzene	0	92	95
Poly-*n*-isopropylstyrene	0	0	5

All studied isopropylbenzenes participate in the reaction of ionic hydrogenation and form a sequence: poly-n-isopropylstyrene < isopropylbenzene < 1,4-methylisopropylbenzene $\approx$ 1,3,5-triisopropylbenzene.

The reaction of ionic hydrogenation of polyisobutylene is often accompanied by secondary processes, first of all, alkylation of isopropylbenzene (polyisobutylene). Direction and depth of the reaction are also significantly affected by the nature of applied catalyst (Table 6).

Table 6. Activity of Various Catalytic Systems in the Reaction of Ionic Hydrogenation of Polyisobutylene with MM = 112 in the Presence of 1,4-methylisopropylbenzene (Polyisobutylene:$C_3H_7C_6H_4CH_3$ = 1:4, 298 K, 1 Hour)

Catalytic system	Ratio of catalyst and polyiso-butylene, mol/mol	Conver-sion of polyiso-butylene, mass%	Yield of hydroge-nation products, mass%	Yield of C_8H_{18}, mol/mol	Dispro-portiona-tion degree of $C_3H_7C_6$-H_4CH_3, %
$HCl \cdot 2AlCl_3 \cdot 3C_6H_3(CH_3)_3$	0.01	93	94	25 (50)	0.8
$HCl \cdot 2AlCl_3 \cdot 3C_6H_5CH_3$	0.01	95	80	20 (40)	1.0
$C_2H_5AlCl_2 \cdot nH_2O^*$	1.0	100	70	0.5 (500)	0.7
$C_2H_5AlCl_2 \cdot nH_2O^*$	0.1	60	—	3.0 (3000)	—
$(C_2H_5)_2AlCl \cdot tert\text{-}C_4H_9Cl$	1.0	100	80	0.3 (—)	—
$C_2H_5AlCl_2 - KU\text{-}2\text{-}8^\dagger$	0.2	100	95	2.5 (0.5)	—
$HF^\ddagger$	8.0	33	34	— (0.1)	—

* Contains 10^{-3} mole of H_2O per mole of Al.

† Prepared under conditions shown in Table 1; medium — 1,4-methyliso- propylbenzene.

‡ According to [43].

Complex catalysts exceed HF in activity and selectivity of action (yield of hydrogenation products). The low efficiency of a heterogeneous catalyst (at the highest selectivity) calculated for SO_3H-group is explained by inertness of an inactivated cation. At the same time, efficiency of the $C_2H_5AlCl_2 \cdot H_2O$ aquacomplex points out that initiation of the reaction can proceed at low concentration of a proton-donor.

Kinetically, ionic hydrogenation of polyisobutylene under the effect of homogeneous catalysts represents a comparatively fast reactions (Figure 4, curves 3 and 4).

General scheme of the process is shown in Scheme 3 (see below).

The main direction can be separated in Scheme 3 – the reaction of ionic hydrogenation (1, 2, 3), and transformations of carbocations of the first (I) and the second (II) generation connected to it (Ia and IIa — main products; Ib and IIb — secondary products).

A specific feature of the reaction of ionic hydrogenation of polyisobutylene under the effect of the acidic catalyst–isopropylbenzenes system is attachment of initial olefin to carbocation II (Scheme 3, reaction 5). Notice that, independent of the type of diisobutylene isomer, 1,1,3,5-tetramethyl-3-neopentylindane is formed only instead of two supposed indane structures Ia. Probably, 2,4,4-trimethylpentene-2 is isomerized into the associated α-isomer. This way of forming indane structures differs from the traditional one, which supposes cyclic dimerization of 1-methyl-4- isopropylbenzene, intermediately formed from cation II.

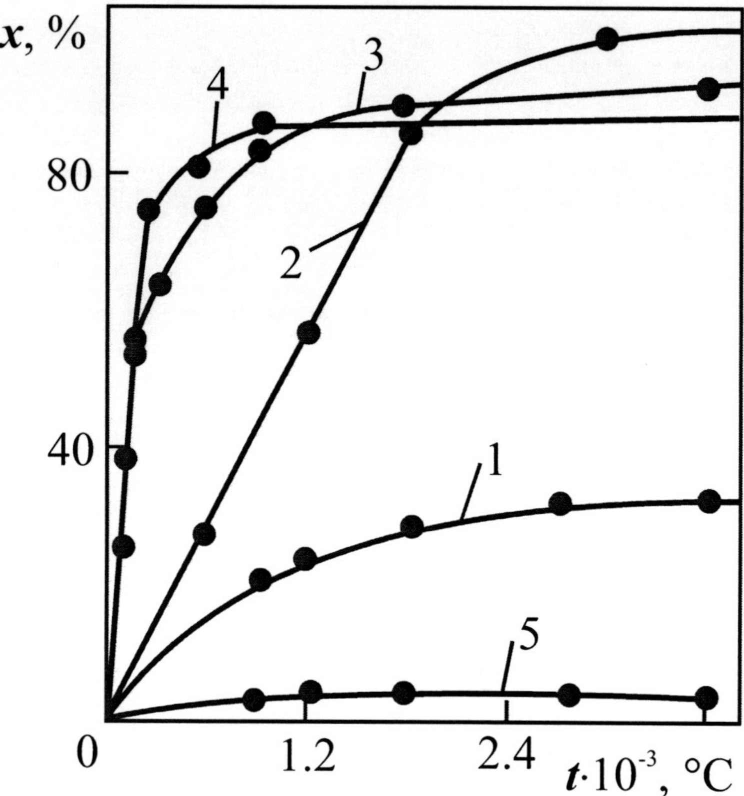

Figure 4. Dependence of diisobutylene conversion on hydrogenation time (298 K) in the presence of isopropylbenzene (1), 1,4-methylisopropylbenzene (2, 3), triisopropylbenzene (4, 5) and catalysts $C_2H_5AlCl_2$ — KU-2-8 (1, 2, 5) and $3C_6H_3(CH_3)_3 \cdot HCl \cdot Al_2Cl_6$ (3, 4) under different molar ratio of substrate:hydrogenating agent:catalyst — 1:5:0.2 (1, 2, 5); 1:5:0.1 (3, 4).

Formation of 2,2-ditolylpropane during hydrogenation (reaction 6, Scheme 3) is the result of conjugated alkylation of toluene by carbocation II of the second generation. Reactions 4-6 in Scheme 3 possess regeneration of the catalyst, which predetermines the catalytic character of ionic hydrogenation.

When the molecular mass of polyisobutylene grows, an excess of hydrogenating agent must be taken: approximately 40-fold for polyisobutylene with MM = 450 and 50-fold for polyisobutylene with MM = 900.

Necessity of a great excess of hydrogenating agent is apparently bound to low concentration of C=C double bonds and its decrease with growth of the molecular mass of polyisobutylene. That is why there is a "critical" molecular mass below which samples of polyisobutylene barely participate in the reaction of ionic hydrogenation (1,500 for heterogeneous catalyst).

Scheme 3

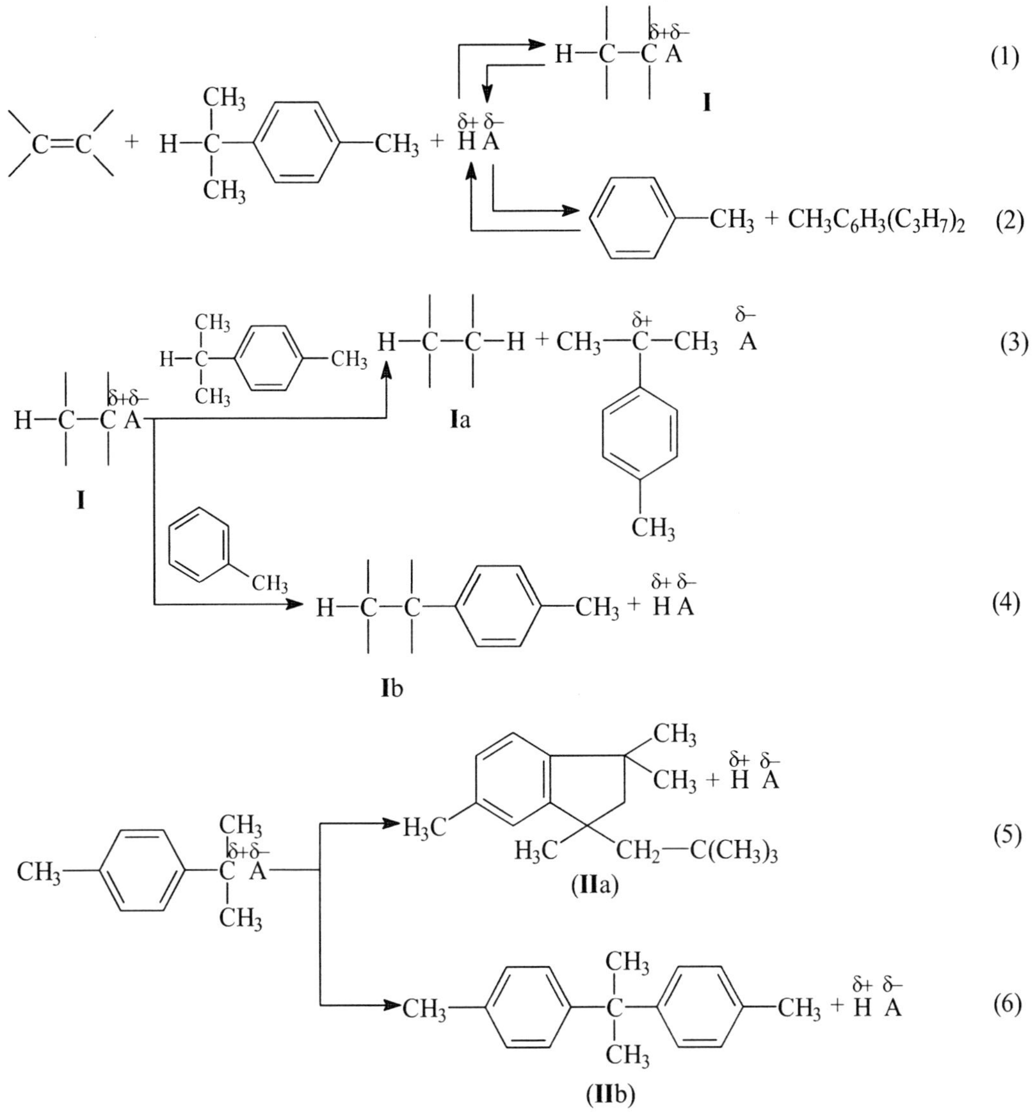

$$(CH_3)_2C=C(CH_3)_2 + H-C(CH_3)_2-C_6H_4-CH_3 + \overset{\delta+}{H}\,\overset{\delta-}{A}$$

(1) $\longrightarrow H-C-\overset{\delta+\ \delta-}{C\ A}$ (**I**)

(2) $\longrightarrow C_6H_5-CH_3 + CH_3C_6H_3(C_3H_7)_2$

$$H-C-\overset{\delta+\delta-}{C\ A}\quad (\mathbf{I})$$

(3) $\longrightarrow H-\overset{CH_3}{\underset{CH_3}{C}}-C_6H_4-CH_3 + H-C-C-H + CH_3-\overset{\delta+}{C}(C_6H_4CH_3)-CH_3\ \overset{\delta-}{A}$ (**Ia**)

(4) $\longrightarrow C_6H_5-CH_3 \to H-C-C-C_6H_4-CH_3 + \overset{\delta+}{H}\,\overset{\delta-}{A}$ (**Ib**)

$$CH_3-C_6H_4-\overset{CH_3}{\underset{CH_3}{\overset{\delta+\delta-}{C\ A}}}$$

(5) $\longrightarrow$ (**IIa**) $+ \overset{\delta+}{H}\,\overset{\delta-}{A}$

(6) $\longrightarrow CH_3-C_6H_4-C(CH_3)_2-C_6H_4-CH_3 + \overset{\delta+}{H}\,\overset{\delta-}{A}$ (**IIb**)

Rise of the ratio $C_3H_7C_6H_4CH_3$:polyisobutylene is insufficient for total high conversion of iso-olefin, through it increases yield of iso-octane and selectivity of the main reaction 3, Scheme 3, at the sacrifice of limiting the secondary reaction 4 (Table 7).

Of interest is one more feature of ionic hydrogenation of polyisobutylene in the presence of a solid catalyst — the absence of temperature influence in the range of, at least, 240–335 K on depth and rate of the process. Because of the technological advantages of this system (for example, ease of release of reaction products), the process of non-degradative ionic hydrogenation of polyisobutylene seems to have considerable potential as a route to obtain synthetic oils, resistant to oxidation by O_2 and O_3.

**Table 7. Influence of the Ratio Hydrogenating Agent:
Polyisobutylene (C_8H_{16}) on the Depth and Selectivity of Ionic Hydrogenation
(298 K, 1 Hour, C_8H_{16} – $1.6 \cdot 10^{-3}$ kg, Catalyst - $3 \cdot 10^{-3}$ kg)**

$C_3H_7C_6H_4CH_3$:polyiso-butylene	Reaction products, mass%			Conversion of polyiso-butylene, mass%	Selectivity by yield of C_8H_{18}, %
	C_8H_{18}	Ib[†]	IIb[†]		
0:1[*]	0	100	0	100	0
0.5:1	15.5	40.6	44.1	100	23
1:1	18.7	19	62.3	76	27
2:1	40.7	15.7	43.6	100	73
4:1	40.0	0	62	100	100

[*] In the absence of *n*-isopropyltoluene, molar ratio of polyisobutylene and toluene is 1:1.
[†] Products of toluene alkylation by carbocations I and II (reactions 4 and 6).

Polyisobutylene Reactions Involving the Main Chain

Substitution reactions proceeding by the free-radical mechanism are typical for polyisobutylene, and are quite often accompanied by simultaneous degradation of macromolecules.

Chlorine derivatives of polyisobutylene are formed at chlorination in solution of dichloroethane, chloroform or tetrachlorocarbon with catalysis by UV-light, I_2 or $FeCl_2$. Chlorination is favored by the presence of HNO_3 or O_3. The process proceeds faster if liquid chlorine is used [44, p. 238]. Attendant sulfochlorides are formed at simultaneous effect of Cl_2 and SO_2 on polyisobutylene in CCl_4 at UV-radiation.

If polyisobutylene is reacted with oxalyl chloride or phosgene, chloroformyl groups can be introduced into the composition of the molecules [14, p. 239]:

$$\sim C(CH_3)_2 - CH_2 \sim \xrightarrow{Cl\bullet} \sim C(CH_3)_2 - \overset{\bullet}{C}H \sim + HCl$$

$$\sim C(CH_3)_2 - \overset{\bullet}{C}H\sim + Cl\overset{O}{\overset{\|}{C}}-\overset{O}{\overset{\|}{C}}Cl \xrightarrow{-\overset{\bullet}{C}OCl} \sim (CH_3)_2 - C(H)C\overset{\nearrow O}{\underset{\searrow Cl}{}}$$

Reaction of polyisobutylene phosphorylation in the presence of O_2 hardly proceeds and generally leads to polymer degradation:

$$\sim C(CH_3)_2-\overset{\displaystyle\curvearrowright}{\underset{\displaystyle\centerdot}{C}}(H)PCl_3 \xrightarrow[-PCl_3]{O_2} \sim C(CH_3)_2-\overset{\displaystyle\curvearrowright}{C}(H)P(O)Cl_2 + POCl_3$$

$$PCl_3 \nearrow\kern-1.2em\swarrow$$

$$\sim C(CH_3)_2-\overset{\centerdot}{C}H\sim$$

$$O_2 \searrow$$

$$\sim C(CH_3)_2-\overset{\displaystyle\centerdot O-O}{\underset{\displaystyle|}{C}}H\sim \xrightarrow{PCl_3} \sim C(CH_3)_2-C(H)OP(O)Cl_2 + Cl\centerdot$$

Diminution of the molecular mass of the polymer and accumulation of oxygen-containing functional groups (acids, ketones, peroxides, etc.) is observed at ozone effect on solutions of polyisobutylene, also including temperatures below 273 K.

$$\underset{\displaystyle\sim CH_2 + O_3}{\overset{\displaystyle\sim C(CH_3)_2}{|}} \longrightarrow \underset{\displaystyle\sim CH-OO\centerdot}{\overset{\displaystyle\sim C(CH_3)_2}{|}}$$

$$\nearrow \quad \sim C\overset{O}{\underset{OH}{\diagup\!\!\diagdown}} + \sim\overset{\centerdot}{C}(CH_3)_2$$

$$\rightarrow \quad \sim C\overset{O}{\underset{C(CH_3)_2\sim}{\diagup\!\!\diagdown}} + \overset{\centerdot}{O}H$$

$$\searrow \quad \underset{\displaystyle\sim CH\sim + R\centerdot}{\overset{\displaystyle OOH}{|}}$$

The rate constant for O_3 reaction with polyisobutylene (293 K, CCl_4) equals $K_{8.1} = 0.012$ l/(mol·s) with 0.05 breaks per single act of the reaction [45]. Obviously, in the sequence of polymers with a saturated hydrocarbon chain, polyisobutylene is the most resistant to ozone. The method of ozonation is suitable for application to obtain grafted copolymers, for example, with styrene [46, p. 122]. Hydroperoxide groups, stable under temperatures up to 380–390 K, are formed at oxidation of polyisobutylene (340–365 K).

A noticeable decrease of molecular mass of polyisobutylene is observed in the presence of peroxides; in particular, at application of dicumyl peroxide at 413 K (1 hour) it was 40-fold decreased and no formation of isobutylene was observed [14, p. 223]. Degradation of polyisobutylene also proceeded in the presence of triazene (378 K, $RN_2NHC_6H_5$). The reactivities of free radicals $R^\centerdot$ form a sequence: $\overset{\centerdot}{C}H_3 > \overset{\centerdot}{C}_2H_5 > \overset{\centerdot}{C}_3H_7 > tert-\overset{\centerdot}{C}_4H_9$.

Of importance from scientific and practical point of view are intramolecular transformations of polyisobutylene, proceeding by the law of accident and end groups, as methods for regulating decrease of the molecular mass of polyisobutylene and MMD width (degradation of macromolecules) and obtaining pure monomer (depolymerization) [12, 47–62].

Table 8. Yield and Composition of Gas Products at Thermal and Catalytic Degradation of Polyisobutylene
$(C_{cat} = 5 \cdot 10^{-4}$ mol/g of Polyisobutylene)

Molecular mass of polyiso-butylene, M_n^0	Catalyst	Reaction temperature, K	Yield of gas products, mass%	Composition of gas products of the reaction, mass %						
				C_1-C_3	iso-C_4H_{10}	n-C_4H_{10}	α-C_4H_8	β-C_4H_8	iso-C_4H_8	iso-C_4H_8 + iso-C_4H_{10}
1100	Without catalyst	653	75.9	1.9	2.1	—	0.1	0.9	16	18.1
1100	AlCl$_3$	643	62.7	1.5	2.1	—	Traces	1.0	19.1	21.2
1100	Li[AlCl$_4$]	583	91.8	1.1	23.1	Traces	0.3	3.1	72.3	95.5
1100	Li[AlCl$_4$]	543	61.1	1.1	40.2	0.1	0.2	3.0	55.4	95.6
1100	Li[AlCl$_4$]	523	52.4	1.4	45.6	0.2	0.2	3.7	49.5	95.1
1100	Li[AlCl$_4$]	503	38.2	1.5	53.1	0.1	0.2	3.0	42.4	95.5
1100	Li[AlCl$_4$]	473	21.5	0.4	63.3	0.2	Traces	1.8	31.4	97.7
575000	Li[AlCl$_4$]	583	10.6	9.5	38.7	1.4	4.7	5.3	40.4	79.1
1100	Na[AlCl$_4$]	623	62.8	1.3	3.9	Traces	0.3	1.7	92.5	96.4
1100	Na[AlCl$_4$]	603	55.0	1.2	6.4	0.1	0.3	2.2	89.8	96.2
1100	Na[AlCl$_4$]	583	47.9	1.1	7.8	Traces	0.3	2.0	88.8	96.6
1100	Na[AlCl$_4$]	553	35.4	0.6	13.5	0.1	0.3	6.6	78.8	92.3
1100	Na[AlCl$_4$]	523	21.7	0.8	32.7	0.2	0.2	4.5	61.5	94.2
575000	Na[AlCl$_4$]	623	53.3	6.2	22.4	1.4	9.4	2.5	58.0	80.4
575000	Na[AlCl$_4$]	603	19.6	16.1	34.3	2.9	0.8	11.0	33.9	68.3
575000	Na[AlCl$_4$]	583	10.8	10.9	48.3	1.7	0.6	10.5	26.7	75.0
1100	K[AlCl$_4$]	603	16.3	2.3	5.0	0.1	4.0	1.6	87.1	92.1
1100	K[AlCl$_4$]	583	6.3	2.1	5.6	Traces	0.5	6.2	85.7	91.2
1100	MgCl$_2$AlCl$_3$	573	93.8	1.4	24.9	0.1	0.1	0.1	73.4	98.3

Note: Hydrogen traces were identified in gas products.

The thermodynamic parameters of polyisobutylene make electrophilic depolymerization easy (Q_{polym} = 53 kJ/mol; T_{border} = 323 K, $E_{depolym}$ = 217 kJ/mol). However, the temperature at which significant polymer decay begins is about 610 K, and thereafter the monomer yield does not exceed 20 mass%.

As can be seen, temperature range of intensive decay of polyisobutylene independent of the molecular mass lies much higher than the limiting temperature of isobutylene polymerization, which is determined by decay of polyisobutylene by a free-radical, but not carbocationic mechanism. The yield and distribution of low-molecular products of free-radical degradation of polyisobutylene are determined by two typical reactions: degradation of macromolecules by free-radical mechanism and radical substitution reactions (chain transmission to the polymer), which generally leads to formation of hydrocarbons of isomeric structure (Table 8) from C_1 to C_{30} [52]. In this case, the idea of selectivity of the process becomes meaningless.

Selectivity and efficiency of the process grows sharply in catalytic degradation of polyisobutylene in the presence of electrophilic catalysts, especially with lower acidity compared with $AlCl_2$ (Table 5.8) [12, 53, 56, 58, 60–62].

The mechanisms of thermal and thermocatalytic degradation of poly-isobutylene are different. Thermal degradation of polymeric products proceeds pre-dominantly by the free-radical mechanism [57] and, at polyisobutylene degradation in the presence of electrophilic catalysts (H_2O as cocatalyst), by the ionic mechanism [58–60], possibly, with participation of free-radical processes [63].

In the general case, depolymerization of polyisobutylene, dependent on the nature of the catalyst, is initiated by both the law of accident (so that the possibility of non-catalytic rupture of internal C–C bonds in macromolecules must be taken into account) and the law of end groups (Scheme 4):

Scheme 4

a) Polyisobutylene degradation by the law of accident:

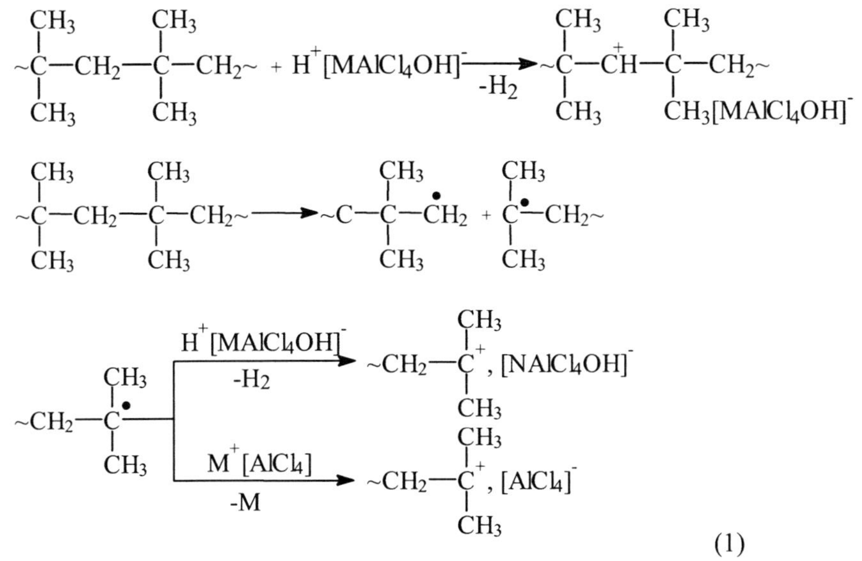

$$\sim\underset{\underset{CH_3}{|}}{\overset{\overset{CH_3}{|}}{C}}-CH_2-\underset{\underset{CH_3}{|}}{\overset{\overset{CH_3}{|}}{C}}-CH_2\sim \;+\; H^+[MAlCl_4OH]^- \xrightarrow[-H_2]{} \sim\underset{\underset{CH_3}{|}}{\overset{\overset{CH_3}{|}}{C}}-\overset{+}{CH}-\underset{\underset{CH_3[MAlCl_4OH]^-}{|}}{\overset{\overset{CH_3}{|}}{C}}-CH_2\sim$$

$$\sim\underset{\underset{CH_3}{|}}{\overset{\overset{CH_3}{|}}{C}}-CH_2-\underset{\underset{CH_3}{|}}{\overset{\overset{CH_3}{|}}{C}}-CH_2\sim \longrightarrow \sim\underset{\underset{CH_3}{|}}{\overset{\overset{CH_3}{|}}{C}}-\overset{\bullet}{C}H_2 \;+\; \overset{\bullet}{\underset{\underset{CH_3}{|}}{C}}-CH_2\sim$$

$$\sim CH_2-\overset{\bullet}{\underset{\underset{CH_3}{|}}{C}} \begin{cases} \xrightarrow[-H_2]{H^+[MAlCl_4OH]^-} \sim CH_2-\underset{\underset{CH_3}{|}}{\overset{\overset{CH_3}{|}}{\overset{+}{C}}},\,[NAlCl_4OH]^- \\[2em] \xrightarrow[-M]{M^+[AlCl_4]^-} \sim CH_2-\underset{\underset{CH_3}{|}}{\overset{\overset{CH_3}{|}}{\overset{+}{C}}},\,[AlCl_4]^- \end{cases}$$

$$\tag{1}$$

and

b) Polyisobutylene decomposition by the law of end groups:

$$\sim CH_2-\underset{\underset{CH_3}{|}}{\overset{\overset{CH_3}{|}}{C}}-CH=\underset{\underset{CH_3}{|}}{\overset{\overset{CH_3}{|}}{C}} + H^+, [MAlCl_4OH]^- \longrightarrow CH_2-\underset{\underset{CH_3}{|}}{\overset{\overset{CH_3}{|}}{C}}-CH_2-\overset{\overset{CH_3}{|}}{\underset{\underset{CH_3}{|}}{\overset{+}{C}}}, [MAlCl_4OH]^- \quad (2)$$

Macrocations formed at temperature higher than T_{border} are β-degraded with isobutylene elimination.

$$\sim \left[CH_2-\underset{\underset{CH_3}{|}}{\overset{\overset{CH_3}{|}}{C}} \right]_n -CH_2-\overset{\overset{CH_3}{|}}{\underset{\underset{CH_3}{|}}{\overset{+}{C}}}, [MAlCl_4OH]^- \longrightarrow \sim \left[CH_2-\underset{\underset{CH_3}{|}}{\overset{\overset{CH_3}{|}}{C}} \right]_{n-m} -CH_2-\overset{\overset{CH_3}{|}}{\underset{\underset{CH_3}{|}}{\overset{+}{C}}}, [MAlCl_4OH]^- + m\cdot iso\text{-}C_4H_8 \quad (3)$$

Diminution of E_a of the process from 217 ± 9 kJ/mol at thermal degradation of polyisobutylene down to 93 ± 4 kJ/mol (reaction 1, Scheme 4) and 70 ± 9 kJ/mol (reaction 2, Scheme 4), and simultaneous elongation of kinetic Z (zip) degradation of the chain from 4 to 20-35 monomeric units at introduction of electrophilic catalysts indicate substitution of the free-radical mechanism of polyisobutylene degradation by a predominantly cationic one. Consequently, the monomer yield grows noticeably. According to these data, seeming noncorrespondence, displayed in the literature, between low yield of the monomer at thermal degradation of polyisobutylene and low values of polymerization heats, and, as a consequence, the upper limit of formation temperature of the polymer, is explained by difference in the mechanism of polyisobutylene degradation (cationic or free-radical mechanism at thermal effect). The possibility of providing degradation of the polymer according to the scheme of cationic depolymerization eliminates this noncorrespondence.

The chemical nature of the catalyst determines the rate and sequence of the reactions of polyisobutylene degradation.

In the general case [53, 58, 60], the following kinetic scheme of the process proceeding can be assumed:

$$\left.\begin{aligned}
P &\xrightarrow{W_P} 2R^\bullet \\
P + cat &\xrightarrow{k} 2R^\bullet \\
R^\bullet + cat &\xrightarrow{k'_{in}} R^+ \\
P + cat &\xrightarrow{k'_{in}} R^+
\end{aligned}\right\} \quad \text{(the law of accident)} \qquad (1)$$

$$P + cat \xrightarrow{k'_{in}} R^+ \quad \text{(the law of end groups)}$$

$$R_n^+ \xrightarrow{k'_{depol}} R_{n-z}^+ + {}_zM \, ; \quad R_n^\bullet \xrightarrow{k_{depol}} R_{n-z}^\bullet + {}_zM$$

$$R^\bullet \xrightarrow{k_{\deg r}} R_n^\bullet + R_m^\bullet \ ; \ R^+ \xrightarrow{k'_{\deg r}} R_n^+ + R_m^+ \qquad (2)$$

$$R^\bullet + P \xrightarrow{k_{pol}} RH + P^\bullet \ ; \ R^+ + P \xrightarrow{k_{pol}} RH + P^+$$

$$R^+ \xrightarrow{k_{ter}} P$$

Here k'_{in}, k_{in}, k'_{depol}, k_{depol}, k'_{degr}, k_{degr}, k'_{pol}, k_{pol}, k_{ter} are rate constants of initiation, depolymerization and degradation of macromolecules under the effect of ionic and radical centers and chain termination, respectively.

For degradation of polyisobutylene by the law of accident, the change of concentration of macrocations $[R]^+$ is described by the equation:

$$\frac{d\left|R^+\right|}{dt} = k_{in}[Cat]\left[R^\bullet\right] - k_{ter}\left[R^+\right]. \qquad (3)$$

In the stationary state

$$\left[R^+\right] = \frac{k_{in}[Cat]\left|R^\bullet\right|}{k_{ter}},$$

where

$$\left[R^\bullet\right] = -\frac{k_{in}[Cat]}{2k_1} + \sqrt{\left(\frac{k_{in}[Cat]}{2k_1}\right)^2 + \frac{W_p}{k_1}} = \frac{k_{in}[Cat]}{2k_1}\sqrt{\frac{4k_1\left(W_p + k_1[Cat]\right)}{\left(k_{in}[Cat]\right)^2} + 1} - 1, \qquad (4)$$

with $W_p = W_T + k_1[Cat]$, the rate of formation of free radicals as a result of thermal termination of the polymeric chain under the effect of catalyst.

In the case of application of small amounts of catalyst, when only a small part of the free radicals formed are transformed into polymeric carbonium ions, the expression for the rate of statistical decay of the polymer by the law of accident is the following:

$$V_1 = \left(k'_{depol} + k'_{degr}\right)\frac{k_{in}[Cat]}{k_{ter}}\sqrt{\frac{W_T + k_1[Cat]}{k_1}} + \left(k_{depol} + k_{degr}\right)\sqrt{\frac{W_T + k_1[Cat]}{k_1}}. \qquad (5)$$

In this case, the limiting stage of the process is the reaction of catalyst interaction with free radical.

If small amounts of catalyst are used (all resulting radicals transform into carbocations), the rate of the process is described by the equation:

$$V_1 = \left(k'_{depol} + k'_{degr}\right)\frac{W_T + 2k_1[Cat]}{k_{ter}}. \tag{6}$$

For degradation of polyisobutylene by the law of end groups, the equation for the rate of depolymerization becomes:

$$V_2 = \frac{k_p}{\left[R^+\right]} = \frac{\left(k'_{depol} + k'_{degr}\right)k'_{in}[Cat]C_p}{k_{ter}} = \frac{\left(k'_{depol} + k'_{degr}\right)k'_{in}A_0[Cat]}{k_{ter}\overline{P}_n}. \tag{7}$$

Here, $C_p = A_0/\overline{P}_n$ is the content of double bonds in polyisobutylene; A_0 is the polymer concentration; $\overline{P}_n$ is the degree of polymerization of polyisobutylene. Taking into account that concentration of active centers $[R^+]$ is determined by the correlation: $\alpha\left[R^+\right]/dt = k'_{in}[Cat]C_n - k'_{ter}\left[R^+\right]$, in the general case, the rate of polyisobutylene depolymerization is described by the equation:

$$\begin{aligned}
V = V_1 + V_2 &= \left(k'_{depol} + k'_{degr}\right)\frac{k_{in}[Cat]}{k_{ter}}\sqrt{\frac{W_T + k_1[Cat]}{k_1}} + \\
&+ \left(k_{depol} + k_{degr}\right)\sqrt{\frac{W_T + k_1[Cat]}{k_1}} + \frac{\left(k'_{depol} + k'_{degr}\right)k'_{in}A_0[Cat]}{k'_{ter}\overline{P}_n}
\end{aligned} \tag{8}$$

The equation (8) allows for reactions of degradation and depolymerization of polyisobutylene on ionic and radical active centers occurring simultaneously, but does not allow for any judgements about the reasons for the difference in selectivity of the process by the monomer yield in the presence of a catalyst. Selectivity of action of catalysts can be quantitatively estimated by value Z – the length of the kinetic chain, which characterizes the amount of monomer yielded on average for each single scission of the macromolecule:

$$Z = \frac{k'_{depol}\left[R^+\right] + k_{degr}\left[R^+\right] + k_{degr}\left[R^\bullet\right]}{W_T + k'[Cat] + k_p\left[R^+\right] + k'_p\left[R^+\right]}. \tag{9}$$

The value of Z and the gross rate of degradation of polymeric products grow with the catalyst activity in the sequence $LiAlCl_4 > NaAlCl_4 > KAlCl_4$ (Table 9).

Length of the kinetic chain Z of polyisobutylene depolymerization decreases with growth of the reaction temperature, whereas the rate constant of polyisobutylene polymerization increases. This is connected with growth of contribution of the thermal component, which increases probability of statistic (by the law of accident) degradation of macromolecules, as well as with reactions of chain transmission and termination.

Table 9. Gross Parameters of Thermocatalytic Degradation of Polyisobutylene
($C_{cat} = 1\cdot10^{-3}$ mol/g, Reaction Time is 1 Hour)

Catalyst	Reaction temperature, K	$Z \pm 10\%$	$k'_{in} \cdot 10^7 + 5\%$, s^{-1}	$E_a \pm 5\%$, kJ/mol
LiAlCl$_4$	583	10	0.6	100
	603	8	1.7	
	623	6	4.3	
NaAlCl$_4$	583	10	0.3	105
	603	7	0.8	
	623	6	2.4	
KAlCl$_4$	583	4	0.4	140
	603	3	1.1	
	623	3	3.7	

The process of electrophilic catalytic degradation of polyolefins can be described by a general equation taking into account contribution of the reactions by both the law of accident and the law of end groups [56]:

$$-\frac{dP}{dt} = z_{eg}k_{in-eg}C_{eb} + k_{in-acc}z_{acc}P. \tag{10}$$

Here P is diminution of the polymer (in parts of initial content); k_{in-acc} and k_{in-eg} are rate constants of initiation of polymer degradation by the law of accident and the law of end groups, respectively; z_{acc} and z_{eg} are the length of kinetic chain at degradation by the law of accident and the law of end groups, respectively; C_{eb} is the concentration of end double bonds. Supposing that end groups able to initiate depolymerization are consumed during degradation by the exponential law $C_{eb} = C_{eb0}\exp(-k_{in-eg}t)$ (C_{eb0} is the initial content of end double bonds), which is typical for polyisobutylene, in particular, we obtain that

$$P = \exp\left(-k_{in-acc}z_{acc}t\right) + \frac{k_{in-eg}z_{eg}C_{eb0}}{k_{in-eg} - k_{in-acc}z_{acc}}\left[\exp\left(-k_{in-eg}t\right) - \exp\left(-k_{in-acc}z_{acc}t\right)\right]. \tag{11}$$

Numerical values of k_{in-eg}, z_{eg}, k_{in-acc} and z_{acc} can be estimated for the initial stages of the process, when $P \ll 1$, $t \ll 1/k_{in-eg}$ and $t \ll 1/k_{in-acc}z_{acc}$ (experimentally, this is possible, if the process is carried out under low-temperature conditions — 200–300°C, when the rate of burnout of chains is rather low). Then diminution of the polymer (in parts of initial content) is described by the equation:

$$P = 1 - k_{in-acc}z_{acc}t - k_{in-eg}z_{eg}tC_{eb0}. \tag{12}$$

Changes in the molecular mass of polyisobutylene can be determined with regard to accumulation of chain breaks. The amount of macromolecules is equal to $n = n_0 + k_{in-acc}t$, and their molecular mass is determined by the expression:

$$\overline{M}_n = \frac{P}{n} \cong \frac{1 - k_{in-acc}z_{acc}t - k_{in-eg}z_{eg}C_{eb0}t}{n_0\left(1 + k_{in-acc}T/n_0\right)} \cong$$

$$\cong \overline{M}_n^0\left(1 - k_{in-acc}z_{acc}t - k_{in-eg}z_{eg}C_{eb0}t - \frac{k_{in-acc}t}{n_0}\right)$$

(13)

Here $\overline{M}_n^0 = 1/n_0 = 1/C_{eb0}$, which corresponds to the presence of one end double bond in each macromolecule.

The expression for relative degree of polymerization of polyisobutylene during degradation is [53]:

$$\frac{d\left(\overline{P}_n/d\overline{P}_n^0\right)}{d\alpha} = -\frac{k_{in-acc}\overline{P}_n^0 + k_{in-eg}z_{eg}/\overline{P}_n^0}{k_{in-acc}z_{acc} + k_{in-eg}z_{eg}/\overline{P}_n^0} .$$

(14)

To calculate kinetic parameters, it is necessary to consider a number of limiting cases for various ranges of the molecular mass of the polymer:

1. At low contribution of reactions proceeding by the law of end groups, i.e. at $k_{in}z_{eg}/\overline{P}_n^0 \ll k_{in-acc}z_{acc} \ll k_{in-acc}\overline{P}_n^0$, the expression (14) for reduced relative degree of polymerization will be the following:

$$\frac{d\left(\overline{P}_n/d\overline{P}_n^0\right)}{d\alpha} = -\frac{\overline{P}_n^0}{z_{acc}} .$$

(15)

2. At small contribution of reactions proceeding by the law of accident, when degradation of polyisobutylene is determined by the law of end groups (low values of), it is obtained:

$$\frac{d\alpha}{dt} \approx \frac{k_{in-acc}z_{eg}}{\overline{P}_n^0} .$$

(16)

In this process, the value of $\overline{P}_n^0$ for initial, most probable MMD $\left(\overline{P}_w/\overline{P}_n \approx 2\right)$ is practically unchanged, i.e. $d\left(\overline{P}_n/\overline{P}_n^0\right)/d\alpha \approx 0$. This conforms to the experimental data (Figure 5, curves 1–3). In this case, application of equation (5.14) is meaningless, because "burnout" of macromolecules at the sacrifice of their carbocationic depolymerization was not taken into account in its deduction, which was taken into consideration in [53].

3. When the main role in changing of the number of chains is played by the law of accident, and in depolymerization then the law of end groups, i.e. at $k_{in-acc}\overline{P}_n^0 \gg k_{in}z_{eg}/\overline{P}_n^0 \gg k_{in-acc}z_{acc}$:

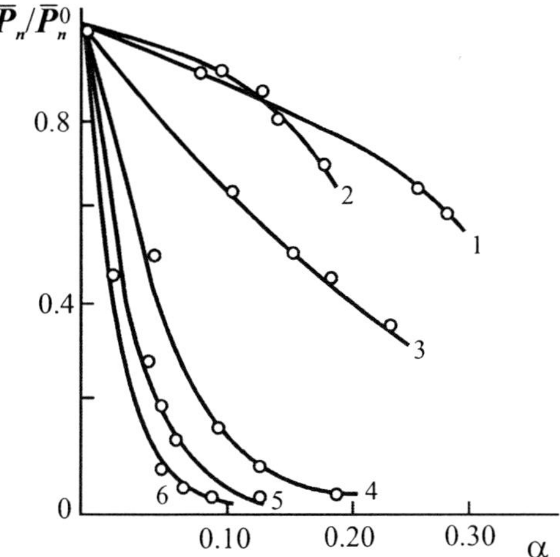

Figure 5. Dependence of the relative degree of polymerization on the depth of transformation (α) during thermocatalytic degradation of polyiso-butylene (PIB) at 573 K in the presence of H[NaAlCl$_4$OH]. Catalyst concentration is $1 \cdot 10^{-3}$ mol/g of polyisobutylene. The initial degree of polyisobutylene polymerization: 1 – 17; 2 – 20; 3 – 49; 4 – 360; 5 – 1790; 6 – 3570.

$$\frac{d\left(\overline{P}_n/\overline{P}_n^0\right)}{dt} = -\frac{k_{in-acc}\left(\overline{P}_n^0\right)^2}{k_{in-eg}z_{eg}}. \tag{17}$$

The three above-mentioned variants correspond to three parts on curves of the dependence of calculated values $d\left(\overline{P}_n/\overline{P}_n^0\right)/d\alpha$ on $\overline{P}_n^0$, obtained by tangents of appropriate experimental curves (Figure 5). Each of these three possible variants – b, c, or a, respectively, is consequently realized with increase in the degree of polymerization (Figure 6).

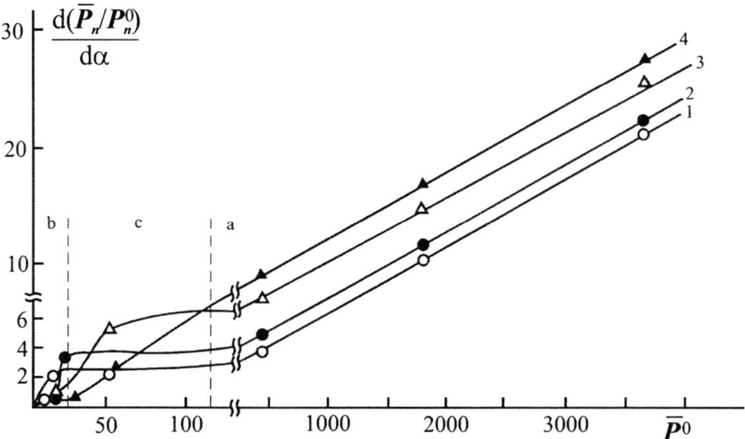

Figure 6. Change of $d\left(\overline{P}_n/\overline{P}_n^0\right)/d\alpha$ in degradation of polyisobutylene on the catalyst NaAlCl$_4$ under various temperatures. Temperature, K: 1 – 473, 2 – 493, 3 – 513, 4 – 533.

Values of z_{acc} and $k_{in\text{-}acc}/k_{in\text{-}eg}z_{eg}$, respectively, can be calculated by tangents for parts a) and c) of the curve of $d\left(\overline{P}_n/\overline{P}_n^0\right)/d\alpha$ against $\overline{P}_n^0$ (Figure 6). The value of $k_{in\text{-}acc}$ can be calculated taking into account that the main contribution to reduction of molecular mass of the polymer is made by the law of accident. This correlates with the dependence $d\left(\overline{P}_n/\overline{P}_n^0\right)/dt = -k_{in-acc}\overline{P}_n^0$, which allows estimation of the $k_{in\text{-}acc}$ value.

Values of z_{acc} and $k_{in\text{-}acc}$, obtained at treatment of experimental data, can be used for estimation of numerical values of $k_{in\text{-}eg}$ and z_{eg} by equation (11). The values of kinetic parameters obtained — the rate constant of initiation and length of kinetic chain of depolymerization of polymers (for polyisobutylene degradation in the presence of catalysts with moderate relative acidity) — are shown in Table 10. They indicate the prevalence of reactions of initiation of catalytic degradation by the law of end groups under low-temperature conditions. The ratio of $k_{in\text{-}eg}$ and $k_{in\text{-}acc}$ decreases with rise of temperature of degradation of polyolefins, and at 260°C their values become commensurable.

Table 10. Main Kinetic Parameters of Catalytic Degradation of Polyisobutylene

Catalyst	T, K	$K_{in\text{-}acc}\cdot10^6$, s^{-1}	$k_{in\text{-}eg}\cdot10^6$, s^{-1}	Z_{acc}	z_{eg}	$k_{in\text{-}eg}/k_{in\text{-}acc}$
H[Na(C$_2$H$_5$)AlCl$_3$OH]	473	0.45	3.50	8	8	7.8
	493	0.48	2.50	7	16	5.2
	513	1.12	2.10	4	26	1.9
	533	1.50	1.90	3	37	1.3
H[NaAlCl$_4$OH]	473	0.53	3.50	7	4	6.6
	493	0.62	3.30	5	9	5.3
	513	0.80	2.90	5	15	3.6
	533	1.20	2.50	4	20	2.1
H[Mg(C$_2$H$_5$)AlCl$_4$OH]	473	0.36	4.90	12	10	13.6
	493	0.45	3.10	11	21	6.9
	513	0.69	2.40	9	33	3.5
	533	0.89	2.20	8	42	2.5
2MgCl$_2$·AlCl$_3$·H$_2$O	473	0.54	4.60	9	7	8.5
	493	0.77	2.70	7	17	3.5
	513	1.07	2.20	6	28	2.1
	533	1.17	2.00	6	38	1.7

Note: Conditions of the reaction conduction: [Cat] = $1\cdot10^{-3}$ mol/g of polyisobutylene. Error of k determination is ±3%, z – ±5%.

Independent of chemical structure of metal-complex catalysts, numerical values of both $k_{in\text{-}eg}$ and $k_{in\text{-}acc}$ hardly changes (within the range of the experimental error). This confirms (with regard to the total yield of gas products, as well as to content of *iso*-C$_4$-hydrocarbons) the conclusion about connection between activity and selectivity of electrophilic catalysts, in particular, of complexes based on metal chlorides with their indexes of relative acidity (Table 11). The ratio of $k_{in\text{-}eg}$ and $k_{in\text{-}acc}$ (Table 10) indicates that initiation of depolymerization of macromolecules at relatively low temperatures of degradation of polymeric products by the law of end groups is preferred. The contribution of reactions proceeding by the law of accident grows with temperature of the process conduction. However, in this case, shortening

Table 11. Acidity Indexes for Electrophilic Catalysts basEd on Metal Chlorides

Catalyst	RAI, mg-eq/g of catalyst (± 0.1)	pK_a (± 0.1)	q_{H^+} (± 0.02)
$H[AlCl_3OH]$	5.0	-22	0.45
$H[AlCl_4]$	4.4	-16	—
$H[FeCl_2OH]$	3.5	-10	0.34
$H[BiCl_3OH]$	2.8	-4.2	0.29
$H[InCl_3OH]$	2.6	-2.5	0.27
$H[MgInCl_5OH]$	1.9	$+3.5$	0.21
$H[CaInCl_5OH]$	1.6	$+4.5$	0.22
$H[MgCl_2OH]$	1.5	$+7.1$	0.19
$H[MgCl_2OH]\cdot H_2O$	1.5	$+7.7$	0.18
$H[Mg(C_4H_9)ClOC_3H_7]$	—	$+7.7$	0.19
$H[AlCl_3OC_2H_5]$	1.5	$+7.7$	0.19
$H[Mg(C_2H_5)ClOH]$	—	$+8.1$	0.17
$H[Mg(C_3H_7)ClOC_3H_7]$	—	$+8.5$	0.18
$H[MgCl_2OH]\cdot 3H_2O$	1.0	$+10.5$	0.16
$H[CaCl_2OH]$	1.0	$+10.5$	0.16
$H[MnCl_2OH]$	1.0	$+10.5$	0.16
$H[BaCl_2OH]$	0.9	$+11.6$	0.15
$H[SnCl_2OH]$	0.6	$+13.8$	0.13
$H[NiCl_2OH]$	0.5	$+14.5$	0.12
$H[ZnCl_2OH]$	0.2	$+17.0$	0.10
$H[CdCl_2OH]$	0.1	$+17.8$	0.10
$H[Al(C_2H_5)Cl_3]$	4.3	-16	—
$H[MgAlCl_5OH]$	2.3	0	0.24
$H[MgAl(C_2H_5)Cl_4OH]$	2.2	$+0.9$	0.25
$H[CaAlCl_5OH]$	2.2	$+0.9$	0.25
$H[CaAl(C_2H_5)Cl_4OH]$	1.8	$+4.0$	0.21
$H[LiAl(C_2H_5)Cl_3OH]$	1.8	$+4.1$	0.18
$H[Al_2Cl_4(O)OH]$	—	$+4.4$	0.21
$H[NaAlCl_4OH]$	1.6	$+4.5$	0.23
$H[SrAlCl_5OH]$	1.6	$+5.0$	0.21
$H[BaAl(C_2H_5)Cl_4OH]$	1.4	$+7.3$	0.18
$H[NaAl(C_2H_5)Cl_3OH]$	1.4	$+7.2$	0.19
$H[Al(C_2H_5)Cl_2OH]$	—	$+7.5$	0.20
$H[Al(C_2H_5)_2ClOH]$	—	$+8.5$	0.18
$H[Al(OH)Cl_2OH]$	—	$+8.5$	0.18
$H[Al_2(CH_3)_3Cl(O)OH]$	—	$+8.5$	0.18
$H[ZnAlCl_5OH]$	1.1	$+9.8$	0.16
$H[CdAlCl_5OH]$	1.1	$+9.8$	0.16
$H[KAlCl_4OH]$	0.9	$+11.6$	0.13
$H[KAl(C_2H_5)Cl_3OH]$	0.8	$+12.0$	0.13

Note: RAI is the relative acidity index (experiment) [1]; pKa is the universal acidity index [2, 3];

of the length of the kinetic chain of decomposition of polyolefins at the initiation of the process by the law of accident (z_{acc}) comparing with the length of kinetic chain of the process initiated by the law of end groups (z_{eg}) are, probably, only apparent. The real growth of amount of monomeric chains, formed at the process initiation by the law of accident, also takes place because in this case the energy of β-decomposition of the carbonium ion decreases but, on the background of sharp increase of the number of breaks in macrochains, no degradation of macromolecules is displayed.

q_{H^+} is the charge on hydrogen atom of the catalyst (quantum-mechanical calculation) [2–5].

At low degree of polymerization $\left(\overline{P}_n^0\right)$, values of z_{eg} are equal to the length of the chain of depolymerization of macromolecules. At high values of $\overline{P}_n^0$, the value of z_{eg} becomes constant, which, obviously, is determined by the ratio of reactions of chain propagation and termination on ionic carbonium active centers, resulting from end groups [8, 9].

Existence of a tertiary carbon atom and especially aliquot C=C bonds (both end and internal ones) in the composition of macromolecules significantly affects the thermocatalytic degradation of polyolefins. In particular, initiation of the process preferably by the law of end groups is typical for low-molecular samples of polyisobutylene with high specific weight of unsaturated end C=C-bonds (Figure 5), because little change of the molecular mass of polymeric products (in the initial stages of the process $\overline{P}_n/\overline{P}_n^0 \approx 1$) is observed during thermocatalytic degradation of the polymer. The ratio $\overline{P}_n/\overline{P}_n^0$ decreases only at greater degrees of transformation. For high-molecular samples of polyisobutylene, isobutylene copolymers with isoprene, containing labile bonds in connection knots of isobutylene clusters and isoprenyl chain with C=C-bond, as well as for polyethylene (independent of the molecular mass) containing internal vinylidene and *trans*-vinylidene C=C-bonds, initiation of the gross-process of degradation of macromolecules proceeds simultaneously by both the law of accident and the law of end groups, but with different contribution of these mechanisms to the total process (Figure 7). The possibility of initiation of degradation of macromolecules by end vinyl groups also cannot be excluded.

$MgCl_2 \cdot AlCl_3$ is characterized by somewhat higher acidity compared with $NaAlCl_4$. This results in higher activity and selectivity of $MgCl_2 \cdot AlCl_3$ compared with $NaAlCl_4$ in the process of polyisobutylene degradation: total yield of gas products and yield of isobutylene itself grow simultaneously. Rise of selectivity of the process is bound to growth of the role of polyisobutylene depolymerization (degradation of macromolecules by the law of end groups). The index of the ratio of rate constants of initiation by the law of accident and the law of end groups in polyisobutylene degradation in the presence of $MgCl_2 \cdot AlCl_3$ is higher than in the presence of $NaAlCl_4$ (Table 10). There is a tendency to reduction of the ratio with temperature rise.

Cationic selective degradation of polyisobutylene is practically interesting for processing of unusual and other polymers down to a monomer [64].

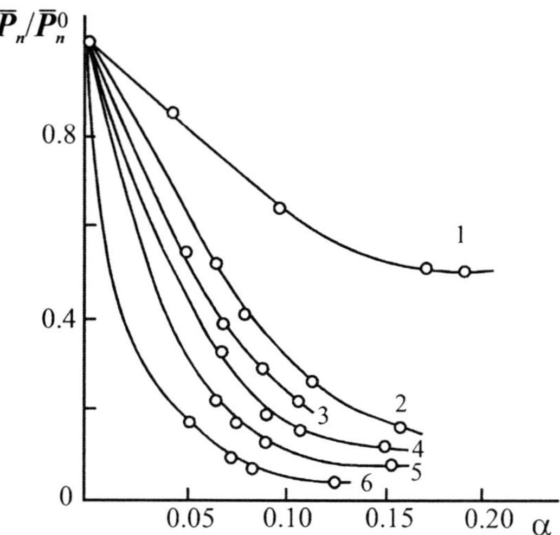

Figure 7. Dependence of the relative degree of polymerization on the transformation depth (α) of low density polyethylene (LDPE) in thermocatalytic degradation at 633 K in the presence of H[NaAlCl₄OH]. Catalyst concentration is $1 \cdot 10^{-3}$ mol/g of LDPE. The initial degree of polymerization of LDPE: 1 - $2 \cdot 10^4$; 2 - $6 \cdot 10^4$; 3 – $1.45 \cdot 10^5$; 4 - $2.5 \cdot 10^5$; 5 - $2.7 \cdot 10^5$; 6 - $8.7 \cdot 10^5$.

Let us point out one more case of decomposition of polyisobutylene polymeric molecules, accompanied by modification of polymer properties — ultrasonic degradation of polyisobutylene in aromatic solvents. In this case, two processes proceed in parallel: degradation of macromolecules and polymerization of arene [65, 66].

The former is characterized by regularities of ultrasonic decomposition of polymers in solvents and proceeds up to reaching some limiting molecular mass, which depends on conditions and method of polymer processing (solution viscosity, thermodynamic quality of the solvent, etc.) [67].

As follows from Table 12, the above factors influence polyisobutylene degradation, but they are not of major significance.

Table 12. Rate Constants of Polyisobutylene Degradation in Various Solvents

Solvent	ΔH_i^{297}, kJ/mol	Higgins constant χ_1	Rate constant of degradation and viscosity of solutions for times τ			
			$0–1.2 \cdot 10^4$, s		$1.2 \cdot 10^4–3.6 \cdot 10^4$, s	
			$k \cdot 10^4$, g/l·min	v_{30}, cSt	$k \cdot 10^4$, g/l·min	v_{30}, cSt
n-Heptane	36.7	0.468	5.81–1.64	21.8–1.00	1.64–0.64	1.00–0.85
Toluene	38.2	0.488	3.45–1.74	17.9–1.15	1.74–0.76	1.15–0.97
Isopropylbenzene	45.0	—	4.02–1.99	28.5–1.70	1.99–0.69	1.70–1.36

Note: Polyisobutylene – $\overline{M}_n = 5.4 \cdot 10^5$, $\overline{M}_w / \overline{M}_n = 3.5$; radiation conditions – 22 kHz, 294 K, 100 W power output, exponential radiator; rate constants of ultrasonic degradation, calculated by the Schmidt equation in the integral form [68]:

$$\frac{M_e}{M_\tau} + \ln\left(1 - \frac{M_e}{M_\tau}\right) = -\frac{K}{C\left(M_e/M_0\right)^2 \tau} + \frac{M_e}{M_i} + \ln\left(1 - \frac{M_e}{M_i}\right),$$

where M_e, M_i, M_τ are values of the molecular mass of polyisobutylene: the limit ($5 \cdot 10^3$), initial and current, respectively.

In all cases, processes are characterized by a non-stationary constant (during the initial period), and then proceed at a constant rate ($k = (0.7\pm0.06)\cdot10^{-9}$ g/l·min). However, physicochemical properties of polyisobutylene after sonication in aromatic solvents differ from those of polyisobutylene, degraded in an inert solvent (heptane), and properties of the initial polymer. First of all, there is a stable yellow-brown color and increased thermostability of products. They may be bound to a change of end aromatic fragments of polymer chains (wide absorption in the range of 220–270 nm in UV-spectra, an exothermic peak on DTA curves in the range of 673–677 K, typical for linear polyphenylene structures).

Transformations of aromatic hydrocarbons under conditions of cavitational influence on polyisobutylene are accompanied by formation of low-molecular polymers, which proceeds via intermediate cation-radical centers (this was proved by the method of selective inhibition of free-radical and cationic centers), the latter being presented as products with polyquinoid structure (ozone absorption) (Table 13).

Table 13. Characteristics of Transformation Products of Aromatic Compounds under Ultrasonic Effect

Initial compound	Element composition, mass%			$\overline{M}_\eta$	UV-spectrum λ, nm (ε, $m^{-3}kg^{-1}m^{-1}$)	Temperature of decay initiation, K		Absorption of $O_3\cdot10^3$, kmol/kg
	C	H	X(Cl, NO$_2$)			Liquid	Powder	
C_6H_6	92	8	—	—	250	—	—	1.40
$C_6H_5CH_3$[*]	88.3	11.7	—	340	223 (1860)	—	—	1.51
$C_6H_5C_3H_7$[†]	92.2	7.8	—	410	246 (2140)	365	—	1.31
$C_6H_5C_3H_7$	94.6	5.4	—	—	—	—	710	—
C_6H_5Cl	75.9	4.2	19.9	—	254 (5640)	365	—	1.10
$C_6H_5NO_2$[*]	84.0	5.6	10.4	296	246 (4210)	361	—	1.50
$C_6H_5NO_2$[†]	81.6	5.3	13.1	—	—	—	730	—

[*] elates to the soluble part of the product.
[†] relates to insoluble product.

At heating or long-term exposure in air, the reaction products transform into insoluble powders, approaching polyolefins in properties (thermal stability above 700 K, concentration of unpaired electrons of 150–$200\cdot10^{18}$ spin/mol). Obviously, polymerization of arenes under ultrasonic effects proceeds similar to catalytic polymerization in systems of the Lewis acid ($AlCl_3$)–oxidant ($CuCl_2$) type [69], but represents a more complex process.

Interaction (recombination) of intermediate products of polyisobutylene decomposition and polymerization of arenes at sonication of solutions causes formation of fragmentary block-copolymers of AB type with end polyphenylene grouping (3–5 units):

$$\sim CH_2\text{-}C(CH_3)_2\text{-}CH_2\text{-}C(CH_3)_2\sim \rightarrow CH_2C^{\bullet}(CH_3)_2 + CH_2\text{-}C^{\bullet}(CH_3)_2\sim$$

Owing to suppression of polyisobutylene decomposition by aromatic fragments from the end of the chain and decrease of the total rate of decomposition, the products obtained are of interest as a basis for synthetic lubricants and additives with improved performance [46].

ANALYTICAL CHARACTERISTICS OF POLYISOBUTYLENE

The IR-spectrum of polyisobutylene is comparatively simple [70]. The band at 1485 cm^{-1} is appropriate for a deformation oscillation of the methyl group. Compared with IR-spectra of paraffin hydrocarbons, this band is shifted by 25 cm^{-1} towards higher frequencies. The doublet at 1370 and 1400 cm^{-1} is appropriate for symmetric deformational oscillations of both methyl groups. In the range of valence oscillations, absorption of methyl groups is displayed at 2880 and 2960 cm^{-1}. Absorption in the area of 1250 cm^{-1} is due to the presence of dimethyl groups in macromolecules. The doublet of absorption bands of equal intensity at 920 and 950 cm^{-1} is also typical. End groups, specially displayed in IR-spectra of low-molecular products, are represented by three-substituted ethylene bonds $\sim CH_2–C(CH_3)_2–CH=C(CH_3)_2$ (absorption at 830 and 1660 cm^{-1}) and vinylidene double bonds $\sim CH_2–C(CH_3)_2–CH_2–C(CH_3)_2=CH_2$ (absorption at 890 and 1640 cm^{-1}). The relation between exo- and endoforms of olefins depend on the type of catalyst, used in the polymer synthesis, and is, for example, 30:70 for application of organoaluminum haloid compounds and 20:80 for Brönsted acids. Vinyl double bonds are absent [71, 72].

IR-spectra of industrial polyisobutylenes, especially of low-molecular samples (octoles), are much more complex, which is associated with influence of additives, inconstancy of the temperature regime and other factors in the polymerization stage. It specially affects the type of double bonds. Beside end vinyliden and three-substituted double bonds, the following ones were observed in IR- and ESR-spectra [73]: internal vinyliden double $\sim CH_2C(=CH_2)–CH_2–C(CH_3)_3$ (1), three-substituted double $\sim C(CH_3)_2–C(CH_3)=CH–CH_3$ (2) and tetra-substituted ethylene $\sim C(CH_3)=C(CH_3)–CH(CH_3)_2$ (3) bonds. Their formation is ascribed to processes of isomerization, skeleton regroupings, in particular, 1,3-methide shear (1), concert 1,2-hydride-2,3-methide shear (3), attachment by the "head-to-head" type with 1,2-methide shear (2). The relation of groupings depends on the nature of the applied catalyst and conditions of synthesis. For example, in the case of BF_3, concentration of vinylidene groupings in macromolecules decreases with time due to their protonation with simultaneous rise of content of three-substituted ethylene bonds. Formation of tetra-substituted ethylene bonds, which demands an unlikely attachment of monomers by the "head-to-head" type, is doubtful [74]. Admittedly, the argument proposed for confirmation — observance of analogously composed fragments in products of thermal decomposition of a high-molecular polymer [74] — is not convincing, because thermal decompo-sition proceeds according to the free-radical mechanism of polymer degradation, and occurrence of various products could be the consequence of secondary reactions of polymeric radicals.

More complicated spectra are observed for copolymers of isobutylene with small amounts of butenes or diene monomers. Besides isobutylene blocks, the following fragments are identified in them [75]: end and methyl groups $CH_3–CH_2–$, $CH_3–CH_2–CH_2–$, $CH_3–CH(CH_3)–CH_2–$, isopropyl groups $–CH(CH_3)_2$, di- and polymethyl groups $–(CH_3)_2C–(CH_2)_n–$

C(CH$_3$)$_2$, internal double bonds, appropriate to 1,4-polymerization of diene hydrocarbons at $n \geq 2$

$$\sim CH_2-CH_2-\underset{\underset{CH_3}{|}}{C}=CH-CH_2-C\underset{CH_3}{\overset{CH_3}{\diagup}}\sim$$

Existence of double bonds, different in reactivity, must be taken into account in the study and application of oligoisobutylenes, for example, as additives to oils, lacquers, etc., which supposes their functionalization by places of double bonds.

NMR-spectrum of H (o-dichlorobenzene) is: δ_{CH_2} = 1.4 m.p., δ_{CH_3} = 1.2 m.p.

NMR-spectrum of ^{13}C (o-dichlorobenzene, 353 K) is: δ_{CH_3} = 31.6 m.p. (quadruplet),

δ_{CH_2} = 60.0 m.p. (triplet), $\delta_{>C<}$ = 38.5 m.p. (singlet).

Molecular masses of moderate- and high-molecular polyisobutylenes are convenient to estimate by the data of characteristic viscosity of samples by the Mark–Hauvink equation $[\eta] = K_{\eta}M^{\alpha}$. Constants K_{η} and α for polyisobutylene–solvent systems are shown in Table 14 (concentration is expressed in g/100 cm^3) [76].

Table 14. Constants K_{η} and α in the Equation $[\eta] = K_{\eta}$ and M^{α}

Solvent	$K_{\eta}\times10^4$	α	Temperature, K	Method[*]	$M\times10^{-4}$
Polyisobutylene					
Isoamylisovalerate	11.4	0.5	295.1	L	16–470
n-Heptane	1.58	0.69	298	L	16–470
Iso-octane	6.75	0.576	293	L	0.8–3
Benzene	11.0	0.5	298	—	—
Cyclohexane	1.35	0.74	298	—	—
Cyclohexanone	2.63	0.69	303	—	—
Isobutylene copolymer with isoprene (BC-2045)					
Carbon tetrachloride	1.07	0.78	298	O	11–40
Toluene	6.6	0.60	298	O	11–40

[*] O – osmometry; L – light scattering.

Typically, $[\eta]$ values in good solvents do not practically depend on temperature (Table 15) [76].

For oligomers of isobutylene and low-molecular polymeric products, in determination of average molecular masses $\overline{M}_n$, it is expedient to use the method of measurement of heat condensation effects (MHCE) [77], as well as the ozonolysis method, more accurate and fasten with high selectivity of unsaturated >C=C<-bonds, using ADS device of the ICPh RAS system (selectivity of the method is 10^{-8} g/l against 10^{-4} g/l for MCHE) [78, 79]. The method is based on the quantitative interaction of ozone with double C=C-bonds in macromolecules of polyisobutylene. Values of $\overline{M}_n$ are calculated from the equation:

$$\overline{M}_n = \frac{1}{D}, \text{ where } D = \frac{C_{st} V^{st}_{border} S_{samp} V_{solv}}{S_{st} V^{samp}_{border} q_{samp}}.$$

Here D is the non-saturation degree; C_{st} is the stilbene (the standard substance) concentration; V^{st}_{border}, V_{solv} and V^{samp}_{pr} are volumes of probes of stilbene, solvent and sample, respectively; S_{samp} and S_{st} are squares of peaks on ozonograms of samples of polyisobutylene and stilbene; q_{samp} is the length of pendant probe.

Table 15. Characteristic Viscosity [η] of Polyisobutylene with $\overline{M}_\eta$ = 1,300,000 in Good and Bad Solvents

Solvent	Characteristic viscosity [η] of polyisobutylene at temperatures			
	298 K	303 K	313 K	333 K
Diisobutylene	2.98	2.90	3.03	2.95
Carbon tetrachloride	4.25	4.30	4.30	4.38
Cyclohexane	4.78	4.75	4.81	4.83
Benzene	1.45	1.77	2.20	2.87

Table 16. Comparative Analysis of Polyisobutylene Samples

$\overline{M}_n$			Content of C=C bonds, mass%	
Ebullioscopy	MCHE	ADC	Iodometry	ADC
500	510	500	7.1	7.0
540	560	550	6.8	6.9
630	700	690	5.2	5.1
800	820	820	4.5	4.4
820	840	840	4.2	4.2
920	940	930	3.9	3.8
940	950	950	3.9	3.6
980	1000	1000	3.8	3.5
1150	1210	1200	3.0	2.9
1250	1250	1210	2.9	2.5
1450	1500	1480	2.5	2.3

Comparative analysis in estimation of molecular masses of low-molecular polyisobutylene can also be used as the method of fine-layer adsorption chromatography (standard "Silufol" plates, the probe of 1–2 mg, elutriator cyclohexane-chloroform in relation 7:1, developer – iodine, determination time 5–10 min). The method is based on measuring the path length of a polymer probe patch and determining values of the molecular mass of polyisobutylene by the calibration curve, and enables molecular masses of polyisobutylene in the composition of polymerizates to be determined, which is important for analytical control in modern high yield processes of polyisobutylene industrial production.

Traditional methods of determination of polyisobutylene MMD: fractionation of polymeric products, gel-chromatography and the universal method based on combination of liquid chromatograph with a specific detector of unsaturated compounds with the ADC method.

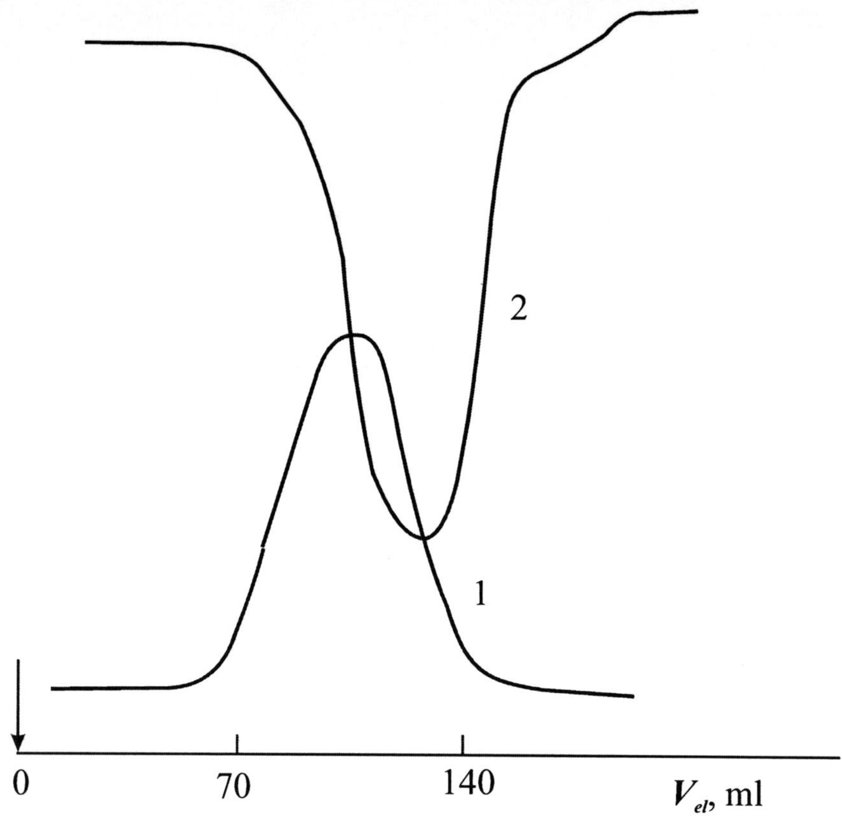

Figure 8. Gel- (1) and ozonochromatograms (2) of polyisobutylene with $\overline{M}_n$ = 970. The arrow marks the probe injection.

For polyisobutylene, ADC-chromatograms differ from chromatograms, because the part of nonsaturation in the low-molecular range is higher, than in the high-molecular one (Figure 8). As practically every macromolecule contains a single end bond, application of a specific detector enables a calibration curve to be composed for isobutylene oligomers, which connects values of average molecular masses to elutriator volume, by combined treating of two chromatograms. Calibration is performed on fractions of isobutylene oligomers with $\overline{M}_n$ = 600-1200 by dividing mass parts of the polymer by the appropriate number of double bonds and determination of values of molecular masses, related to the elutriator volume. Semi-logarithmic anamorphose of $\overline{M}_n$ values on the elutriator volume V_{el} is depicted in Figure 9.

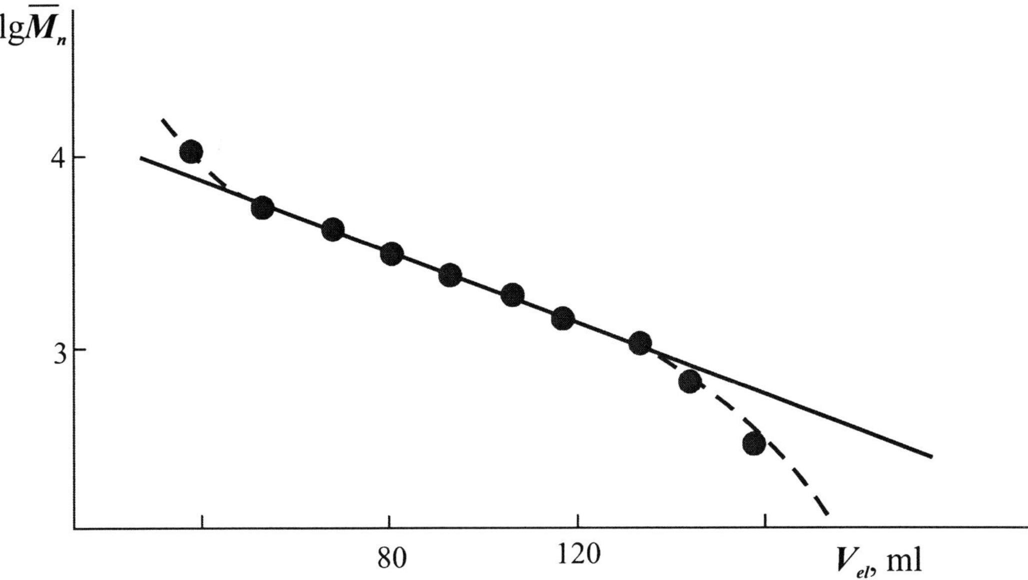

Figure 9. Dependence of logarithm of the average molecular mass of polyisobutylene on the elutriant volume.

Table 17. Relation between Some Molecular Characteristics of ollgoisobutylenes[*]

Molecular mass $\overline{M}_n$	Content of C=C bonds, mass%	Viscosity (373 K), m^2/s
500	7.1	0.59
600	5.9	1.00
660	5.5	1.12
700	5.2	1.50
740	5.0	1.70
800	4.6	2.00
830	4.3	2.08
875	4.1	2.68
900	4.0	3.00
960	3.9	4.80
1000	3.8	4.90
1050	3.6	5.10
1200	3.0	9.00
1420	2.6	12.20

[*] MMD of polymer fractions is not wider than 1.5.

Comparison of MMD of polyisobutylene samples, obtained by ozono- and gel-chromatograms, based on the calibration dependence and calibration on narrow fractions of the polymer, verifies the high accuracy of the universal method. Application of the specific method to decode MMD curves allows fast and reliable determination of $\overline{M}_n$ value of polyisobutylene in every point of the chromatogram. Consequently, the universal method gives information about MMD in connection with the nonsaturation degree of the sample dependent on conditions of synthesis.

Values of molecular mass and kinematic viscosity of oligoisobutylene are in clear dependence (Table 17).

It can be easily predicted that widening of MMD, keeping $\overline{M}_n$ and the iodine number of oligoisobutylenes constant, will lead to a significant growth of viscosity.

REFERENCES

[1]   Huivink, R. and Staverman, A., *Chemistry and Technology of Polymers,* (Translation from German), 1965, v. 1, 675 p., Moscow–Leningrad, Khimia. (Rus)

[2]   Suter, U.W., Saiz, E., and Flori, P.J., *Macromolecules*, 1983, v. 16(8), pp. 1317–1328.

[3]   Allegra, G., Benedetti, E., and Pedone, C., *Macromolecules*, 1970, v. 3(6), pp. 727–785.

[4]   Boyd, R.H. and Breitling, S.M., *Macromolecules*, 1972, v. 5(1), pp. 1–7.

[5]   Bartos, J. and Hlonskova, Z., *Eur. Polym. J.*, 1989, v. 25(1), pp. 21–24.

[6]   Hayashi, H., Flori, P.J., and Wignall, G.D., *Macromolecules,* 1983, v. 16(8), pp. 1318–1325.

[7]   De Boll, L.C. and Suter, W.W., *Macromolecules,* 1987, v. 20(6), pp. 1424–1425.

[8]   Dejean de Ca Batle, R., Zauprete, F., and Monnerie, L., *Macromolecules*, 1989, v. 22(6), pp. 2617–2622.

[9]   Boyer, Raymond F. and Enns, John B., *J. Appl. Polym. Sci.*, 1986, v. 32(3), pp. 4075–4107.

[10]   Raficov, S.R., Budtov, V.P., and Monakov, Yu.B., *Introduction into Physicochemistry of Polymers*, 1978, 328 p., Moscow, Nauka. (Rus)

[11]   Malanga, M. and Vogl, O., *Polym. Eng. Sci.,* 1983, v. 23(10), pp. 597–600.

[12]   Sangalov, Yu.A., Minsker, K.S., and Berlin, Al.Al., *Highmolec. Comp.*, 1983, Ser. A, v. 25(7), pp. 1451–1457. (Rus)

[13]   *Chemical Reactions of Polymers*, Ed. By E.M. Fettes, (Translation from English), 1967, v. 2, 536 p., Moscow, Mir. (Rus)

[14]   *Chemical Reactions of Polymers*, Ed. By E.M. Fettes, (Translation from English), 1967, v. 1, 503 p., Moscow, Mir. (Rus)

[15]   Chandra, R., *Progr. Polym. Sci.*, 1985, v. 11(1–2), pp. 1–22.

[16]   Bartos, J., *Collect. Czech. Chem. Commun.*, 1985, v. 50(8), pp. 1699–1713.

[17]   Baramboim, N.K., *Mechanochemistry Of High–molecular Compounds*, 1978, 383 p., Moscow, Khimia. (Rus)

[18]   Cazale, A. And Porter, R., *Reactions of Polymers Under Tensions*, 1983, 441 p., Leningrad, Khimia. (Rus)

[19]   Sangalov, Yu.A. and Yasman, Yu.B., *Uspekhi Khimii*, 1985, v. 54(7), pp. 1206–1229. (Rus)

[20]   Sangalov, Yu.A., Yasman, Yu.B., Minsker, K.S., and Prokofiev, K.V., *Chemistry and Technology of Petroleum and Oil,* 1984, No. 6, pp. 30–32. (Rus)

[21]   Shamaeva, Z.G., Popova, V.V., Monakov, Yu.B., and Tolstikov, G.A., In: *Thes. of VI Conf. on "Ageing and Stabilization of Polymers"*, Ufa, 1983, p. 166. (Rus)

[22]   Shamaeva, Z.G., Monakov, Yu.B., Galin, F.Z., Kovardakova, V.V., and Tolstikov, G.A., *Acta Polym.*, 1984, v. 35(8), pp. 561–564.

[23] Chang, V.S.C. and Kennedy, J.P., *Polymer Bull.*, 1981, v. 5(7), pp. 379–384.

[24] Macosko, C.W. and Saam, J.G., *Polymer Bull.*, 1987, v. 18(5), pp. 463–471.

[25] Sangalov, Yu.A., Yasman, Yu.B., Khudaiberdina, Z.I., and Minsker, K.S., *Highmolec. Comp.*, 1981, Ser. A, v. 23(12), pp. 2652–2656. (Rus)

[26] Kasianov, V.V. and Naverro, E.A., Izv. VUZov. *Khimia i Khim. Tekhnol.*, 1977, v. 20(11), pp. 1623–1626. (Rus)

[27] Kolyavina, S.I., *Neftepererabotka i Neftekhimia*, 1981, No. 9, pp. 16–18. (Rus)

[28] Berkly, E.Yu., Novikov, D.D., and Entelis, S.G., *Highmolec. Comp.*, 1967, v. A9(12), pp. 2754–2763. (Rus)

[29] Kuliev, A.M., *Chemistry and Technology of Additives to Oil and Petroleum*, 1972, 358 p., Moscow, Khimia. (Rus)

[30] Belov, P.S., *Netepererabotka i Neftekhimia*, 1991, No. 9, pp. 14–16. (Rus)

[31] Biglova, R.Z., Malinskaya, V.P., Sagitdinova, Kh.F., and Minsker, K.S., *Highmolec. Comp.*, 1994, v. A36(8), pp. 1276–1280. (Rus)

[32] Biglova, R.Z., Malinskaya, V.P., Sagitdinova, Kh.F., and Minsker, K.S., *Highmolec. Comp.*, 1995, v. A37(9), pp. 1474–1481. (Rus)

[33] Biglova, R.Z., Zaikov, G.E., and Minsker, K.S., *Intern. J. Polymeric Mater.*, 1998, v. 42, pp. 219–248.

[34] Mukmeneva, N.A., Cherezova, E.N., Biglova, R.Z., Malinskaya, V.P., and Minsker, K.S., *Highmolec. Comp.*, 1997, v. A39(6), pp. 953–959. (Rus)

[35] Sangalov, Yu.A., Biglova, R.Z., Kamalova, N.G., and Minsker, K.S., *Bashkir. Khim. Zh.*, 1998, v. 5(4), pp. 37–41. (Rus)

[36] Minsker, K.S., Biglova, R.Z., Kamalova, N.G., and Zaikov, G.E., Russian Polymer News, 1997, v. 2(3), pp. 4–10.

[37] *General Organic Chemistry*, Ed. by N.K. Kochetkov, 1982, v. 2, p. 239, Moscow, Khimia. (Rus)

[38] Gumerova, E.F., Ivanova, S.R., Ponomareva, E.D., Budtov, V.P., Berlin. Al.Al., and Minsker, K.S., *Highmolec. Comp.*, 1989, v. 31(8), pp. 607–611. (Rus)

[39] Boone, D.E., Eisenbraum, P.W., Flanagan, P.W., and Grigsby, R.D., *J. Org. Chem.*, 1971, v. 36(15), pp. 2042–2048.

[40] Yasman, Yu.B., Sangalov, Yu.A., Prokofiev, K.V., Khudaiberdina, Z.I., and Nelkenbaum, E.M., *Neftekhimia*, 1983, v. 23(4), pp. 500–507. (Rus)

[41] Kursanov, D.N., Parnes, Z.N., Kalinkin, M.I., and Loim, N.M., *Ionic Hydrogenation*, 1979, 192 p., Moscow, Khimia. (Rus)

[42] Sangalov, Yu.A., Yasman, Yu.B., Gladkikh, I.F., and Minsker, K.S., *Doklady AN SSSR*, 1981, v. 261(2), pp. 428–432. (Rus)

[43] Ipatief, V.N., Pines, H., and Olberg, R.L., *J. Amer. Chem. Soc.*, 1978, v. 70(6), pp. 2123–2128.

[44] Guterbock, G., *Polyisobutylene and Isobutylene Copolymers*, 1962, 363 p., Leningrad, Gostoptekhizdat. (Rus)

[45] Razumovsky, S.D. and Zaikov, G.E., *Ozone and Its Reactions with Organic Compounds*, 1974, 322 p., Moscow, Nauka. (Rus)

[46] Minsker, K.S. and Sangalov, Yu.A., *Isobutylene and Its Polymers*, 1986, 224 p., Moscow, Khimia. (Rus)

[47] Grassi, N., *Chemistry of Degradation of Polymers*, 1959, 252 p., Moscow, Izdinlit. (Rus)

[48] Jamaguchi, T., Kamiguchi, T., and Ito, T., *J. Chem. Soc. Jap. Chem. Ind. Chem.*, 1976, pp. 1171–1174.

[49] Kodaira, J., Osawa, Z., and Ando, H., *J. Chem. Soc. Jap. Chem. Ind. Chem.*, 1977, pp. 1892–1898.

[50] Sangalov, Yu.A., Prochukhan, Yu.A., and Minsker, K.S., *Highmolec. Comp.*, 1978, v. B20(3), pp. 686–688. (Rus)

[51] Seger, M. and Cantow, H.I., *Polym. Bull.*, 1979, v. 5, pp. 347–354.

[52] Minsker, K.S., Ivanova, S.R., and Biglova, R.Z., *Russian Chemical Reviews*, 1995, v. 64(5), pp. 462–479.

[53] Berlin, Al.Al., Minsker, K.S., Sangalov, Yu.A., and Prochukhan, Yu.A., *Highmolec. Comp.*, 1983, v. A25(7), pp. 1458–1466. (Rus)

[54] Vasile, C. and Onu, P., *Acta Polym.*, 1985, v. 36, pp. 543–550.

[55] Ivanova, S.R., Ponedelkina, I.Yu., Berlin, Al.Al., and Minsker, K.S., *Highmolec. Comp.*, 1986, v. A28(2), pp. 266–271.

[56] Berlin, Al.Al., Gumerova, E.F., Ivanova, S.R., Minsker, K.S., and Karpasas, M.M., *Highmolec. Comp.*, 1987, v. B29(8), pp. 604–607. (Rus)

[57] Grassi, N. and Scott, J., *Degradation and Stabilization of Polymers*, 1988, 248 p., Moscow, Mir. (Rus)

[58] Ivanova, S.R., Gumerova, E.F., Minsker, K.S., Zaikov, G.E., and Berlin, Al.Al., *Prog. Polym. Sci.*, 1990, v. 15, pp. 193–215.

[59] Ivanova, S.R., Gumerova, E.F., Berlin, Al.Al., and Minsker, K.S., *Highmolec. Comp.*, 1991, v. A33(2), pp. 342–349. (Rus)

[60] Ivanova, S.R., Gumerova, E.F., Berlin, Al.Al., and Minsker, K.S., *Highmolec. Comp.*, 1991, v. A33(2), pp. 430–448. (Rus)

[61] Ivanova, S.R., Ponedelkina, I.Yu., Romanenko, T.V., Minsker, K.S., Karpasas, M.M., and Minsker, K.S., *Highmolec. Sangalov, Comp.*, 1986, v. A28(6), pp. 1217–1221. (Rus)

[62] Nanbu, H., Yumiko, Y., and Ikemura, T., *Polym. J.*, 1986, v. 18(11), pp. 871–875.

[63] Khalafov, F.R., Novrusova, F.A., Ismailov, E.G., and Krentsel, B.A., *Acta Polym.*, 1990, v. 41(4), pp. 226–229.

[64] Sangalov, Yu.A. and Prokofiev, K.V., *Chemistry and Technology of Petroleum and Oils*, 1985, No. 3, pp. 2–4. (Rus)

[65] Nelkenbaum, Yu.Ya., Prokofiev, I.K., and Sangalov, Yu.A., *Highmolec. Comp.*, 1986, v. A28(5), pp. 1058–1063. (Rus)

[66] Sangalov, Yu.A., Prokofiev, I.K., Polovinkina, G.M., and Zorin, Bashkir. *Khim. Zh.*, 1995, v. 2(2), pp. 34–38. (Rus)

[67] Baramboin, N.K., *Mechanochemistry of High–molecular Compounds*, 1978, 384 p., Moscow, Khimia. (Rus)

[68] Basedow, A.M., Ebert, K., *Adv. Polym. Sci.*, 1977, v. 22, pp. 83–148.

[69] Engstrom, G.G. and Kovacic, P., *J. Polym. Sci., Polymer Chem. Ed.*, 1977, v. 15(10), pp. 2453–2468.

[70] Thompson, H.W. and Torkington, P., *Trans. Faraday Soc.*, 1945, v. 41(3), pp. 246–260.

[71] Siriuk, A.G., *Chemistry and Technology of Petroleum and Oils*, 1972, No. 6, pp. 52–57. (Rus)

[72] Puskas, I., Banas, E.M., and Nerheim, A.G., *J. Polym. Sci., Symp.*, 1976, No. 56, pp. 191–202.

[73] Sangalov, Yu.A., Khudaiberdina, Z.I., and Yasman, Yu.B., In: *Thes. VI Conf. on Aging and Stabilization of Polymers*, Ufa, 1983, p. 8, UNI. (Rus)

[74] Slobodin, V.M. and Matusevitch, N.I., *Highmolec. Comp.*, 1963, v. 5, pp. 774–776. (Rus)

[75] Kichkin, G.I., Kupreev, A.I., and Lakhshi, V.A., *Chemistry and Technology of Petroleum and Oils*, 1973, No. 7, pp. 47–49. (Rus)

[76] Fetters, L.J., Hadjichristidis, N., J.S. Linder, Mays, J.W., and Wilson, W.W., *Macromolecules*, 1991, v. 24(11), pp. 3127–3135.

[77] Bekli, E.Yu., Novikov, D.D., and Entelis, S.G., *Highmolec. Comp.*, 1967, Ser. A, v. 9(12), pp. 2754–2763. (Rus)

[78] Pozniak, T.I., Lisitsin, D.M., Novikov, D.D., Berlin, Al.Al., Diachkovsky, F.S., Prochukhan, Yu.A., Sangalov, Yu.A., and Minsker, K.S., *Highmolec. Comp.*, 1980, Ser. A, v. 22(6), pp. 1424–1427. (Rus)

[79] Pozniak, T.I., Lisitsin, D.N., Novikov, D.D., and Diachkovsky, F.S., *Highmolec. Comp.*, 1977, Ser. A, v. 19(5), pp. 1168–1170. (Rus)

In: Physical Organic Chemistry: Theory and Practice
Eds: A. D'Amore and G. E. Zaikov, pp. 81-91

ISBN 1-59454-275-9
© 2005 Nova Science Publishers, Inc.

Chapter 4

THE CONDENSATION OF DIETHILMETHYLTHIENYLSILANES WITH THE POLYCYCLIC BISPHENOLES

*L. M. Khananashvili, E. G. Markarashvili, Z. Sh. Lomtatidze,
Ts. N. Vardosanidze, E. I. Khubulava and D. A. Girgvliani*
Tbilisi State University, Tbilisi, Georgia

ABSTRACT

In the present paper the reaction of etherification of dichlorsilanes with the extent of ethanol and the partial hydrolitic condensation of methylthienyldietoxisilane and the reaction of polycondensation of methylthienylsiloxanes with polycyclic bisphenoles 4,4-(2-norborniliden)diphenol (B-1) and hexahydro-(4,7-methyleniden-5,5imiden)dyphen (B-2) in the presence of catalyst sodium butilat is studied.

Key words: etherification, hydrolysis, poly-condensation, bactericides, thermo mechanics.

INTRODUCTION

The synthesis of organic silanes and siloxanes containing the thienyl group with the silicon atom together wit organic radicals is of big interest. Such compounds are interesting for the production of oil and petroleum resistant materials [1] as well as biologically active substances [2]. At the same time with the addition of polycyclic compounds to the main chain of thienylcontaining organosiloxanes the thermal resistance of the polymers is improved [1]. That's why the synthesis of thienyl-containing polylilesters with the cyclic fragments in the chain is of great interest.

In the present paper the reaction of etherification of dichlorosilanes with the extent of ethanol and the partial hydrolytic condensation of methylthienyldietoxilane with the various concentration of the water and the reaction of polycondensation of methylthienylsiloxanes with polycyclic bisphenoles.

EXPERIMENTAL

IR spectrum of the synthesized compounds where obtained on the UR-10 spectrometer in the liquid state.

PMR spectrum was registered using TESLA spectrometer.

Chromatographic analysis of the initial compounds was performed on LHM-80, model 2, column 3000x4 mm, nozzle Chromosorb-W, phase 5 mass% silicon organic polymer SE-30, gas carrier Helium.

Thermo gravimetric analysis was performed by the Paulic Paulic Erday derivator MOM-102, with the heating speed of 10 centigrade per minute.

THE SYNTHESIS OF THE COMPOUNDS I AND II

100 ml of solid toluene, 2M absolute ethyl alcohol and 2M pyridine are placed in the 4 neck chemical flask. From the drop funnel the mixture of 1M chlorsilane and 100ml absolute toluene is added during the 4 hours. After the complete addition of chlorsilane the mixture is stirred during 18 hours at the temperature of 80-100°C, then separating the sediment by filtration. The volatile products are removed from the liquid mass. After the distillation in the vacuum the compounds I and II are separated. The characteristics of the compounds I and II are given in table 1.

PARTIAL HYDROLYSIS OF THE DIETOXIMETHYLTHIENYLSILANE

0.1M compound I and the different amount of 90% ethyl alcohol are placed in the round-bottomed chemical flask with the reverse refrigerator (see table 2). The content of the chemical flask is stirred permanently and is heated to 80-90 °C. The duration of the reaction is varied from 24 to 54 hours depending on the polymerization of – diethoximethylthienylsiloxanes. After the alcohol and reaction mass is driven away at the pressure of 1 mm mercury to receive the compounds III-V and in the case of compounds VI-X is vacuumed.

Characteristics of compounds I-X are given in table 1.

Table 1. Characteristics of the compounds I-X

No Compound	Extent %	Boiling Temp (mm. mercury	N_D^{20}	d_4^{20}	MR_D		M		OH_2H_5, %	
					obtained	calculated	obtained	calculated	obtained	calculated
I	92.9	136/I	1.542	1.166	76.66	77.36	282.5	284	31.37	31.69
II	80	138-140/2	1.5072	1.1029	58.26	58.48	215	216	42.56	42.55
III	70.4	148-151°/1	1.5293	1.1809	89.51	89.52	357	358	25.12	25.13
IV	58.1	162-167°/1	1.5512	1.2589	120.58	120.56	501.8	500	18.12	18.0
V	57.4	186-197°/1	1.5735	1.3369	151.59	151.6	641.9	642	14.14	14.01
VI	51.9	-	-	-	-	-	783	784	11.52	11.47
VII	50.8	-	-	-	-	-	1064	1068	8.49	8.42
VIII	61.9	-	-	-	-	-	1350	1352	6.67	6.66
IX	59.7	-	-	-	-	-	1493	1494	6.015	6.02
X	50.5	-	-	-	-	-	2060	2062	4.29	4.36

Element Composition %								
Obtained				Formula	Calculated			
C	H	S	Si		C	H	S	Si
50.69	5.54	22.49	9.91	$C_{12}H_{16}O_2S_2Si$	50.70	5.63	22.59	9.85
50.12	7.39	14.78	12.97	$C_9H_{16}O_2SSi$	50	7.40	14.8	12.96
46.90	6.13	17.86	15.65	$C_{14}H_{22}O_3S_2Si_2$	46.92	6.14	17.87	15.09
45.81	5.59	19.22	16.92	$C_{19}H_{28}O_4S_3Si_3$	45.6	5.6	19.2	16.8
44.83	5.27	19.84	17.46	$C_{24}H_{34}O_5S_4Si_4$	44.85	5.29	19.93	17.44
44.39	5.09	20.41	17.83	$C_{29}H_{40}O_6S_5Si_5$	44.38	5.10	20.40	17.8
43.81	4.85	20.99	18.34	$C_{39}H_{52}O_8S_7Si_7$	43.82	4.86	20.97	18.35
43.48	4.73	21.29	18.64	$C_{49}H_{64}O_{10}S_9Si_9$	43.49	4.73	21.30	18.63
43.38	4.685	21.43	18.73	$C_{54}H_{70}O_{11}S_{10}Si_{10}$	43.37	4.68	21.41	18.74
43.095	4.56	21.71	19.04	$C_{74}H_{94}O_{15}S_{14}Si_{14}$	43.06	4.55	21.72	19.0

Table 2. The Ratio of the Initial Compounds and the Duration of the Reaction during the Partial Hydrolysis of the Methylthienildiethoxisilane

Methylthienil diethoxisilane	Amount of alcohol, gram	Amount of water, ml	Duration of the reaction	Polymerization degree in diethoxisilane
0,1	9	0,9	24	2
	12	1,2	24	3
	13,2	1,32	24	4
	13,5	1,35	30	5
	14,4	1,44	36	7
	17,7	1,77	42	9
	19,9	1,99	48	10
	20,1	2,01	54	14

THE POLYCONDENSATION OF DIETOXIMETHYLTHIENYLSILANES AND SILOXANES WITH BIS-PHENOLS

0.5 gram of bis-phenol, compound VI and 0.02 ml of sodium butilat is placed in the 3 neck chemical flask equipped with the mixer, thermometer, cooler and connected to the collector of the condensed matter. The reaction mixture is permanently mixed and heated. The evolved ethyl alcohol is separated at the temperature of 160 ^{o}C. Then temperature is raised up to 180 ^{o}C. Condensation is finished in 4 hours, after 75% of ethylene alcohol is evolved. At the end the polymer is dropped by the petrolein ester from chloroform solution, is drained in the vacuum at the temperature of 70 degrees during 7 hours and till obtaining the permanent weight.

The synthesis of the rest of the polymers is carried out similarly.

Table 3. The Influence of Thienylethoxisilanes I, II and Siloxanes III and IV on the Growth of Some Microbes

Test objects	Control	compounds							
		I		II		III		IV	
		0,1 g/l	0,01 g/l	0,1 g/l	0,01 g/l	0,1 g/l	0,01 g/l	0,1 g/l	0,01 g/l
		The value of depression zones of test objects, mm							
Bacterium tumefaciens	0	0,5	0,5	1,5	0,5	1	0,5	0,5	0,5
Xanthomonas campestiis	0	1	1	1	0,5	1	0,5	0,5	0,5
Pectobacterium aroideae	0	2	0,5	1	0	0,5	0,5	0,5	0,5
Streptomyus spp;	0	1,5	0	0	0	0	0	0	0
Nocardiophsis spp;	0	1	0,5	0	0	0	0	0	0

DISCUSSION OF THE RESULTS

It is known that the thienylclorosilanes are very unstable during the hydrolysis in the strong acid area as the splitting of the thienyl-silicon chains occurs. That's why to obtain the thienylclorosilanes we have performed the reaction

$$RR'SiCl_2 + 2\ C_2H_5OH \xrightarrow[-2Py*HCl]{+2Py} RR'Si(OC_2H_5)_2$$

where R=R'= (I); R' = (II).

The dietoxisilanes synthesized are transparent substances with light specific smell, well dissolved in polar and non-polar organic solvents.

The α,ω-bis(etoxi)oligomethylthienylsiloxanes where synthesized by the partial hydrolytic condensation of the compound I with the different amount of water:

$$n\ Me(Th)Si(OC_2H_5)_2 + (n-1)H_2O \longrightarrow C_2H_5O\left[\begin{array}{c} Me \\ | \\ Si\ O \\ | \\ Th \end{array}\right]_n Si\ OC_2H_5$$

where : n=1 (III), 2(IV), 3(V), 4(VI), 6(VII), 8(VIII), 9(IX), 4(X).

Compounds III-V are non viscous liquids, while the compounds VI-X are viscous compounds. All compounds are well soluble in different organic solvents. The properties of the compounds I-X are given in table 1.

The structure of the compounds I-X are confirmed by the IR and PMR analysis.

In the IR spectrum of the compounds I-X the strong absorption bands are observed at 705 and 1220 cm^{-1} and mid absorption bands at 1500,1520 and 1800 cm^{-1} and weak absorption bands at 3035 cm^{-1}, meaning the presence of C-S bonds of thienyl groups. The strong absorption bands at 1250 cm^{-1} indicate the Si-Me bonds in compounds I-X.

The strong absorption bands at 1075-1100 cm^{-1} mid absorption bands at 945-990 cm^{-1} indicates the Si-O-C bonds. In addition the detection of strong absorption bands at 1020-1070 cm^{-1} for compounds III-X indicates the existence of Si-O-C bonds.

In PMR spectrum of the compounds I-X the peaks at 0.34 m.d. are observed, meaning Si-Me and 7,2-7,7 thienyl groups.

Compounds I-IV where tested for bacterial and actinomicet characteristics. The photo pathogenic bacteria *Xantomonas campestris,Pectobacterium aroideae and Bacterium tumefaciens (*provoking cancer of vine) and actinomicets *Streptomyus spp and Nocardiophsis spp.* The bacterial and actinomicets of the compounds where detected by alveolar method.

Table 4. Physical and Chemical Characteristics and the Element Composition of the Polymers

No	Element section of the polymer	weight	Softening temperature °C	η_{spec}	Obtained				Gross formula	Calculated			
					C	H	Si	S		C	H	Si	S
1	$\left[\begin{array}{c} C_4H_3S \\ -Si-O- \\ OC_2H_5 \end{array} R_1 \right]_4$	950	90-95	0.15	54.38	5.31	11.76	13.45	$(C_{43}H_{50}O_9Si_4S_4)_m$	54.32	5.26	11.79	13.47
2	$\left[\begin{array}{c} C_4H_3S \\ -Si-O- \\ OC_2H_5 \end{array} R_1 \right]_7$	1466	80-85	0.17	49.92	5.03	13.39	15.26	$(C_{61}H_{74}O_{15}Si_{17}S_7)_m$	49.53	5.05	13.37	15.28
3	$\left[\begin{array}{c} C_4H_3S \\ -Si-O- \\ OC_2H_5 \end{array} R \right]_9$	1810	60-65	0.19	48.38	4.99	13.9	15.93	$(C_{73}H_{90}O_{19}Si_9S_9)_m$	48.4	4.97	13.92	15.91
4	$\left[\begin{array}{c} C_4H_3S \\ -Si-O- \\ OC_2H_5 \end{array} R_2 \right]_4$	990	85-90	0.14	55.75	5.44	11.34	12.91	$(C_{46}H_{54}O_9Si_4S_4)_m$	55.76	5.45	11.31	12.93
5	$\left[\begin{array}{c} C_4H_3S \\ -Si-O- \\ OC_2H_5 \end{array} R_2 \right]_7$	1506	75-80	0.16	50.97	5.19	13.03	14.85	$(C_{64}H_{78}O_{15}Si_{17}S_7)_m$	50.99	5.18	13.01	14.87
6	$\left[\begin{array}{c} C_4H_3S \\ -Si-O- \\ OC_2H_5 \end{array} R_2 \right]_9$	1850	55-60	0.18	49.31	5.07	13.61	15.89	$(C_{76}H_{94}O_{19}Si_9S_9)_m$	49.26	5.08	13.62	15.57

Table 4. Physical and Chemical Characteristics and the Element Composition of the Polymers (Continued)

No	Element section of the polymer	weight	Soft. temp. °C	η_{spec}	Obtained				Gross formula	Calculated			
					C	H	Si	S		C	H	Si	S
7	$\left[\begin{array}{c}C_4H_2SCl \\ \mid \\ -Si-O- \\ \mid \\ OC_2H_5\end{array}R_1\right]_4$	1088			47.42	4.24	10.27	13.03	$C_{43}H_{46}O_9Si_4S_4Cl_4$	47.43	4.23	10.29	13.05
8	$\left[\begin{array}{c}C_4H_2SCl \\ \mid \\ -Si-O- \\ \mid \\ OC_2H_5\end{array}R\right]_7$	1707			42.85	3.93	11.47	14.56	$C_{61}H_{67}O_{15}Si_7S_7Cl_7$	42.87	3.92	11.48	14.55
9	$\left[\begin{array}{c}C_4H_2SCl \\ \mid \\ -Si-O- \\ \mid \\ OC_2H_5\end{array}R_1\right]_9$	2121			41.3	3.83	11.87	15.08	$C_{73}H_{50}O_{19}Si_9S_9Cl_9$	41.31	3.82	11.88	15.07
10	$\left[\begin{array}{c}C_4H_2SCl \\ \mid \\ -Si-O- \\ \mid \\ OC_2H_5\end{array}R_2\right]_4$	1128			48.95	4.44	9.93	12.58	$C_{46}H_{50}O_{15}Si_4S_4Cl_4$	48.94	4.43	9.93	12.59
11	$\left[\begin{array}{c}C_4H_2SCl \\ \mid \\ -Si-O- \\ \mid \\ OC_2H_5\end{array}R_2\right]_7$	1748			43.94	4.07	11.23	14.23	$C_{64}H_{71}O_{15}Si_7S_7Cl_7$	43.95	4.06	11.22	14.22
12	$\left[\begin{array}{c}C_4H_2SCl \\ \mid \\ -Si-O- \\ \mid \\ OC_2H_5\end{array}R_2\right]_9$	2161			42.2	3.92	11.67	14.8	$C_{76}H_{85}O_{19}Si_9S_9Cl_9$	42.21	3.93	11.66	14.79

The investigations show, that the compounds I-IV show bacterial activity and suppress the growth of the phitopatogenic bacteria, while only the compound I shows the actinomicetic activity. Compound I is most active phitopatogenic compound.

The poly-condensation reaction of compounds VI, IX and X with polycyclic bisphenoles -4,4'-(2-norborniliden)diphenol (B-1) and hexahydro-(4,7-methyleniden-5-iliden)diphenol (B-2) where investigated at the correspondent temperatures of 170-180 °C in the presence of catalyst sodium butilate at the equimole amounts of the above compounds. The synthesis may be described by the following scheme:

$$m\ C_2H_5O - \left[\begin{array}{c} Me \\ Si\ O \\ Th \end{array}\right]_n - \begin{array}{c} Me \\ Si\ OC_2H_5 \\ Th \end{array} + mHOR'OH \longrightarrow$$

$$\longrightarrow C_2H_5O - \left[\left[\begin{array}{c} Me \\ Si\ O \\ Th \end{array}\right]_n - RO\right]_m - H$$

where: R =   n=4 (XI), n=9(XII), n=14(XIII)

R =   n=4(XIV), n=9(XV), n=14(XVI)

The process of poly-condensation was controlled by the change of specific viscosity (in 10% solution of benzene at 20 °C) as well as by the amount of the discharged alcohol in time (see Figures 1 and 2).

The investigation of the reactions of the compounds VI, IX and X with B-1 and B-2 has shown that the viscosity of the poly-condensed compounds rises and reaches the peak in 1.5 hours for the polymers XI-XIII and in 2.5 hours for the polymers XIV-XVI (see Figure 1). Approximately same time the maximum quantities of alcohol is yielded (see Figure 2).

After the evolving of alcohol is finished the heating was continued for 3 hours at 180°C / 3mm. mercury, resulted with production of cross-linked polymers. After the evaporation of solvent the solid fragile films are left. Polymers are well soluble in solvents. Such films are characterized by the low softening temperature (Table 1.).

The thermal stability of the polymers XI-XVI where investigated via TGA method (thermal gravimetric analysis).

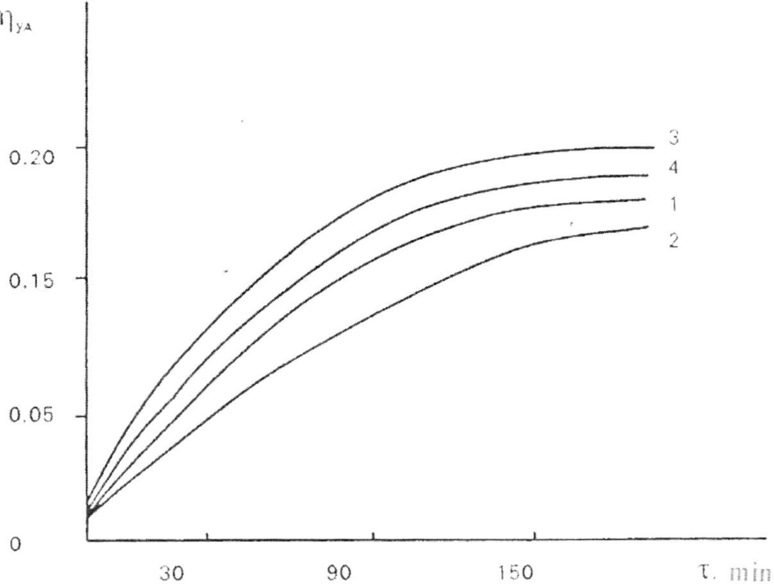

Figure 1.

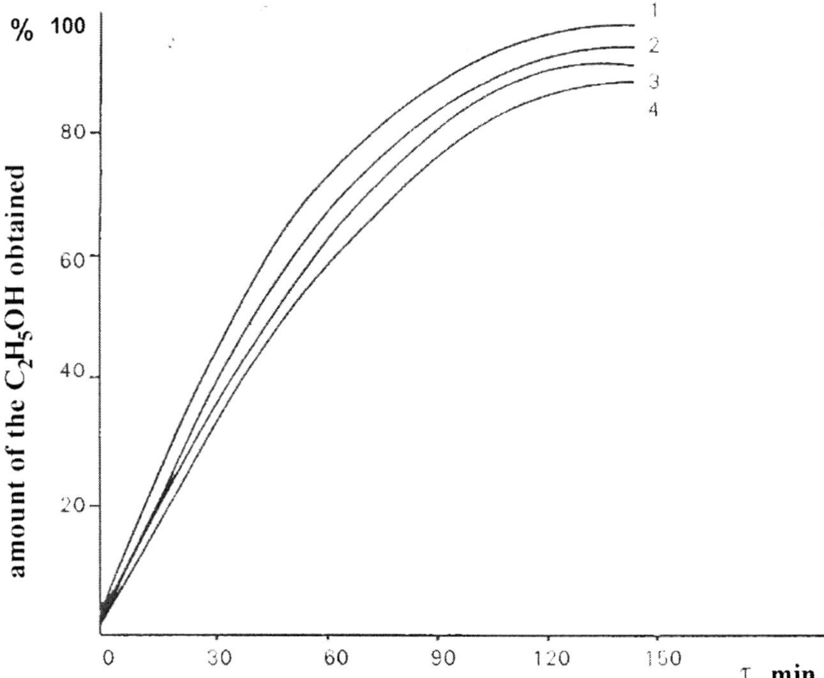

Figure 2.

Polymers XI-XVI lose the 508% of the initial weight within the temperatures 50-450^{0}C. at 450-550^{0}C polymers lose weight intensively. Above the temperature of 550^{0}C only 5% of the initial weight is left (Figure 3).

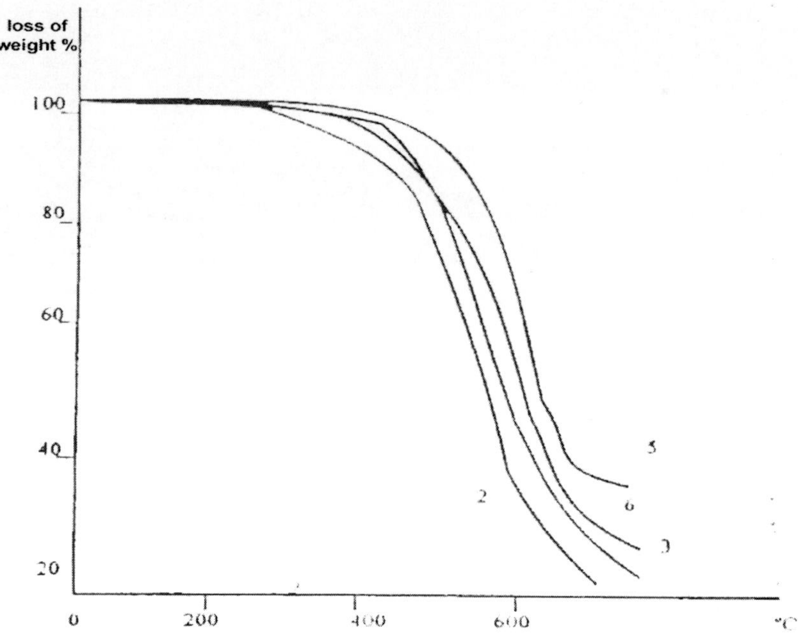

Figure 3.

Thermo-mechanical investigations show (Figure 4.) that the polymers XI-XVI transform from the glassy condition to liquid with the intensive deformation. The increase of the methylthienylsiloxi links (polymers XIII, XVI), as well as leading of B-2 into the mail chain of the polymers (polymers XIV-XVI) thermo-mechanical curve is shifted to the higher temperatures.

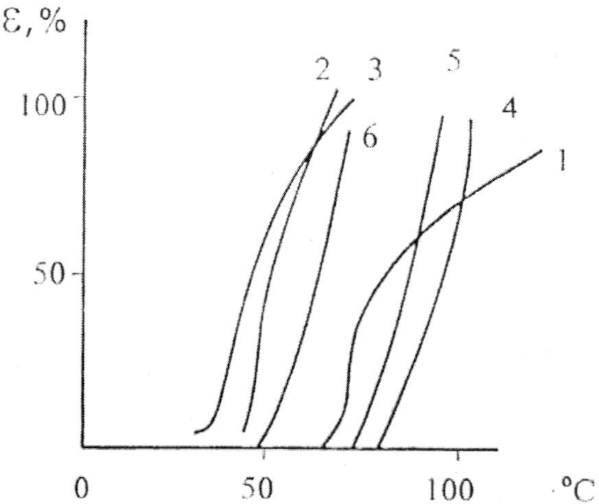

Figure 4.

In the present paper the reaction of the etherification of dichlorsilanes with alcohols was investigated together with the partial hidrolitic condensation of methylthienyldialoxisilanes with the various amount of water. Synthesized methylthienylsilanes and siloxanes were investigated by IR and PMR spectrometers. The policondensation of thienylalcoxisiloxanes with polycyclic bisphenols. It is shown, that the thermal resistance and the temperature of the softening of polymers are determined by the structure of the polycyclic bis-phenols as well as by the length of the poliorganosiloxane oligomer chain.

REFERENCES

[1]   Sobolevskiy M.V., Muzovskaya O.L., Popeleva G.S. The characteristics and range of application of siliconorganic products. *M. Chemistry*, 1975, p.269.

[2]   Voronkov M.G., Zelchan G.I., Tabenko B.M., Savushkina V.I., Chernishev E.A. *ChGS* 1976, 6, 772.

In: Physical Organic Chemistry: Theory and Practice
Eds: A. D'Amore and G. E. Zaikov, pp. 93-110

ISBN 1-59454-275-9
© 2005 Nova Science Publishers, Inc.

Chapter 5

KINETIC STUDY OF POLYPROPYLENE NANOCOMPOSITE THERMAL DEGRADATION

Sergei M. Lomakin and Gennadi E. Zaikov*
N.M. Emanuel Institute of Biochemical Physics, 119991 Kosygin 4, Moscow, Russia
Irina L. Dubnikova and Svetlana M. Berezina
N.N. Semenov Institute of Chemical Physics, Moscow, Russia

ABSTRACT

Polypropylene has wide acceptance for use in many application areas. However, low thermal resistance complicates its overall practice. The new approach in thermal stabilization of PP is based on the synthesis of polypropylene nanocomposites. This paper discussed new advances in the study of the thermal degradation of polypropylene nanocomposite. The observed results are interpreted by a proposal kinetic model, and predominant role of the one-dimensional diffusion reaction type under the thermal degradation of polypropylene nanocomposite. According to provided kinetic analysis, polypropylene nanocomposite demonstrated the transcendent thermal and fireproof properties as compared with the neat polypropylene.

Keywords: charring, diffusion, kinetics, nanocomposite, polypropylene, thermal degradation

INTRODUCTION

Polypropylene (PP) was the major synthetic stereo-regular polymer to achieve industrial importance. Today PP presents the fastest growing material for technical end-uses where high tensile strength coupled with low-cost are essential features. However, relatively low thermal

---

* Correspondence to: Sergei M. Lomakin, N.M. Emanuel Institute of Biochemical Physics of Russian Academy of Sciences, Moscow, Russia, Kosygin 4, 119991, mailto:lomakin@sky.chph.ras.ru

resistance and flammability of PP complicates their usage in many fields of applications. The new approach in thermal stabilization of PP is based on polymer nanotechnology [1-2].

Over the past decade, polymer nanocomposites have received considerable interest as an effective way for developing new composite materials, and they have been studied widely. Because of the larger surface area and surface energy of the additives when individual particles become smaller, it is not an easy task to obtain homogeneously dispersed organic / inorganic composites when the additives are down sized to nano-scale [3]. Melt intercalation has been successful in preparing polymer clay nanocomposites [4]. Recently the thermal degradation behavior of nanocomposites based upon PP-organoclay was studied by Zanetti, Camino *et. al.* using isothermal and dynamic thermogravimetry [5]. They suggested that the oxygen charring action and scavenging effect in the nanocomposite increases as the volatilization proceeds and that in the nanocomposite a catalytic role is played by the intimate polymer-silicate contact that may further favor the oxidative dehydrogenation-crosslinking-charring process.

This silicate morphology may act as an efficient barrier to oxygen diffusion towards the bulk of the polymer. Surface polymer molecules trapped within the silicate are thus brought to a close contact with oxygen to produce the thermally and oxidative stable charred material providing a new char-layered silicate nanocomposite acting as an effective surface shield [5]. In the present study a routine set of TGA analytical data was accomplished to provide the kinetic analysis of thermal degradation in air of PP nanocomposite based on layered organoclay (Cloisite 20A). The results obtained in this study gave an additional evidence of diffusion-controlled character of thermal degradation of PP nanocomposite caused by the catalytic-charring effect of nanosilicate clay.

EXPERIMENTAL

Synthesis

Polypropylene (PP) by Moscow petroleum refinery (MFI = 0.7) and maleic anhydride-modified oligomer (MAPP - Licomont AR 504 by Clariant co.) with $M_n{\sim}2900$; MA-content~4 wt.%. were blended/mixed using a laboratory Brabender mixing chamber for 2 min. at the first stage. Then the 7% wt. of organoclay (Cloisite 20A - Na^+ montmorillonite modified by dimethyl, dihydrogenatedtallow ammonium chloride by Southern Clay Co.) was added to the PP-MAPP - melt at a rotor speed of 60 rpm and set temperature of 190°C (PP-MAPP- Cloisite 20A). 10 min. mixing time was used in all experiments.

Characterization

WAXS analysis of nanocomposite layered structure was carried out with a DRON-2 X-ray Diffractometer with Cu-Kα radiation. Diffraction patterns were collected in reflection-mode geometry from 2° to 10°2θ.

AFM studies were performed with commercial scanning probe microscope Nanoscope IIIA and IV MultiMode (Digital Instruments/Veeco Metrology Group, USA) in tapping mode, at ambient conditions. Conventional etched Si probes (stiffness ~40 N/m, resonant frequency 160-170 kHz) were used. The amplitude of the free-oscillating probe, A_0, was varied in the 10-20 nm range while the set-point amplitude, A_{sp}, ranged from 0.5 to 0.8 A_0. Imaging was conducted on the flat sample surfaces being prepared at –80°C with an ultramicrotome MS-01 (MicroStar Inc., USA) equipped with a diamond knife.

Thermogravimetrical Analysis (TGA) has been performed with a MOM Q 1500 instrument at the heating rates of 3, 5 and 10°/min in air.

Kinetic analysis of PP compositions thermal degradation was carried out using Thermokinetics software by NETZSCH-Gerätebau GmbH.

RESULTS AND DISCUSSION

Structure Characterization

It is always necessary to carefully characterise the polymer structure in order to ensure a sort of dispersion for the nanoclays in polymers. XRD analysis and TEM would provide some information on the nanocomposite structural morphology.

The diffraction patterns for nanoclay and nanocomposites are displayed in Figure 1 (a). The Cloisite 20A itself has a single peak peak at around 3.6° with d-space of 2.4 nm.

The diffraction patterns for nanoclay and nanocomposites are displayed in Figure 1 (a). The Cloisite 20A itself has a single peak at around 3.6° with d-space of 2.4 nm.

The shift of the clay basal spacing d_{001} from 3.6° to 2.2° in PP-MAPP-Cloisite 20A sample suggests the intercalated nanocomposite sample have higher d-space (4.2 nm) than that in the original clay (2.4 nm), it may have some exfoliated structures, and considering the smearing of peak in nanocomposite sample, the nanocomposite structure have not been well exfoliated.

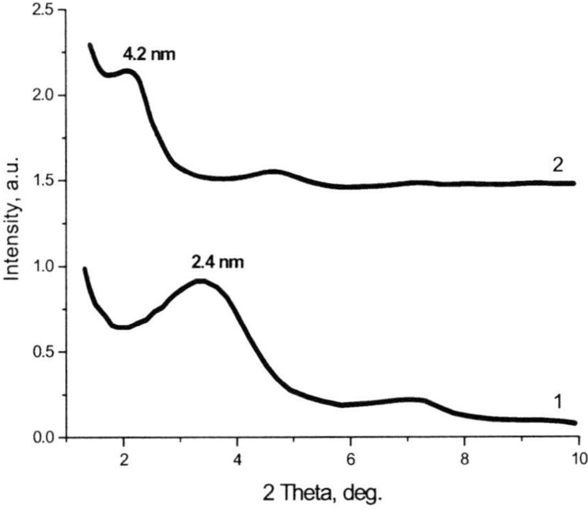

Figure 1. WAXS analysis for Cloisite 20A (1) and PP-MAPP-Cloisite 20A (2).

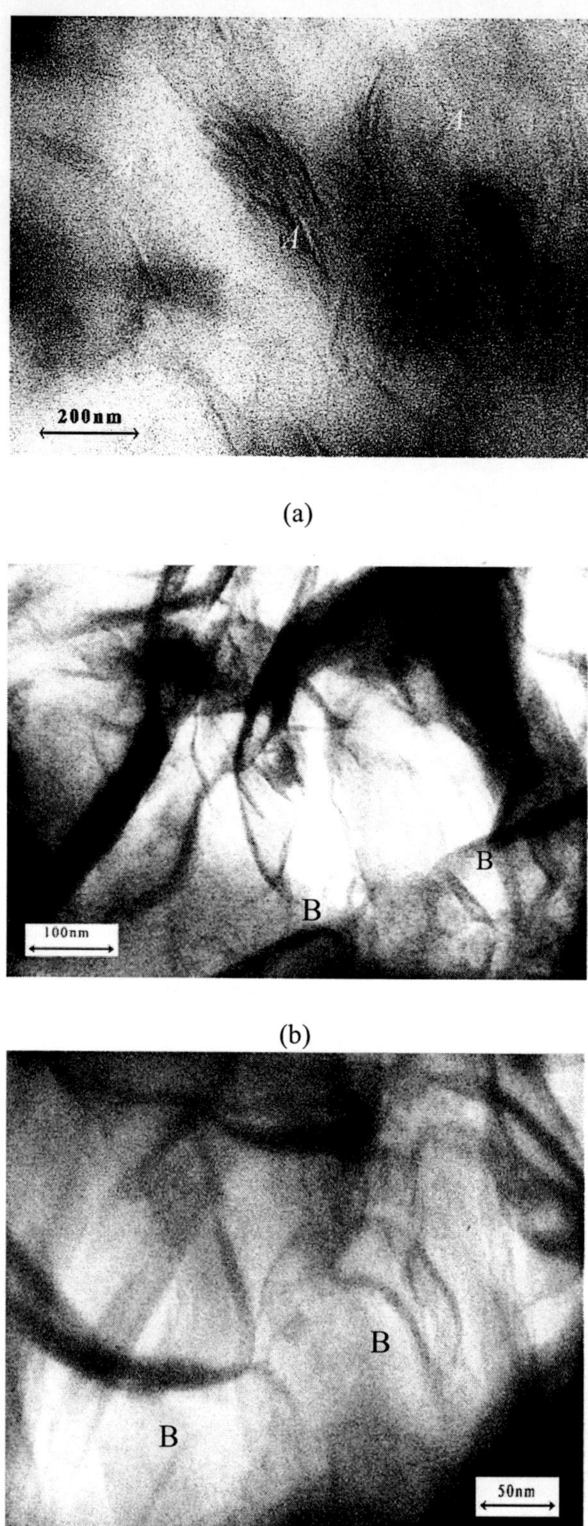

Figure 2.TEM photos with different magnification of PP-MAPP-Cloisite 20A where: *A* - stacks of layers (intercalated tactoids), B - exfoliated monolayers.

However, the XRD can only detect the periodically stacked montmorillonite layers; for all these nanocomposites there also exists a large number of exfoliated layers as well, which can be directly observed by transmission electron microscopy (TEM).

In Figure 2 we present TEM images with different magnification which indicate the presence of intercalated tactoids (A) and exfoliated monolayers (B) coexisting in the nanocomposite structure. The intercalated structures are characterized by a parallel registry that gives rise to the XRD reflection of Figure 1.

AFM studies indicate the similar morphology characteristics for PP-MAPP-Cloisite 20A (Figure 3). Height and phase images were simultaneously recorded on polymer surfaces. Height image presents surface topography, whereas phase images provide a sharp contrast of fine structural features and emphasize differences in sample components. We presume the availability of exfoliated monolayer units (A) as well as more complex multilayered tactoids (B) (Figure 3).

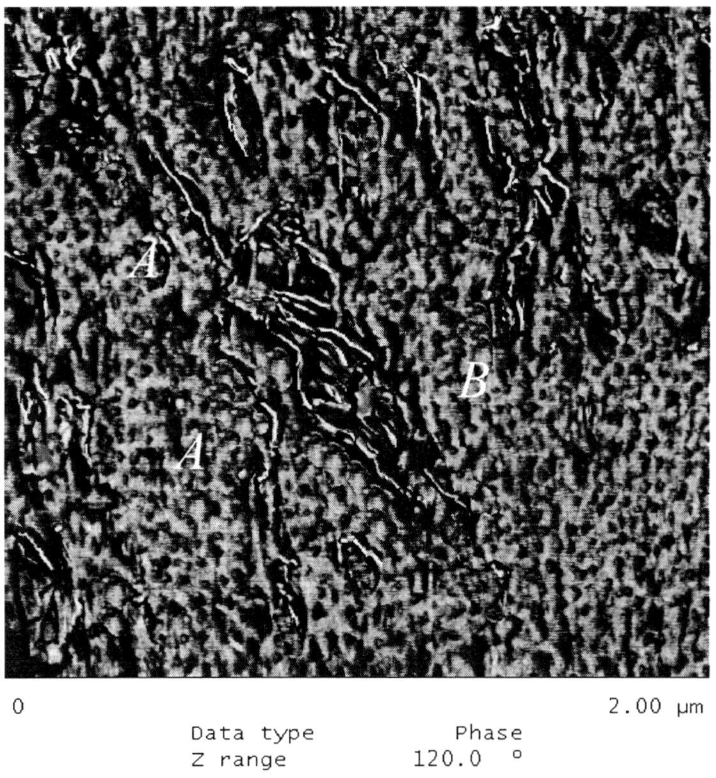

Figure 3. AFM image of PP-mPP with 7% Cloisite where: *A* - exfoliated monolayer units, B - multilayered tactoids.

THERMAL DEGRADATION STUDY

TG analysis of PP and PP nanocomposite (PP-MAPP-Cloisite 20A) show that at heating in air at 10°C/min, PP volatilizes completely, in two steps beginning at about 300°C with maximum rate at 400°C through a radical chain process propagated by carbon centered radicals originated by carbon-carbon bond scission (Figure 4) [6]. Below 200°C, the hydroperoxidation on C-H bonds, in which oxygen addition occurs to the carbon radicals created within the polymer chain by H abstraction initiates radical-chain degradation of PP [7], whereas above 200-250°C, oxidative dehydrogenation of PP takes over [8]. Depolymerization and random scission by direct thermal clevege of carbon-carbon bonds becomes possible in air as in nitrogen, above 300°C.

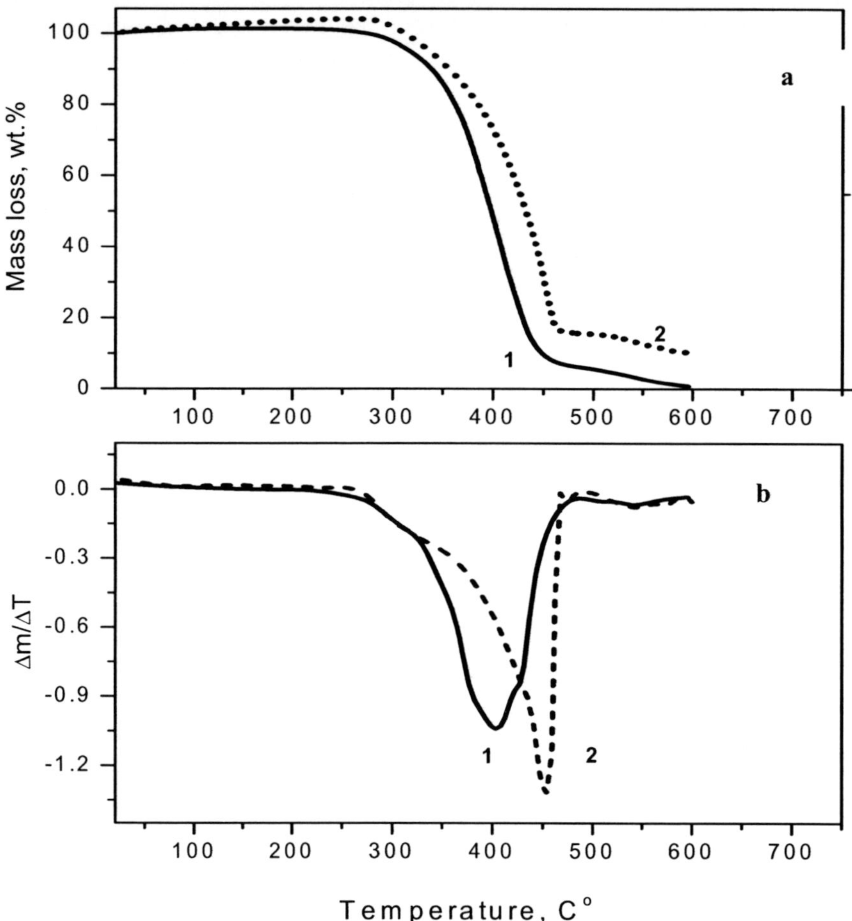

Figure.4. TG – (*a*) and DTG – (*b*) curves of PP (1) and PP-MAPP-Cloisite 20A (2) in air at the heating rate of 10°/min.

Stabilizing effect of Δ50°C (Figure 4 b) of PP-MAPP-Cloisite 20A over neat PP calculated with the maximum rate of mass loss can be explain by means of the barrier effect of the silicate nanolayers which operate in the nanocomposite level against oxygen diffusion, shielding the polymer from its action.

A char residue from neat PP is left at 450°C (5%), due to charring promoted by oxidative dehydrogenation. Then it slowly decomposes on heating up to 600°C in air (Figure 4). On the other hand, thermal degradation of PP-MAPP-Cloisite 20A in air results to much more stable char form which doesn't oxidize even at 600°C (Figure 4 and 5).

Silicate nanostructure executes a role of efficient barrier to oxygen diffusion towards the native polymer. Surface polymer molecules trapped within the silicate are thus brought to a close contact with oxygen and catalytic - silicate layers to produce the thermally and oxidative steady carbonized structures (Scheme on Figure 5).

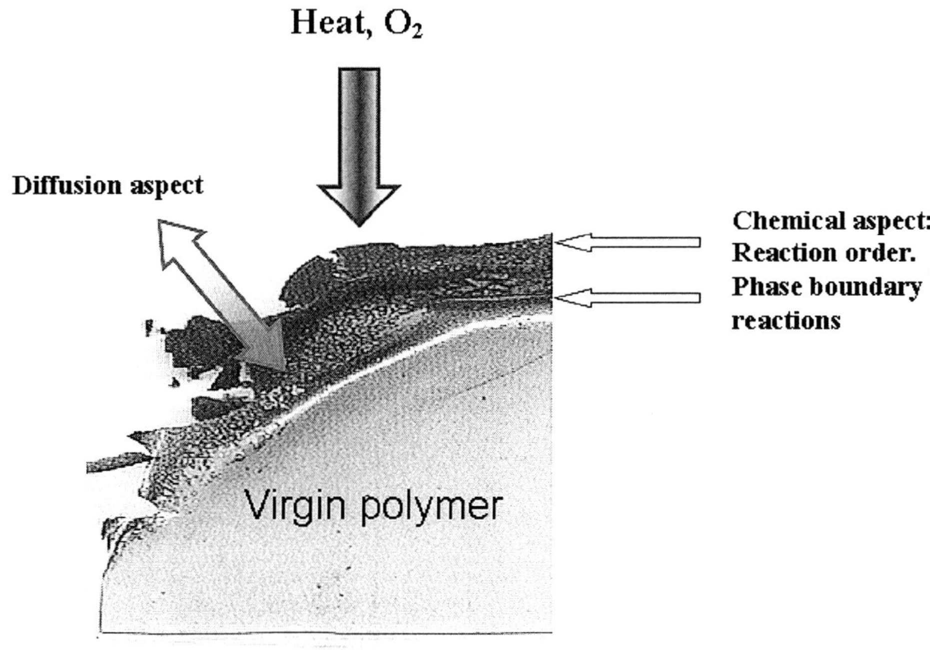

Figure 5. Scheme of the thermo-oxidative degradation of PP nanocomposite (PP-MAPP-Cloisite 20A).

Kinetic Analysis Using TGA Data

Kinetic studies of materials degradation have been carried out for many years using numerous techniques to analyze the data. Most often, TGA is the experimental method of choice and the only technique to be explored here. TGA involves placing a sample of polymer on a microbalance within a furnace and monitoring the weight of the sample during some temperature program. It is generally accepted that materials degradation obeys the basic equation (1) [9]

$$dc/dt = - F(t, T\ c_0\ c_f) \tag{1}$$

where: t - time, T - temperature, c_0 - initial concentration of the reactant, and c_f - concentration of the final product. Equation $F(t,T,c_0,c_f)$ can be described by two separable functions, $k(T)$ and $f(c_0,c_f)$:

$$F(t,T,c_o,c_f) = k(T(t) \cdot f(c_o,c_f) \tag{2}$$

Arrhenius equation (4) will be assumed to be valid for the following:

$$k(T) = A \cdot exp(-E/RT) \tag{3}$$

Therefore,

$$dc/dt = - A \cdot exp(-E/RT) \cdot f(c_o,c_f) \tag{4}$$

A series of reactions types: classic homogeneous reactions and typical solid state reactions, is listed in Table 1 [9].

Table 1. Reaction Types and Corresponding Reaction Equations, $dc/dt = - A \cdot exp(-E/RT) \cdot f(c_o,c_f)$

Name	$f(c_o,c_f)$	Reaction type
F_1	c	first-order reaction
F_2	c^2	second-order reaction
F_n	c^n	n^{th}-order reaction
R_2	$2 \cdot c^{1/2}$	two-dimensional phase boundary reaction
R_3	$3 \cdot c^{2/3}$	three-dimensional phase boundary reaction
D_1	$0.5/(1 - c)$	one-dimensional diffusion
D_2	$-1/ln(c)$	two-dimensional diffusion
D_3	$1.5 \cdot e^{1/3}(c^{-1/3} - 1)$	three-dimensional diffusion (Jander's type)
D_4	$1.5/(c^{-1/3} - 1)$	three-dimensional diffusion (Ginstling-Brounstein type)
B_1	$c_o \cdot c_f$	simple Prout-Tompkins equation
B_{na}	$c_o^n \cdot c_f^a$	expanded Prout-Tompkins equation (na)
$C_{1\text{-}X}$	$c \cdot (1+K_{cat} \cdot X)$	first-order reaction with autocatalysis through the reactants, X. $X = c_f$.
$C_{n\text{-}X}$	$c^n \cdot (1+K_{cat} \cdot X)$	n^{th}-order reaction with autocatalysis through the reactants, X
A_2	$2 \cdot c \cdot (-ln(e))^{1/2}$	two-dimensional nucleation
A_3	$3 \cdot c \cdot (-ln(e))^{2/3}$	three-dimensional nucleation
A_n	$n \cdot c \cdot (-ln(e))^{(n-1)/n}$	n-dimensional nucleation/nucleus growth according to Avrami/Erofeev

The analytical output must fit of measurements with different temperature profiles by means of a common kinetic model.

Kinetic analysis of PP compositions thermal degradation in air at heating rates of 3, 5 and 10°/min. was carried out using NETZSCH Thermokinetics software in order to provide an extra evidence of the diffusion-stabilizing effect of nanoclay structure (Figure 6).

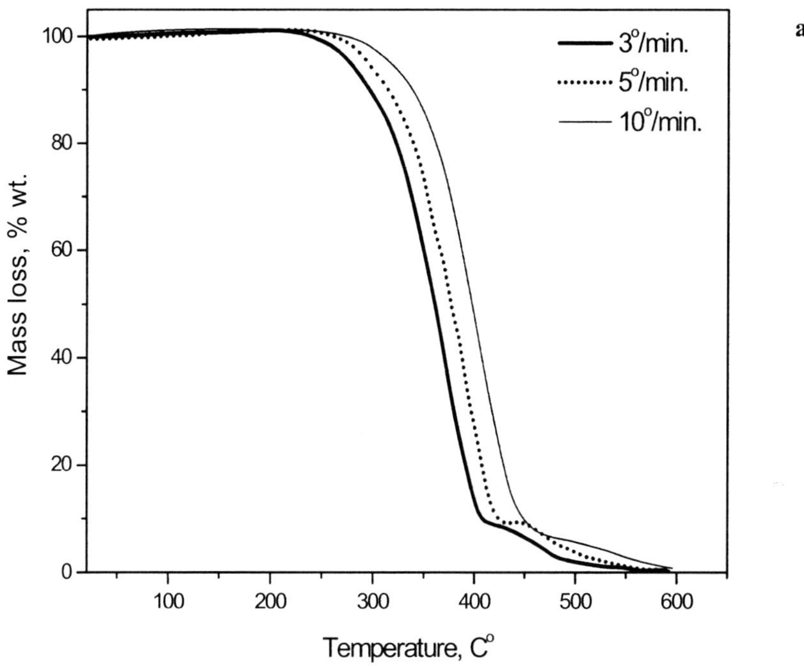

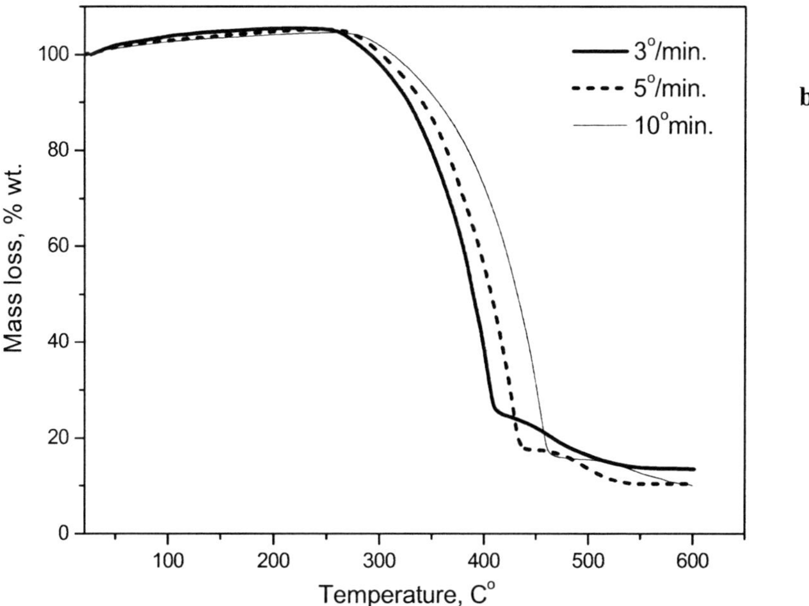

Figure 6. TG curves of PP (a) and PP-MAPP-Cloisite 20A (b) in air at different heating rates.

Model-free methods evaluations were chosen as the starting points in kinetic analysis of neat PP and PP-mPP with 7% Cloisite 20A for determining the activation energy in the development of the model. Figure 7 shows a corresponding Friedman analysis, where the activation energy is a function of partial mass loss change [10].

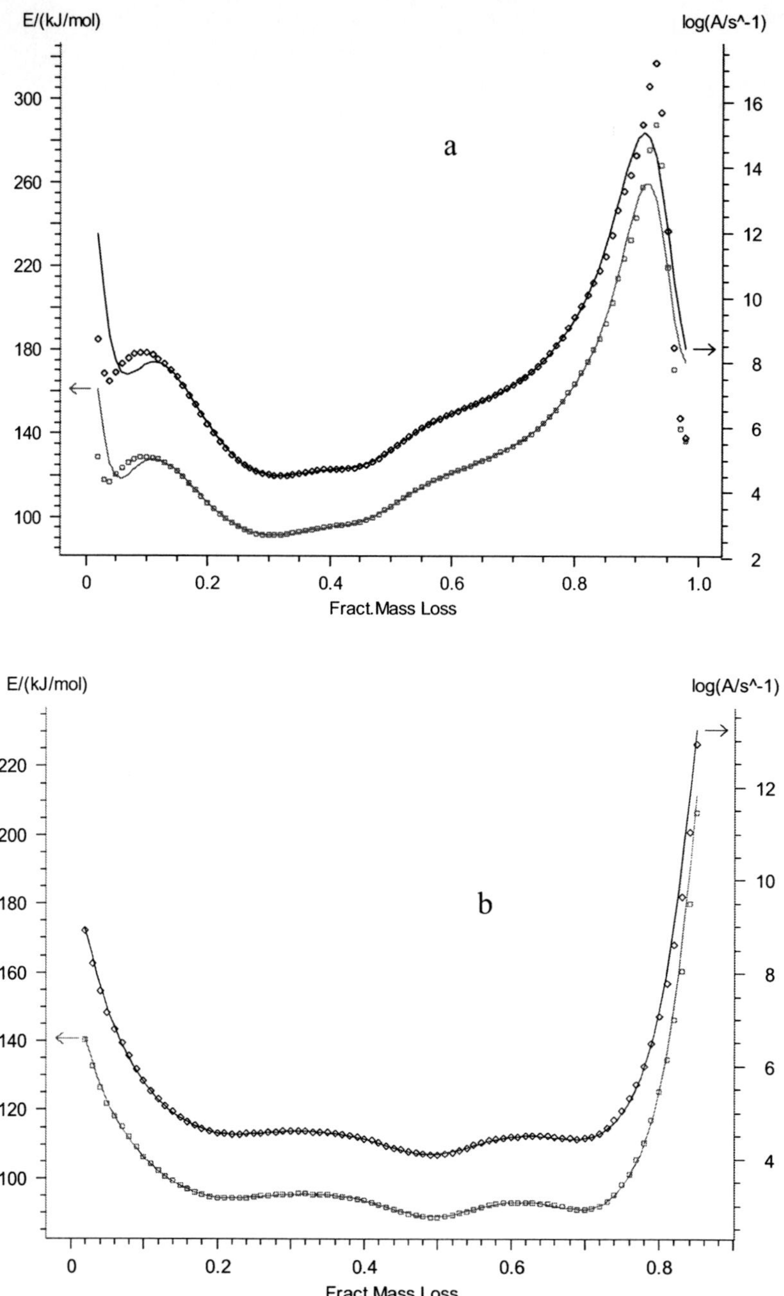

Figure 7. Friedman Analysis of neat PP – (a) and PP-MAPP-Cloisite 20A – (b)

The curves show higher values at the beginning of the sintering process, i.e. at lower partial-length-change values, and considerably higher values at the end of the process (a) and particularly (b). This indicates the presence of a multiple-step process.

First round analysis by Friedman method indicates a complexity of the scheme for neat PP and PP-MAPP-Cloisite 20A thermal degradation in air [10]. Non-linear fitting procedure established the two - stage scheme for neat PP

$$A \rightarrow X_1 \rightarrow B \rightarrow X_2 \rightarrow C \tag{4}$$

and the triple - stage scheme for PP-mPP with 7% Cloisite 20A (Figure 4 a, b) [9,11].

$$A \rightarrow X_1 \rightarrow B \rightarrow X_2 \rightarrow C \rightarrow X_3 \rightarrow D \tag{5}$$

Taking these findings into consideration for neat PP, a fit was attempted using nonlinear regression with model (4), where the nth-order (F_n) reaction type was used for all steps of the reaction (Figure 8, Table 2).

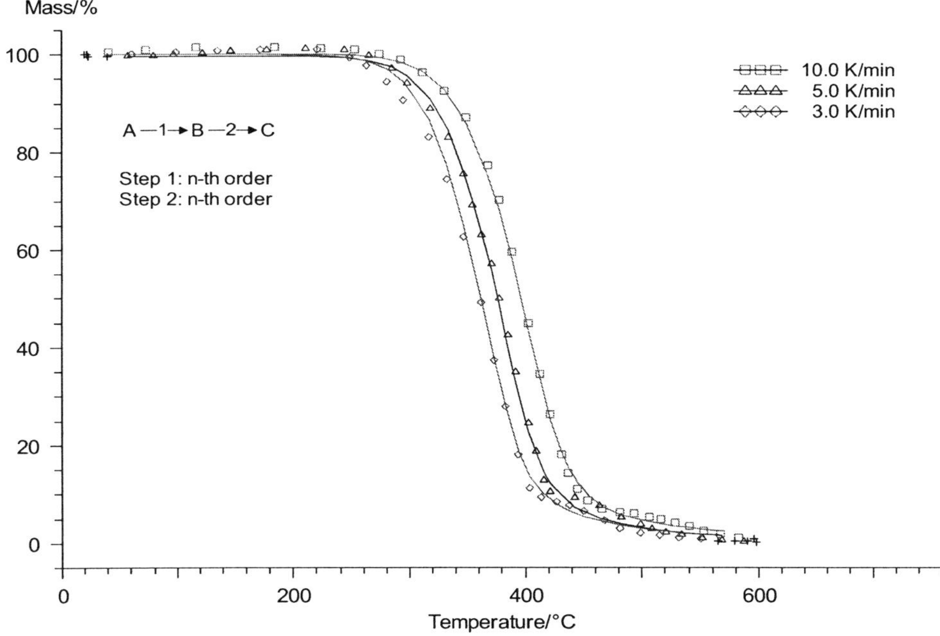

Figure 8. Non-linear kinetic modelling for neat PP.

With this expanded model, an excellent fit is possible for all three measurements. The kinetic parameters are listed in Table 2.

A more sophisticated model (5), based on different reaction types, was chosen for PP-MAPP-Cloisite 20A thermal degradation. The parameters are listed in Table 3.

Table 2. Kinetic Parameters Resulting from Multiple-Curve Analyses (Heating Rates 3, 5 and 10 deg/min) with Reaction Model ($A \rightarrow X_1 \rightarrow B \rightarrow X_2 \rightarrow C$) from TG Measurement of Neat PP

Reaction models (types, Xi)	Parameter	Value	Corr. Coeff.
$F_n \rightarrow F_n$	$\log A_1$, s^{-1}	6.39	0.9994485
	E_1, kJ/mol	110.28	
	N_1	1.13	
	$\log A_2$, s^{-1}	9.85	
	E_2, kJ/mol	151.64	
	N_2	2.59	

Table 3. Kinetic Parameters Resulting from Multiple-Curve Analyses (Heating Rates 3, 5 and 10 K/min) with Different Reaction Models ($A \rightarrow X_1 \rightarrow B \rightarrow X_2 \rightarrow C \rightarrow X_3 \rightarrow D$) from TG Measurement of PP-MAPP-Cloisite 20A

Reaction models (types, Xi)	Parameter	Value	Corr. Coeff.
$F_n \rightarrow F_n \rightarrow F_n$	$\log A_1$, s^{-1}	6.33	0.997415
	E_1, kJ/mol	113.40	
	n_1	1.16	
	$\log A_2$, s^{-1}	8.82	
	E_2, kJ/mol	150.97	
	n_2	2.46	
	$\log A_3$, s^{-1}	11.56	
	E_3, kJ/mol	188.55	
	n_3	0.78	
$F_n \rightarrow D_1 \rightarrow F_n$	$\log A_1$, s^{-1}	6.90	0.998841
	E_1, kJ/mol	113.42	
	n_1	1.21	
	$\log A_2$, s^{-1}	4.72	
	E_2, kJ/mol	100.02	
	$\log A_3$, s^{-1}	12.04	
	E_3, kJ/mol	199.85	
	n_3	1.17	
$F_n \rightarrow D_2 \rightarrow F_n$	$\log A_1$, s^{-1}	6.45	0.997409
	E_1, kJ/mol	113.37	
	n_1	1.68	
	$\log A_2$, s^{-1}	6.23	
	E_2, kJ/mol	118.43	
	$\log A_3$, s^{-1}	11.68	
	E_3, kJ/mol	197.24	
	n_3	0.95	
$F_n \rightarrow D_3 \rightarrow F_n$	$\log A_1$, s^{-1}	6.49	0.997280
	E_1, kJ/mol	113.62	
	n_1	2.04	
	$\log A_2$, s^{-1}	8.26	
	E_2, kJ/mol	152.32	
	$\log A_3$, s^{-1}	11.84	
	E_3, kJ/mol	197.06	
	n_3	0.94	

Table 3. Kinetic Parameters Resulting from Multiple-Curve Analyses (Heating Rates 3, 5 and 10 K/min) with Different Reaction Models ($A{\rightarrow}X_1{\rightarrow}B{\rightarrow}X_2{\rightarrow}C{\rightarrow}X_3{\rightarrow}D$) from TG Measurement of PP-MAPP-Cloisite 20A (Continued)

Reaction models (types, Xi)	Parameter	Value	Corr. Coeff.
$F_n{\rightarrow}D_4{\rightarrow}F_n$	$\log A_1$, s^{-1}	6.58	0.997337
	E_1, kJ/mol	113.68	
	n_1	2.01	
	$\log A_2$, s^{-1}	7.23	
	E_2, kJ/mol	138.94	
	$\log A_3$, s^{-1}	11.97	
	E_3, kJ/mol	195.19	
	n_3	0.98	
$F_n{\rightarrow}A_2{\rightarrow}F_n$	$\log A_1$, s^{-1}	6.61	0.997557
	E_1, kJ/mol	114.40	
	n_1	2.10	
	$\log A_2$, s^{-1}	5.60	
	E_2, kJ/mol	105.21	
	$\log A_3$, s^{-1}	12.03	
	E_3, kJ/mol	199.34	
	n_3	1.23	
$F_n{\rightarrow}A_3{\rightarrow}F_n$	$\log A_1$, s^{-1}	6.50	0.997613
	E_1, kJ/mol	114.07	
	n_1	2.02	
	$\log A_2$, s^{-1}	4.81	
	E_2, kJ/mol	95.25	
	$\log A_3$, s^{-1}	12.36	
	E_3, kJ/mol	200.13	
	n_3	1.46	
$F_n{\rightarrow}A_n{\rightarrow}F_n$	$\log A_1$, s^{-1}	6.73	0.998355
	E_1, kJ/mol	114.72	
	n_1	0.97	
	$\log A_2$, s^{-1}	4.75	
	E_2, kJ/mol	97.73	
	$\log A_3$, s^{-1}	11.86	
	E_3, kJ/mol	200.01	
	n_3	1.26	
$F_n{\rightarrow}R_2{\rightarrow}F_n$	$\log A_1$, s^{-1}	7.10	0.997984
	E_1, kJ/mol	114.89	
	n_1	1.27	
	$\log A_2$, s^{-1}	5.10	
	E_2, kJ/mol	105.66	
	$\log A_3$, s^{-1}	12.13	
	E_3, kJ/mol	200.34	
	n_3	1.76	
$F_n{\rightarrow}R_3{\rightarrow}F_n$	$\log A_1$, s^{-1}	7.15	0.997898
	E_1, kJ/mol	114.62	
	n_1	1.10	
	$\log A_2$, s^{-1}	5.22	
	E_2, kJ/mol	108.92	
	$\log A_3$, s^{-1}	11.56	
	E_3, kJ/mol	188.55	
	n_3	0.78	

Taking these fittings for PP-MAPP-Cloisite 20A, a best approximation was attempted using nonlinear regression with model (5), based on the best fit quality (correlation coefficient) (Figure 9), where the one-dimensional diffusion (D_1) reaction type was used for the second step of the reaction (Figure 10, Table 3), whereas the nth-order reaction models were chosen for the first and third steps respectively.

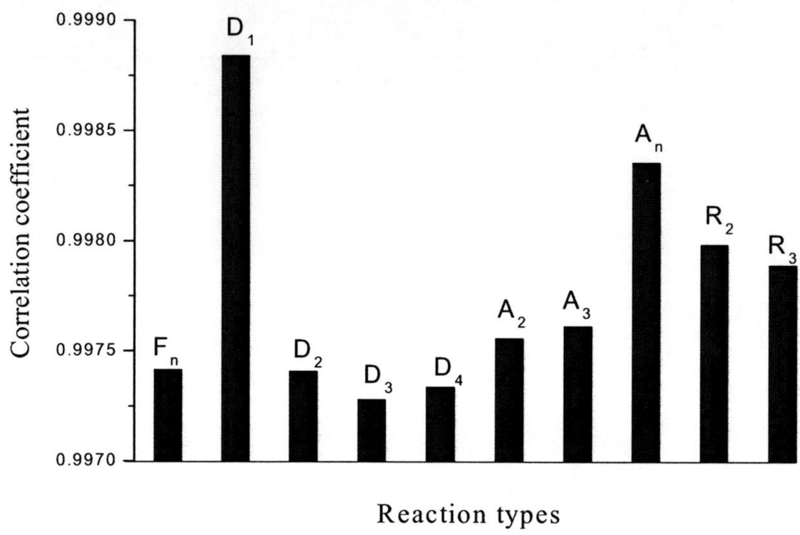

Figure 9. Fit quality between the multiple steps models. Correlation coefficients of different reaction types.

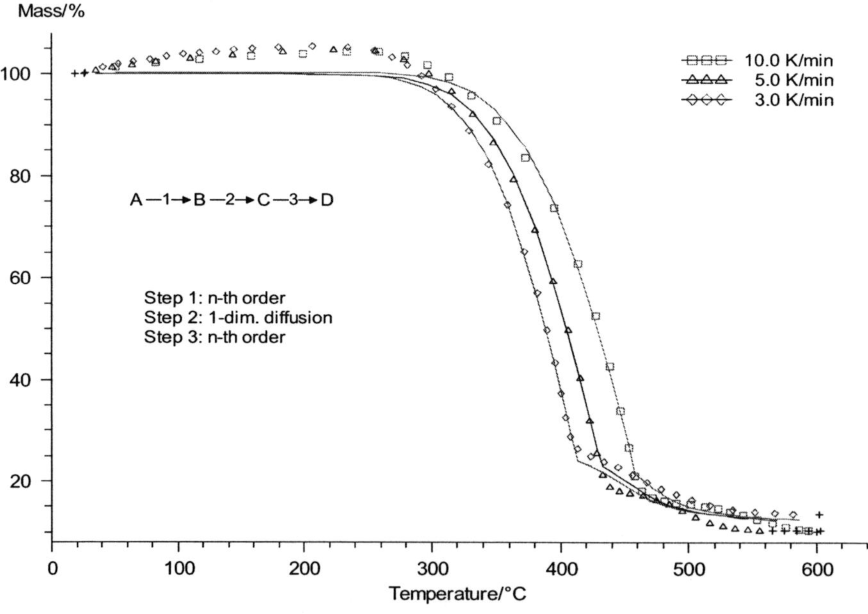

Figure 10. Non-linear kinetic modelling for PP-MAPP-Cloisite 20A.

These results show that the second step in thermal degradation of PP-MAPP-Cloisite 20A is described by one-dimensional diffusion (D_1) reaction type which is liable for the overall process of the carbonization in nanocomposite polypropylene structure.

Flame Resistant Properties

The recent interest in the reported char-promoting functionalized dispersed nanoclays to yield nanocomposite structures having enhanced fire and mechanical properties, when the clays are present only at levels of 2~10%, prompts their investigation as potential fire retardants.

Because of its wholly aliphatic hydrocarbon structure, neat polypropylene by itself burns very rapidly with a relatively smoke-free flame and without leaving a char residue. It has a high self-ignition temperature (570°C), a rapid decomposition rates and hence has a high flammability.

Polypropylene nanocomposites have attracted more and more interest in flame retardant area in recent years due to their improved fire properties [12-14]. It is suggested that the presence of clay can enhance the char formation providing a transient protective barrier and hence slowing down the degradation of the matrix [13,14]. Thus, the study of isothermal flash pyrolysis of PP composition under the temperature higher 400-500°C allows us to forecast their flammability.

The kinetic results of the present study let us the basis to predict the mass loss of material under isothermal pyrolysis conditions using the same thermokinetics software.

Figure 11 shows the partial reaction curves as a function of time with temperature (400 – 600°C) as a parameter.

It is clearly seen that under conditions of polymer ignition and initial surface combustion, the mass loss for PP-MAPP-Cloisite 20A and its rate are noticeably lower then adequate values for the neat PP. An improvement in flame resistance of PP-MAPP-Cloisite 20A over the neat PP happens as a result of the char formation providing a transient protective barrier. In the present study this phenomena was interpreted in terms of isothermal kinetic analysis.

Apart from this information, the graphs of mass loss rates (*dm/dt*) vs. time for neat PP and PP-MAPP-Cloisite 20A indicate the depression of the degradation (fuel) products under the isothermal pyrolysis conditions at 600°C (Figure 12). It is well known that the temperature of 600°C corresponds to an incident heat flux of 35 kW/m^2 which is referred to the full scale fire scenario [15].

CONCLUSION

The kinetic data obtained by dynamic TGA designate thermal stabilization effect of nanoclay structure into a polymer matrix, caused in one-dimensional diffusion process of catalytic-charring throughout the thermal degradation of PP nanocomposition. According to provided kinetic analysis, polypropylene nanocomposite demonstrated the transcendent thermal and fireproof behaviour in relation to neat polypropylene.

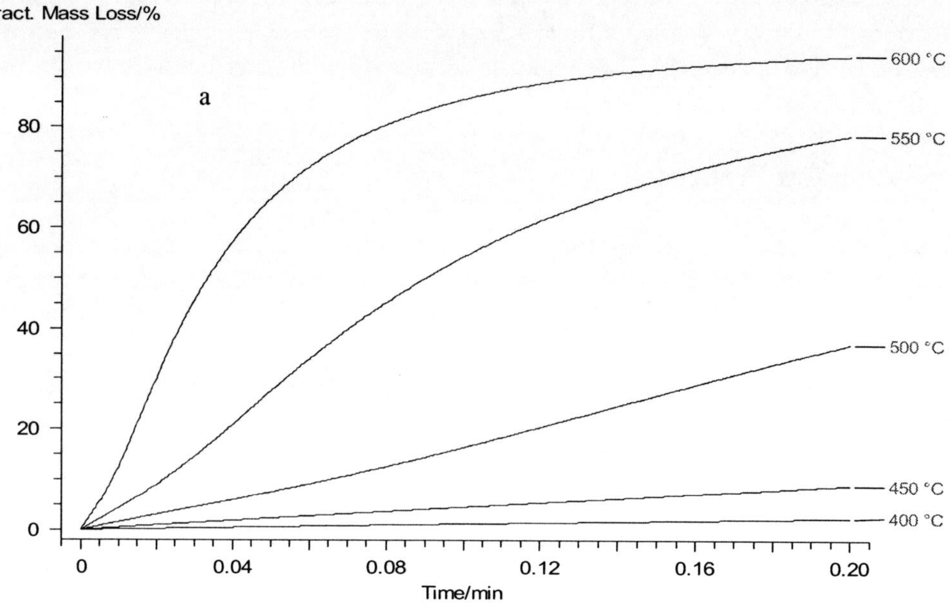

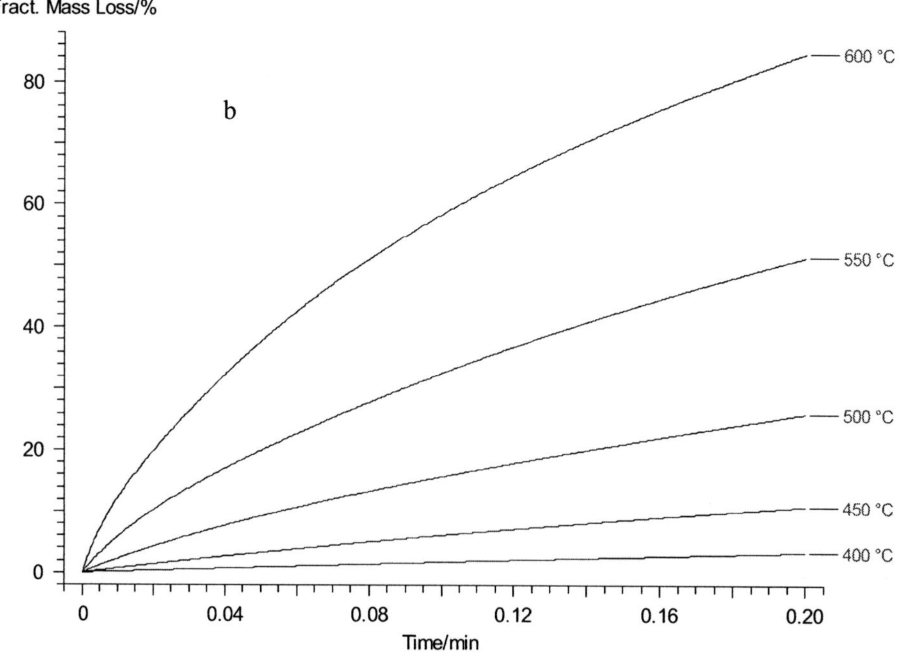

Figure 11. Fractional reaction vs. time (15 sec.) for neat PP (a) and PP-MAPP-Cloisite 20A (b).

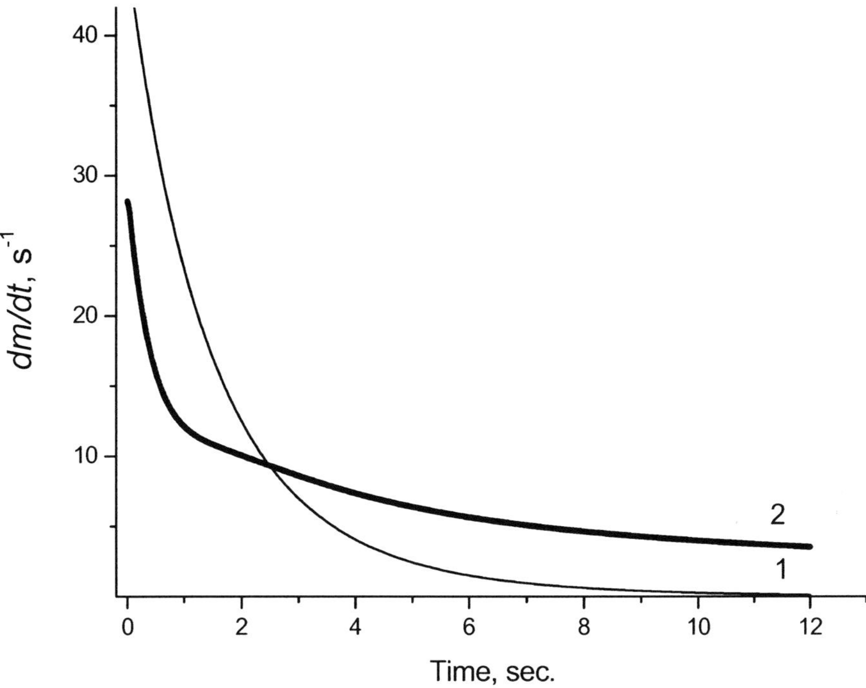

Figure 12. Mass loss rates vs. time for neat PP-MAPP-Cloisite 20A (2) under the isothermal heating condition of 600°C.

ACKNOWLEDGEMENTS

This work was supported by the Russian Foundation for Basic Research, project N 04-03-32052.

The authors are pleased to acknowledge Mrs. Natalia Erina from Veeco Metrology of LLC AFMs, Santa Barbara, Ca for supplying AFM data, and the Moscow division of NETZSCH-Gerätebau GmbH for Thermokinetics software.

REFERENCES

[1]   Giannelis E, *Adv Mater*, *8*:29 (1996).

[2]   Gilman JW., Kashiwagi T, Nyden MR, Brown JET, Jackson C L, Lomakin SM, Giannelis E P, Manias E, *in Chemistry and Technology of Polymer Additives*, Chapter 14, ed by Ak-Malaika S, Golovoy A, Wilkie CA, p 249, Blackwell Science Inc., Malden MA (1999).

[3]   Zanetti M, Lomakin S, Camino G, *Macromol Mater Eng*, 279:1-9 (2000).

[4]   Kojima Y., Usuki A., Kawasumi M., Okada A., Fukushima Y., Kurauchi T., Kamigaito O., *J Mater Res*, 8:1185 (1993).

[5]   Zanetti M, Camino G, Reichert P, Mülhaupt R, *Macromolecular Rapid Communications* 22:176-180 (2001).

[6]   Grassie N, Scott G, *in Polymer degradation and stabilization*, Cambridge University Press, Cambridge, p 275 (1985).

[7]   March J, *in Advanced Organic Chemistry*, McGraw-Hill Kogakusa Ltd, Tokyo, p.367 (1977).

[8]   Benson SW, Nogia PS, *Account of Chemical Research*, 12:233 (1979).

[9]   Opfermann J, *J Thermal Anal Cal*, 60: 641 (2000).

[10]   Friedman H L, *J Polym. Sci.*, C6:175 (1965).

[11]   Opfermann J, Kaisersberger E, *Thermochim Acta*, 11:167 (1992).

[12]   Lomakin SM, Zaikov GE, *in Modern Polymer Flame Retardancy*, VSP Int. Sci. Publ. Utrecht, Boston, p. 272 (2003).

[13]   Gilman J W, *Applied Clay Sci*, 15:31 (1999).

[14]   Gilman G W, Jackson C L., Morgan A B, Harris R H, Manias E, Giannelis E P, Wuthenow M, Hilton D, Phillips S, *Chem. Mater*, 12:1866 (2000).

[15]   Babrauskas V, *Fire and Materials*, 19:243 (1995).

In: Physical Organic Chemistry: Theory and Practice
Eds: A. D'Amore and G. E. Zaikov, pp. 111-120

ISBN 1-59454-275-9
© 2005 Nova Science Publishers, Inc.

Chapter 6

INVESTIGATION OF PHYSICAL, MECHANICAL AND RHEOLOGICAL PROPERTIES OF COMPOSITIONS CONTAINING ORGANIC AND NON-ORGANIC FILLERS

O. Legonkova and A. Bokaref
Moscow State University of Applied Biotechnology, Moscow, Russia
V. Vasiliev*
Institute of Organoelement Compounds, Moscow, Russia

ABSTRACT

The article is devoted to investigation of mechanical and rheological properties of compounded mixes. There were taken different fillers, organic and inorganic. Both of them bring biodegrability to the whole composition.

Key words: fillers, compounded mix, physical and mechanical properties, rheological properties

INTRODUCTION

In the last years polymer compositions, consisting of mixtures of polymers and inorganic substances of different chemical origin, structure and of organic substances in the form of blends and alloys [1] are widely spread nearly in all spheres of human life. This general approach of receiving of compositions opens great possibilities for assembling new and new combinations.

---

* Institute of Organoelement Compounds, 28, Vavilova str., Moscow 117813, Russia; e-mail: ol@te-ka.ru, fax: 7(095) 474-4784, tel: 7 (095) 277- 0713.

At the same time properties of polymers compositions, as a rule, are not diagnosed as a sum of properties of their components, but are determined through differences in chemical and physical processes during interactions of components on the boundary of phases [2].

Recently polymer blends have been finding their application as biodegradable polymers [3-7] that include biodegradable component, i.e.: cellulose, its esters, starch (widely spread), etc. All these components lead to bio-damage of the whole composition. Such types of materials are of great interest from practical point of view.

In literature investigations of physical and mechanical and rheological properties of highly filled materials, containing of polymer matrix and of mixture of organic and inorganic fillers at one time that brings biodegrability of the final composition, are insufficiently paid attention to. And it is of great interest to explore changes in strength properties of the composition, as it is this figure that is the most difficult to be diagnosed, and is peculiarly fixed by probabilistic processes of origin and evolution of structural damages in loaded compositions, by nature of components and their durability properties, and by phase interactions which provide the realization of needed characteristics.

EXPERIMENTAL PART

This work is devoted to discussion of the results of investigation of physical – mechanical and rheological properties of compounded materials. Organic and inorganic fillers were taken. Waste of grain thrashing were chosen as an organic filler, it has particles of 63-240 mkm in dimension, piled density – 350 kg/m^3 , moisture – 4%. Water soluble mineral complex fertilizer of type "Rastvorin-A" was used as an inorganic filler which consists of the necessary microelements needed for growth and evolution of plants as well as the main trace elements, such as: Mn, Cu, Zn, B, Mo, in the form of their salts.

Co-polymer of ethylene and vinyl acetate (CEVA) with different content of vinyl acetate groups; co-polyamide (Co-PA), received by polycondensation of adipinic and sebacic acids and hexamethylendiamine, trade mark – H-005 (Tg = 115-120^{0}C, melting index – 15 g\10min); co-polymer of acrylic acid and styrolene – Lentex A4 were taken as polymer matrix.

The criterion, governing the chose of the said above polymer materials, was the presence of functional groups in all these polymers. Another determining factor was the low temperature of their processing (130-150^{0}C). This factor is rather important from the point of preservation of properties of fillers during getting samples.

Mechanical properties of samples, differing to each other with a wide range of ratio of ingredients, were measured through "Instron" machine, rheological properties were determined through method of capillary viscometry via defining melting index (MI), one of the main factor, characterizing the ability of compounded mix to be processed in specific manufacture with one or another method.

Blending was carried out through dispersion of mixture, laminar melting and pressing. During these processes samples in the form of plates of 500 microns in thickness were gotten.

RESULTS AND DISCUSSION

During dispersion of fillers into CEVA, regardless of the content of vinyl acetate, it was revealed that the increase of content of both organic and inorganic fillers into two phase system: polymer-filler, strength and deformation at break came down, figures 1,2. In case of blending with organic filler (waste of grain thrashing), samples became more rigid. In case of blending with mineral fillers even in great amount (mass.60%) all samples retain relatively high plasticity (deformation at break amounts 400% in highly compounded mix). This behavior may be connected with decrease of polymer in the bulk of composite and with plastifying action of inorganic filler, which is mixture of different salts.

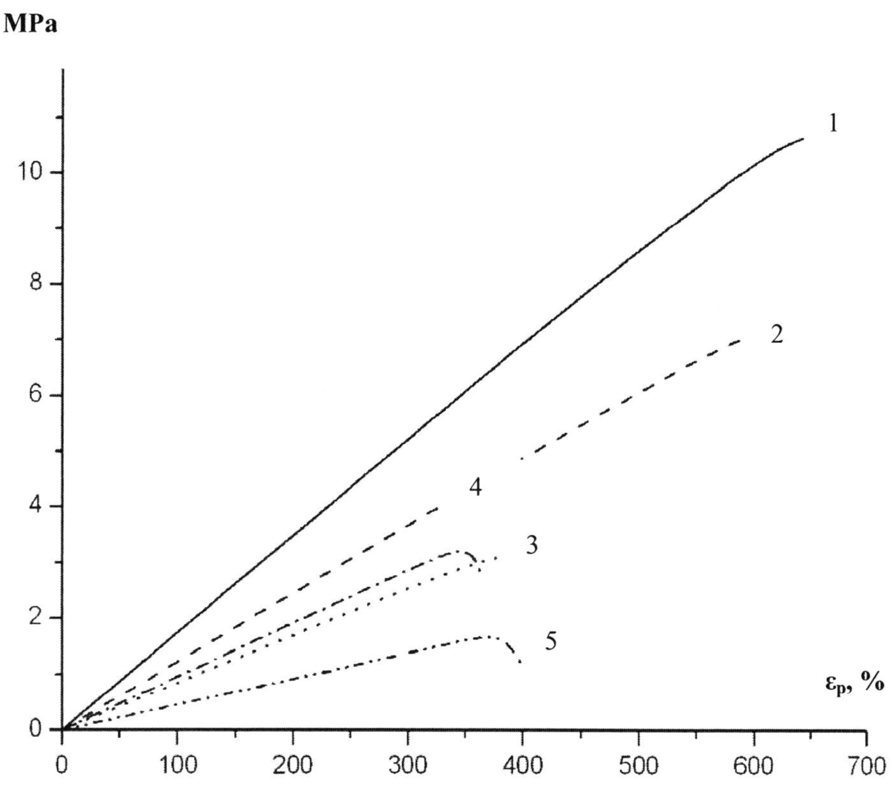

Figure 1. Dependence of strength (σ_p) on deformation (ε_p) at break for two component system CEVA/inorganic filler at different percentage of filler, mass. %: 1 – 100/0; 2 – 80/20; 3 – 60/40; 4 – 50/50; 5 – 40/60.

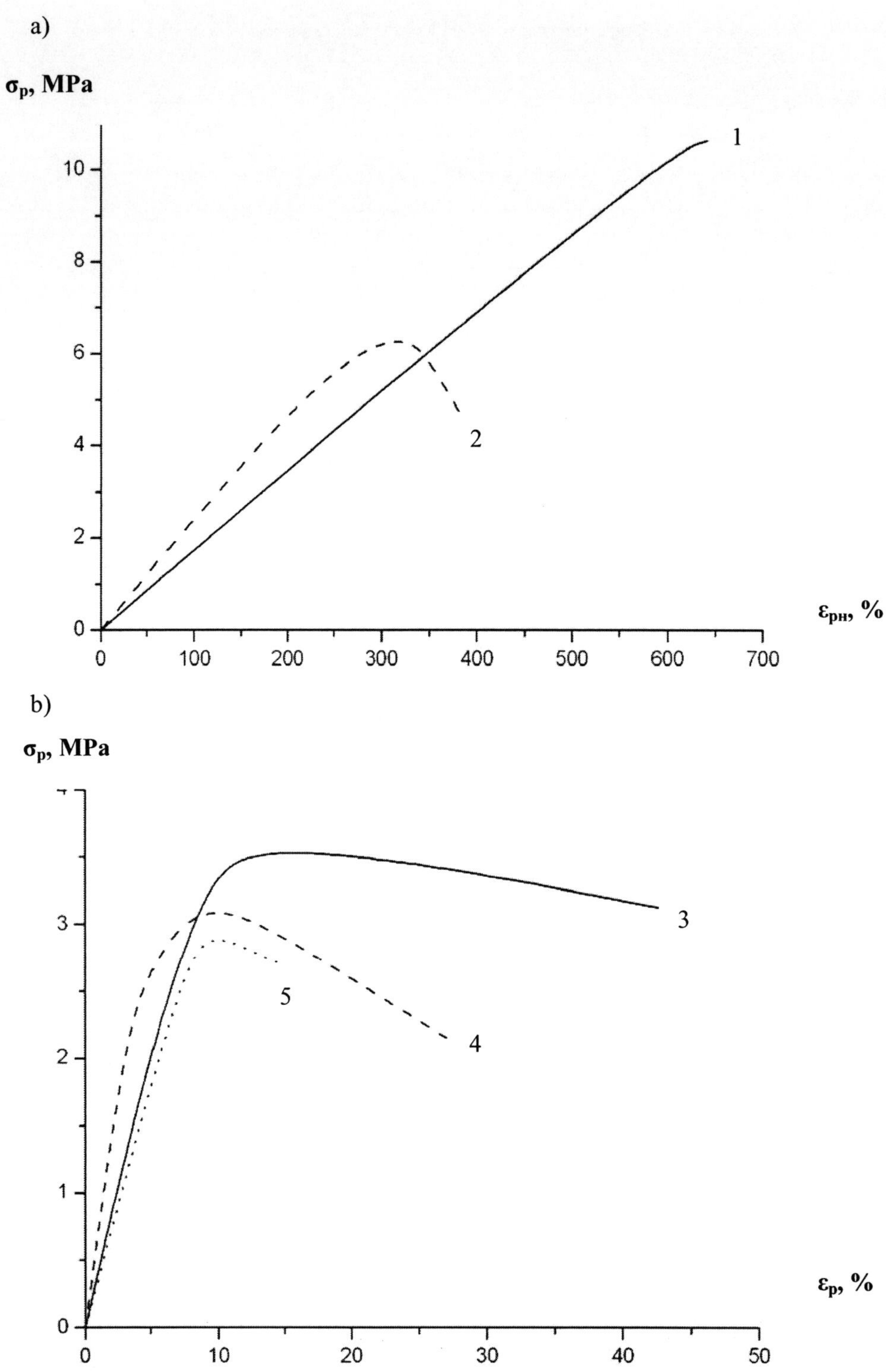

Figure 2. Dependence of strength (σ_p) on deformation (ε_p) at break for two component system CEVA/organic filler at different percentage of filler, mass. %: 1 – 100/0; 2 – 80/20; 3 – 60/40; 4 – 50/50; 5 – 40/60.

Having a look on three component system CEVA-inorganic filler- organic filler, Figure 3., it should be marked that increase of percentage of inorganic filler (mineral fertilizer) makes the final samples more plastic and pliable.

The curves of changes of mechanical properties of samples, based on CEVA dependently on content of vinyl acetate groups in polymer, have similar character.

On the Figure 4 the dependence of ratio $(\sigma_{max})/(\sigma_{рисх})$ on percentage of inorganic filler is shown (where σ_{max} is strength at break for filled with inorganic component samples and $\sigma_{рисх}$ is strength at break for not filled samples). Samples were prepared while having used CoPA polymer as polymer matrix

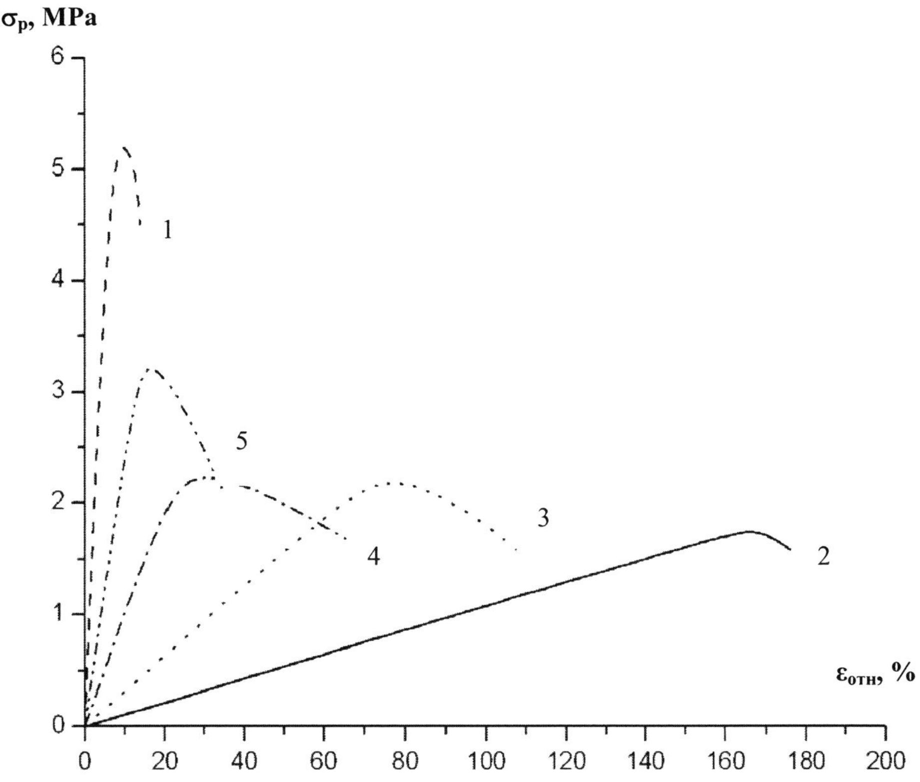

Figure 3. Dependence of strength (σ_p) on deformation (ε_p) at break for system based on CEVA at different percentage of fillers, mass. %: 1 – CEVA/organic filler - 50/50; 2 – CЭVA/inorganic filler - 50/50; 3 – CЭBA/inorganic filler/organic filler - 50/37/13; 4 – CЭVA/inorganic filler/organic filler - 50/25/25; 5 – CЭvA/inorganic filler/organic filler - 50/13/37.

While putting inorganic filler up to 10% into composition, Figure 4, durability at break slightly falls. It might be connected not only with the decrease of amount of polymer in the cross-section, but also with the fact that at small extent of filling even in the presence of adhesion between polymer and filler but in the lack of aggregation of powder particles, the filler behaves as thickener of tensions and potential source of growth of cracks [8]. At expansion of inorganic filler up to 30% the ratio $(\sigma_{max})/(\sigma_{рисх})$ increases more than twice. It could be explained both with increasing amount of molecular interactions between components of the composition and with the occurrence that the increase of content of filler leads to rising of obstacles for growth of the crack which promote braking the processes of demolition [9]. The further reduction of strength is typical for highly compounded mixes with the great amount of fillers and the lowest content of polymer matrix. At concentration of inorganic filler nearby 70% the samples became brittle, have strength much lower then the initial sample. Deformation at break for all samples decreases with increasing amount of fillers, and finally became twice lower.

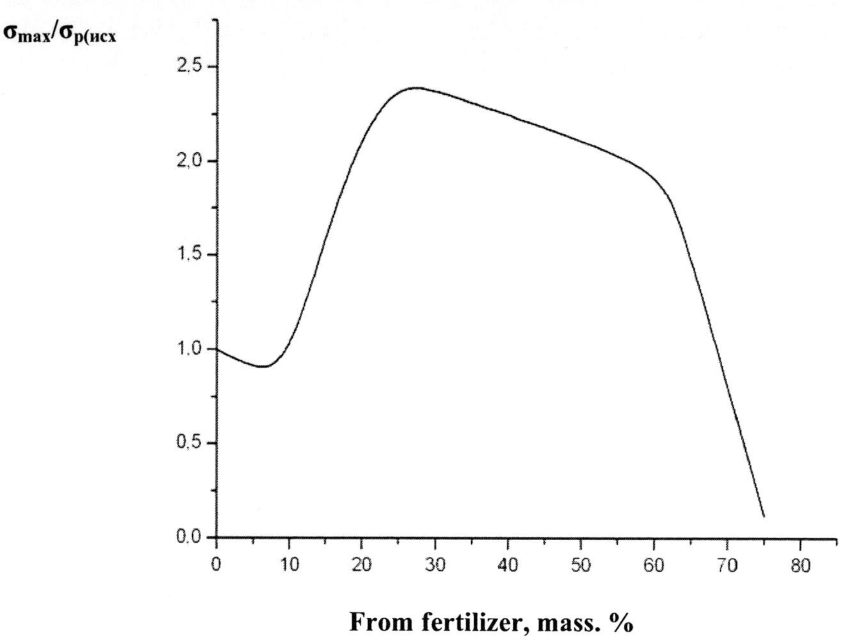

Figure 4. Change of the ratio $\sigma_{max}/\sigma_{рисх}$ depending on content of inorganic filler for two component system CoPA/fertilizer

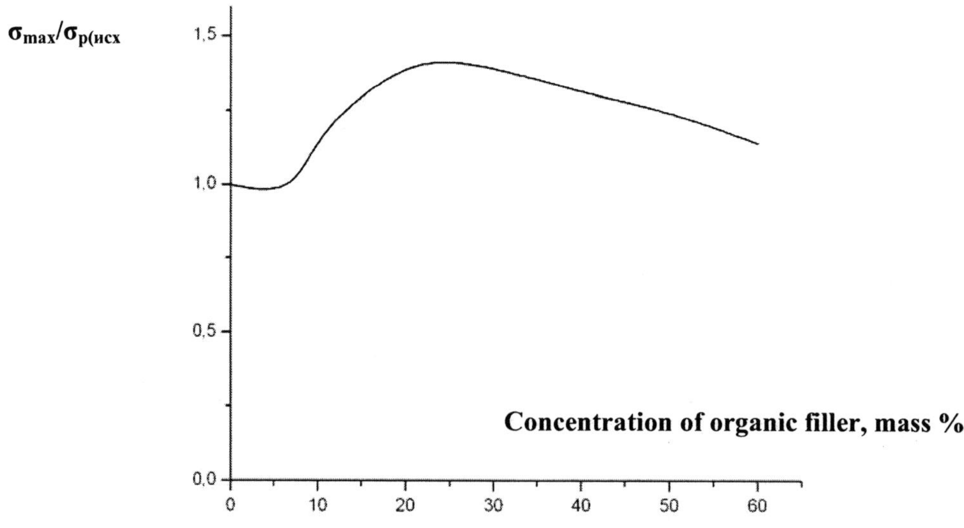

Figure 5. Change of the ratio $\sigma_{max}/\sigma_{рисx}$ depending on content of organic filler for two component system CoPA/waste.

On the figure 5 the similar regularity is shown, i.e. the effect of relative strengthening of the filled with organic compound system is observed. However this result is not so considerable as we can see in two component system with inorganic component.

On the figure 6, changes of durability at break for three component system CoPA/organic filler/inorganic filler depending on content of organic and inorganic fillers are shown. The amount of CoPA is permanent and equals to 50%. It has to be noted once again that increase of the amount of waste brings more rigid and solid samples.

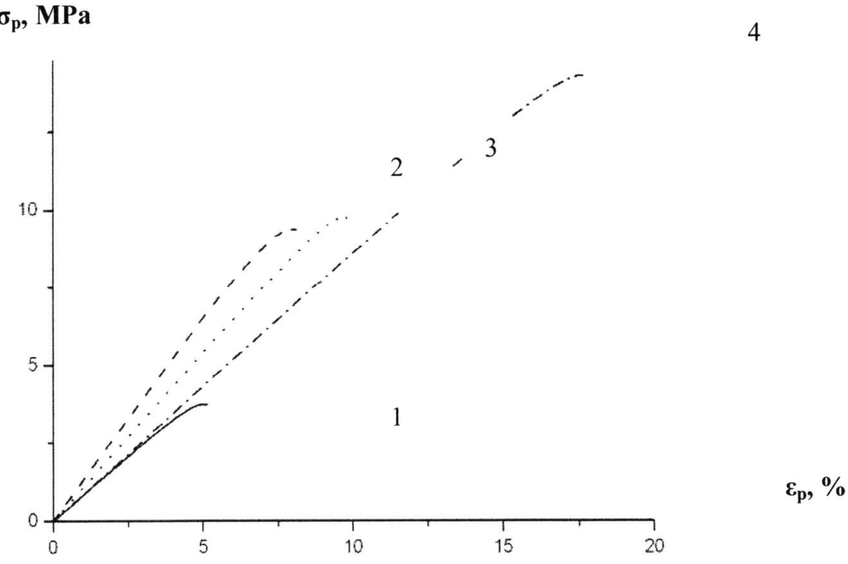

Figure 6. Dependence of strength (σ_p) on deformation (ε_p) at break for three component system – CoPA/organic filler/inorganic filler, mass.%: 1 – 50/10/40; 2 – 50/20/30; 3 – 50/25/25; 4 – 50/40/10.

In case we take co-polymer of acrylic acid and styrolene – Lentex A4 in the amount of 40% , inflate with inorganic filler, then we have the decreased strength (0.9 time smaller). If we inflate organic filler, we have the increased strength (2.7 time bigger).

So, implication of inorganic filler to CoPA up to 10% brings not only to negligible decrease of durability of the compounded mix but to lowering of melting index (MI), figure 7. And then, having inflated the system with filler till 60%, we've got dramatic enlargement of MI in comparison with the value of MI of the original polymer. This effect confirms plastifying activity of inorganic filler in polymer system. After 60% of inflation, the MI reduces. This behavior is connected with glut of the compounded mix with inorganic filler.

The revealed effect is defined by total influence of individual components of inorganic filler on viscosity of composition. To explain this effect, the polymer CoPA was inflated with individual salts constituted the inorganic filler separately. The inflation was carried out in the amount of 1-2%, taking into account that the discussed effect was noted at 25-30% concentration of complex fertilizer, Figure 7, and the content of individual component in inorganic filler "Rastvorin A" is 5-15%. Chloride of potassium, magnesium, oxide of phosphorus, nitrate of potassium, nitrate of ammonium were introduced into polymer. And it was found out that it was nitrogen-containing salts cause this sizeable increase of MI of synthetic polymers, table 1.

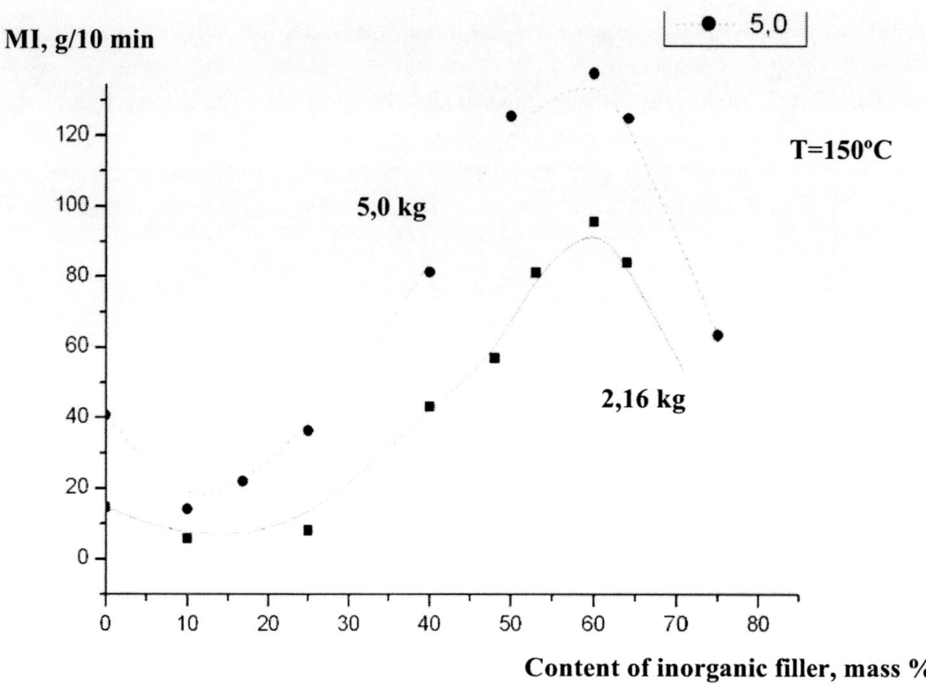

Figure 7. Dependence of MI on concentration of inorganic filler in two compound systerm CoPA/fertilizer.

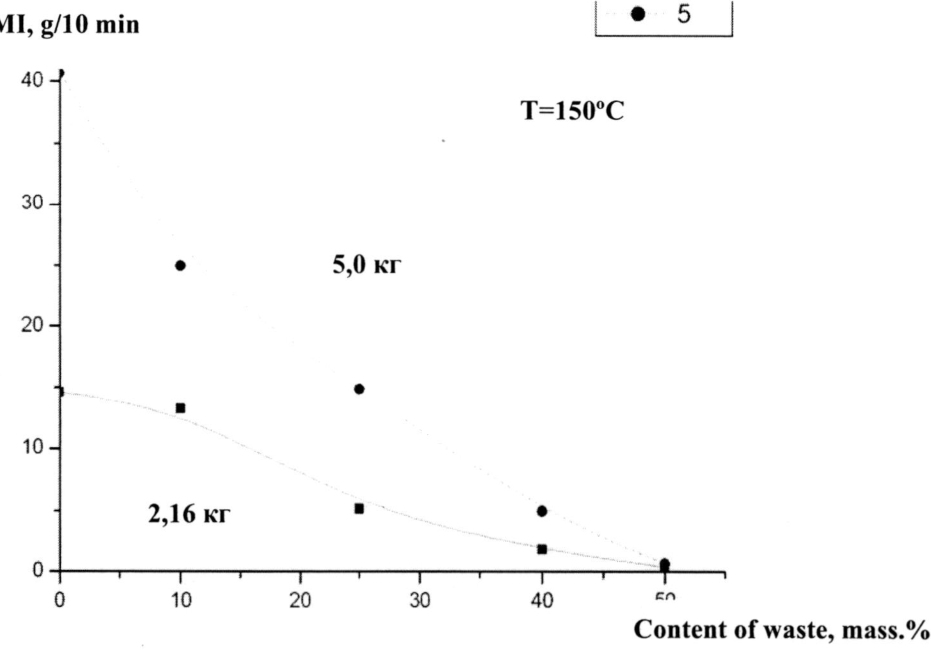

Figure 8. Changes of MI in dependence of content of organic filler in the blend based on CoPA.

Table 1. Change of MI in Dependence on Concentration of Individual Salts

Content, mass.%	Not filled samples	MgCl2	KCl	P2O5	KNO3	NH$_4$NO$_3$
1	7,8	7,5	6,7	8,1	2,9	4,3
2	7,5	7,1	6,3	7,1	Too high to be measured	

Inflation of minor amount of organic filler enlarges viscosity of the system. Mixture containing 50% of organic waste is deprived of fluidity, Figure 8. The received results are typical for highly compounded mix in the absence of interaction between filler and polymer matrix.

CONCLUSIONS

1. Impact of inorganic filler on mechanical properties of polymer compound depends on the nature of polymer matrix: for CEVA, independently on its mark, and for Lentex A4 the increase of content of inorganic filler results in decreasing of strength at break; for CoPA the increase of content of inorganic filler leads to enlarging of durability more than twice.

   However, for all these types of polymers inflation of inorganic filler brings to the lowering of viscosity of the whole composition, that event is connected with chemical interactions between the components of inorganic filler (namely, nitrogen – containing salts) and polymer matrix.

2. For all these types of polymer matrix the inflation of organic filler results in getting rigid and rough system. In case of using CoPA and Lentex A4 as polymer matrix augmentation of strength at break is noticeable.

3. To receive compounded mix it is important to keep the order of introducing of the components, namely: at first, it's necessary to create highly compounded two – component system polymer – inorganic filler, having the lowest viscosity and then the organic filler should be added. At that sequence we can get three component system which is homogeneous and compatible.

REFERENCES

[1]   S.Katz, V. Milewski. *Handbook of fillers and reinforcements for plastics*, Moscow, "Chemistry", 1981, 735 p.

[2]   J.Lipatov. *Physical and chemical principles of polymer fillin*, Moscow, "Chemistry", 1991, 260 p.

[3]   Patent RU 217432 Cl.

[4]   Patent RU 2126427 Cl.

[5]   Patent RU 2126023 Cl.

[6]   Patent RU 2123014 Cl.

[7]   Patent RU 2117016 Cl.

[8]   J. Manson, L.Sperling. *Polymer blends and composites*, Moscow, "Chemistry", 1979, 430 p.

[9]   J.Lipatov. *Colloid chemistry of polymers*, Kiev, "Naukova dumka", 1984, 340p.

In: Physical Organic Chemistry: Theory and Practice
Eds: A. D'Amore and G. E. Zaikov, pp. 121-126

ISBN 1-59454-275-9
© 2005 Nova Science Publishers, Inc.

Chapter 7

Inhibition of Radiation Destruction and Oxidation of Gamma Irradiated Polypropylene

J. N. Aneli,[*] M. A. Donadze, M. S. Kutsia and L. G. Kapanadze
Corrosion Center of Georgian Technical University, Tbilisi, Georgia

Introduction

The stabilization of polymers at influence of different physical and chemical factors is one of most important problems solved mainly by two ways: 1) synthesis of polymers containing such functional groups in macromolecules, which display the protector properties at influence of aggressive medium, 2) introduction of different organic or inorganic substances in macromolecular systems, in view of acceptors of energy of external influences with consequence dissipation of the reaction promoting to abruption of destruction processes in the macromolecular system [1].

The present work is devoted to comprehensive study of physical and chemical processes in polypropylene (PP) stabilized by different phenols and amines at and after irradiation by gamma ray using several physical and chemical methods of structural analysis. The samples on the basis of PP (mm. 390 000) and following stabilizers: 2,6 ditret-buthyh-4-methylphenol (ionol); 3,5-ditret-buthyl-pyrocatechine (BPC); phenyl -naphtylamine (neozon-D); 2,2-methylen-bis-4-methyl-6-tretbuthil phenol (MTBP) were used. These substances were introduced in PP in amount of 0,4 mmol/g via mechanical mixing of components. The samples were irradiated at 293 K and pressure of residual gases 1 Pa in the glass vessel, which did not give a signal of electron spin resonance (ESR) after irradiation. The temperature of the samples at irradiation (radiation source - Co^{60}, dose rate - 0,1 MGR/hour) did not exceed 303-308 K. ESR spectra were registered on Brucker type radio-spectrometer. ESR measurements were made by standard methods. Infrared spectroscopy (IRS) study was

---

[*] Jimsher Aneli – doctor of technical Sci., leading scientific collabortor of the Corrosion Center of the Georgian Technical University, Tbilisi, Georgia. Tel: +(99532) 31-5205; E-mail: jimaneli@posta.ge

carried out by Perkin-Elmer system device. Thermo-gravimetric measurements were curry out by using of differential-thermal analysis (DTA) method. Irradiated in vacuum PP dissolved in organic solvents and the amount of gel-fraction was estimated by extraction of irradiated polymer in o-ksylol. The degree of polymer destruction was estimated by decreasing of melting point by means of DTA method. It was shown that alkyl-derivative phenols are best effective objects, which influence on destruction processes of PP. Thus non-irradiated PP is melted at 428 K, when irradiated in vacuum at 300 K at 16 MGr PP is melted at 396 K and system PP+ionol is melted at 405 K. ESR spectra (singlet type signal) of irradiated both pure and stabilized PP before oxidation are similar and correspond to polyenil radicals [2]. The half life time of these radicals is higher than 1 year at keeping of samples in air (Figure 1,a). The four components of ESR line for ionol and MTBP indicated the formation of radicals with unpaired electron interacting with 3 protons of methyl group of ionol molecules (Figure 1,b). The ESR spectra of irradiated neozon-D and BPC are singlet type lines (Figure 1,c,d). The spectra ESR of polymer radicals are similar both for pure and stabilized PP at start period after irradiation of materials, because the concentration of polyenyl radicals is too higher than stabilizers radicals in irradiated systems PP+stabilizer.

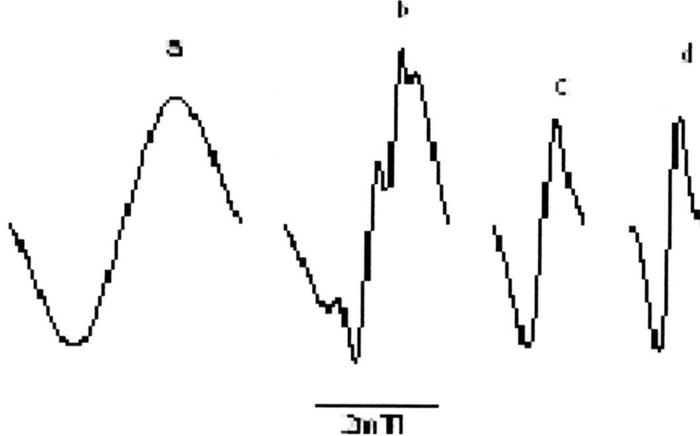

Figure 1. The spectra of gamma irradiated PP without and with stabilizers (a), ionol and MTBP (B), neozon-d(C) and BPC (d)

The concentration of free radicals in irradiated pure PP is higher than in PP+stabilizer. Among stabilizers ionol decreased concentration of polymer radicals in PP more efficiently (Figures 2, 3).

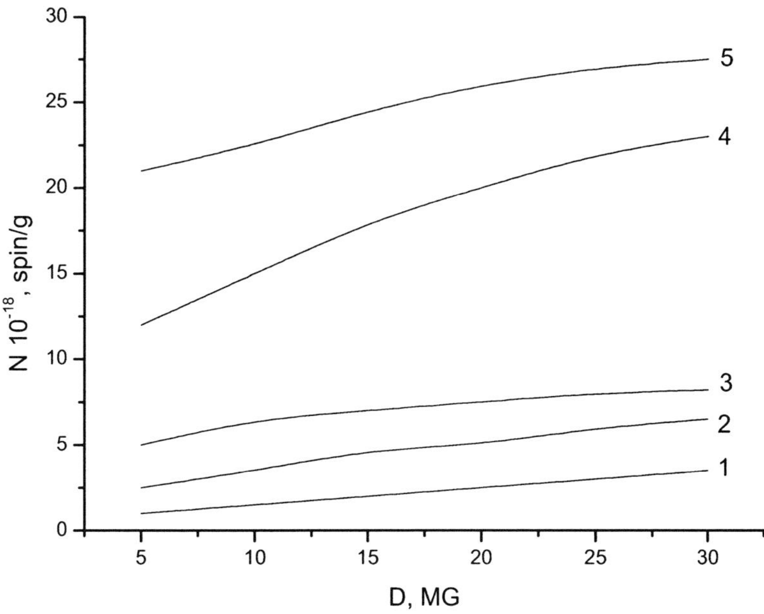

Figure 2. Dependence of the radical concentration in PP+ionol (1), PP+MTBP (2), PP+BPC (3), PP+neozon-d (4), PP (5) on the radiation doses

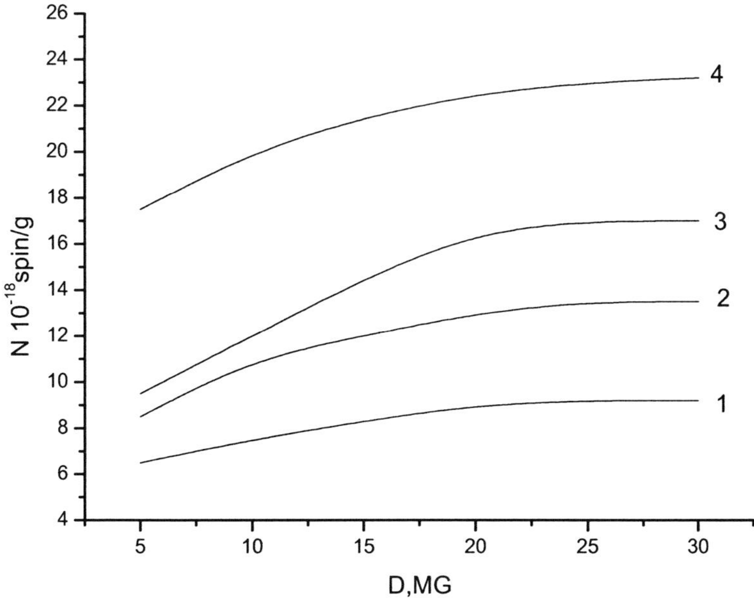

Figure 3. Dependence of radical concentration in neozon-d(1), BPC (2), ionol (3) and MTBP (4) on the radiation doses

It is known that recombination of free polymer radicals often leads to formation of cross-links and consequently-gel [4]. The recombination of these radicals requires much time because of high rigidity and low mobility of PP macromolecular segments. The crystal structure of polymer undergoes to destruction at heat treatment (melting, extraction). It must be noted that irradiation of PP at 77 K forms the alkyl and allyl radicals with multi-component spectra [3], which leads to facilitation of conditions of cross links formation. The amount of gel-fraction allows to calculate the number of free radicals (Table. 1) found by ESR method. Here is supposed that each two radicals give one cross links. However the number of chains is higher, than the sum of measured radical concentration of PP and stabilizers. Apparently the purl of formed radicals is recombined during irradiation [5]. The extraction data allow us to calculate total amount of cross linked molecules, formed in result of radical recombination during irradiation and extraction.

Table 1. Concentration of Free Radicals and Gel-fraction in PP and in PP+Stabilizer (Stabilizer Content in PP is 0,4 mmol/g, Irradiation Dose 16 MGr)

Sample	Ge Content of Gel- Fr fraction, mass. %	Co Content of free radicals, sp spin/g x 10^{18}
PP PP	79. 79. 0	1, 1. 6
PP PP + neozon-D	71. 71. 3	1, 1,.0
PP PP + ionol	0. 0.0	0. 6.0
PP PP + BPC	0. 0.0	0. 0.7
PP PP + MTBP	70. 7 0.1	12 12

In accordance with table 1, the distinction in amounts of gel-fraction in stabilized PP is very high, while the distinction between radical concentrations for these materials is not considerable. Thus, the distinction between radical concentrations for PP+neozon D and PP+ionol is not evidence, while the amount of gel-fraction for first system is 71,3 % and for second one - 0%. Apparently total radical concentration of polymer and stabilizer radicals cannot be only one criterion of effectiveness of stabilizers in irradiated PP.

Comparison of antiradiation properties of ionol and MTBP (molecules of later represent redoubled molecules of ionol) shows that large inhibitor molecules display lower protector properties than small ones.

Free radical recombination, conducted at destruction of crystal structure of PP leads to formation cross links only in non-stabilized PP or in PP+neozon-D. The formation of cross links does not form in PP+phenols. However, in this case the radicals disappeared after heating at extraction. The radical recombination may be passed by another mechanism.

It is very interesting that at extraction of samples of irradiated PP+ionol the stabilizer do not wash from the samples. Probably in this case the compounds of type RI or RI-IR (R – radical, I – inhibitor) are formed in result of reaction between polymer (R) and stabilizer (inhibitor) (I) radicals. These reactions are concerned with reaction R+R $\rightarrow$ R-R, which leads to formation of the crosslinks.

The spectra of ESR of irradiated systems PP + stabilizer show the change of initial spectra at keeping of irradiated samples on air symmetric singlet in pure irradiated PP is gradually transformed with time into nonsymmetrical one, which is correspond to RO$_2$ type peroxide radical [6]. This result is confirmed by ESR data. (Figures 4, 5). The change of radical concentration for PP + inhibitor is decreased to some extent and in case of PP + neozon-D and PP + BPC - is increased. In the last two systems the formation of inhibitor radicals takes place by following reaction: $RO_2 + IH \rightarrow ROOH + I$.

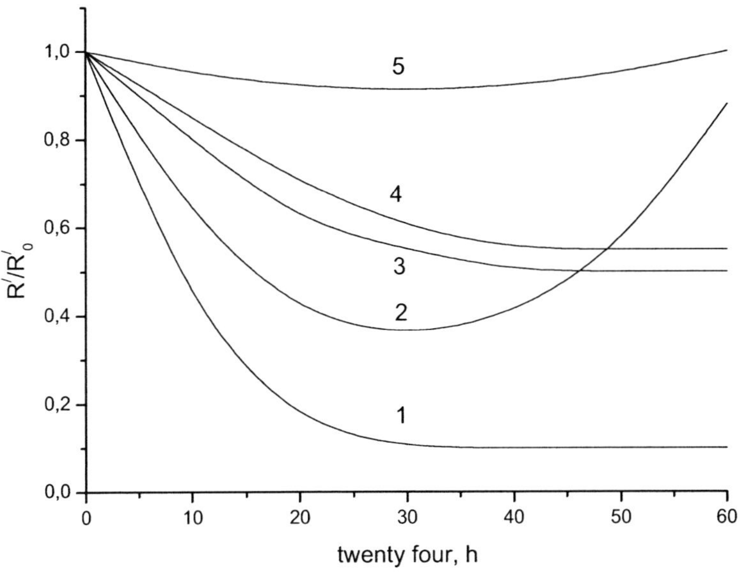

Figure 4. Dependence of radical concentration in PP(1), PP+neozon d (2), PP+ionol (3), PP+MTBP(4), PP+BPC (5)

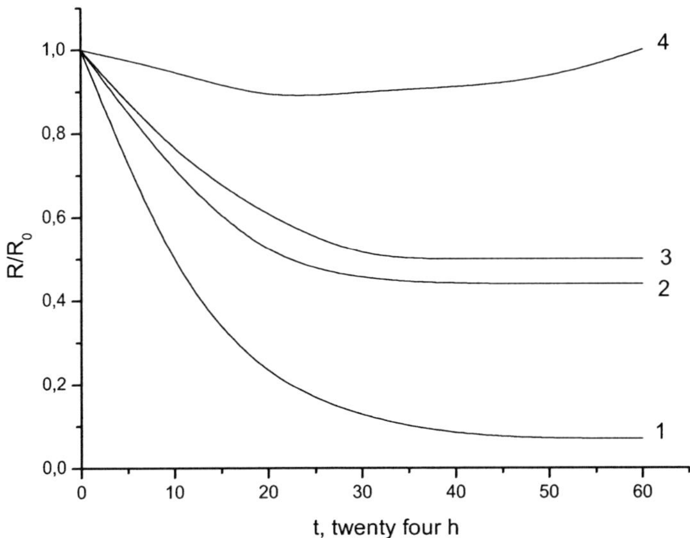

Figure 5. Dependence of radical concentration in ionol(1), MTBP(2), neozon d(3) and BPC(4)

At the investigation of compatibility of PP with different inhibitors in the system PP + ionol the considerable mutual solubility of components was observed. This result is confirmed by decreasing of the melting temperature of irradiated PP. The change of melting points for PP + neozon-d and PP + MTBP was not observed. This fact testifies about bed contacts (compatibility) between components in last systems.

The irradiated PP has low stability at heating in air. Thus, irradiated at 16 MGr PP is oxidized at 400 K. The inhibitors displace the start of oxidation to high temperature range. The amines and phenols display high antioxidant properties.

The experimental data allow us to make a conclusion that all good radiation protectors of PP are simultaneously good antioxidants, but not vice versa.

REFERENCES

[1]   Gross N. and Skott J. *Destruction and Stabilization of Polymers*. Transl. from engl., M., Nauka, 1986, 380 p./Rus./.

[2]   Milinchuk V.K., Klinshpont E.R.., Pshezhetski S. *Macro-radicals*. M.,Khimia, 1980 /Rus./.

[3]   Emanuel N.M. and Buchachenko A.L. *Chemical Physics of Polymer Aging and Stabilization*. M., Nauka. 1982, 359 p./Rus./.

[4]   Nechitailo N.A., Sanin P.I., Polak I.S., Goldenberg A.G., Aneli J.N. *Materials of International Sympozium on Radiation Chemistry of Polymers*. Nauka M.,, 1966, p.272./Rus./

[5]   *ESR of Free Radicals in Radiation Chemistry*. Khimia M., 1972, 389 p./Rus./.

In: Physical Organic Chemistry: Theory and Practice   ISBN 1-59454-275-9
Eds: A. D'Amore and G. E. Zaikov, pp. 127-132   © 2005 Nova Science Publishers, Inc.

Chapter 8

CREATION OF THE CONDUCTING POLYMER COMPOSITES BASED ON SILICON WITH RELAY EFFECT

Jimsher N. Aneli, * *Mamuka S. Kutsia and Medea M. Bolotashvili*
Institute of Machine Mechanics of the Georgian Academy of Sciences, Tbilisi, Georgia

ABSTRACT

The great interest to conducting polymer composites (CPC) is connected with wide possibilities of their technical application. Low (light) specific weight, high corrosion firmness, stable working capabilities under extreme conditions are typical features of CPC, mechanical and thermal properties of which are not less significant than of traditional conducting materials. The presented work is devoted to creation and investigation of some electro-physical properties of new CPC on the basis of some silicon and carbon black fillers. It was established that the properties of CPC essentially depend not only on the selection of CPC components, but also on the technological methods of production of these conducting materials. For obtaining of CPC with desirable conducting properties the method of polymerization filling with simultaneous application of high pressures (up to 1000 MPa) at variation of temperature from range of 150 ... 250 ^{0} C has been used. The obtained CPC materials in view of thin (0,2 ... 2 mm) sheets are characterized by anomalous dependence of electrical resistance on temperature – sharp change of electrical resistance by 3-4 orders. Our investigation shows that this change is reversing. The temperature at which the change takes place depends on pressure, applicable to the samples. It was established also that external electrical pulses might reach analogical change of the sample resistance level. We make conclusion that these CPC adsorb the electromagnetic waves in the range from some millimeters to 3-4 centimeters. Of these materials adsorption coefficients are up to 0,7. The obtained

---

* Main scientific collaborator of the Institute of Machine Mechanics of Georgian Academy of Sciences, doctor of technical sciences, professor. Tel: +(99532) 31-69-67; E-mail: jimaneli@posta.ge

materials are flexible in wide range of temperatures (-140 ... 250 ^{0}C) and their mechanical modules are up to 10 MPa.

INTRODUCTION

Among of CPC based on various polymers and conducting fillers (carbon black, graphite, metallic powders) the composites with silicon attract an attention of researchers by keeping flexible properties at wide range of temperatures [3]. The absent of fundamental theory of conductivity of CPC in general restricts a development of physics, chemistry and technology of these materials. In particular, of the silicon elastomers filled by various conducting fillers.

In presented work the temperature dependence of specific volumetric electrical resistance and electromagnetic wave absorption properties of dimethyl vinyl methyl siloxan (PDMVS) elastomers filled by carbon black of types P357-E, PM-75, P803 and thermally expanded graphite (TEG) have been investigated.

The samples were prepared by using of polymerization filling [4] developed further in work [5] and additive vulcanization methods.

Polymerization filling was conducted using mixture of octamethyl – ciclotetrasiloxane and 1,3,5,7 – tetramethyl – 1,3,5,7 – tetravinyl – cyclotetrasiloxane and 1,3,5,7 – tetramethyl – 1,3,5,7 – tetravinyl – cyclotetrasiloxane in the presence of lamellar carbon black compositions with potassium ($C_{24}K$) or thermal expanded graphite according to reaction considered in [5].

The product obtained was treated by water or 0,1% water of acetic acid, and than diethylaminomethyl trietoxysilane as hardener was added. Further treatment of the mixture was conducted according to the same method, used in additive vulcanization of rubber mixtures [6].

The conductance of samples was measured by 4-electrod method [2]. Determination of the temperature dependence of the samples ρ_v was conducted in the thermostat. In all tests the rate of the temperature change was the same – 1 K per min. Accuracy of ρ_v measurement did not exceed 10%.

Figure 1 shows temperature dependence of ρ_v for composites based on polymerization filled polymethyl – vinylsiloxan (PDMVS) elastomer filled by two types of carbon black, P803 and P357E. The features of these dependences are characterized by significant effect of type and content of the filler on their view.

The composite with P803 type of carbon black displays the ρ_v increase by several orders in the range near 350K. Temperature dependences are reversible. This shows their physical nature. Let us mention that no above-mentioned anomalies in ρ_v dependences on temperature at temperatures far from melting point, or degradation of polymers were found in literature. This phenomenon called by us as the relay effect [7], is generally stipulated by the structural phase transition in the polymer part of the composite. It is known [3] that siloxane polymer chains obtain various conformations depending on temperature, spiral-tangle for example. Such polymorphous transformation is characterized by a jump-like increase of the free volume with temperature and seriously affects technical indexed of rubbers. Mentioned transition in the system of PDMVS composites is observed only in the cases, when vulcanizes were obtained by the additive vulcanization technique and if in this case relatively low structured fillers were used. Thus, the anomalous increasing of ρ_v near 350 K for the composite containing 100 mass parts of P803 carbon black is weakened with the increase of the filler content. In the case of P357E carbon black this anomalous increase of ρ_v disappears. Moreover, the maximum in the ρ_v – T curve shifts to the side of high temperatures with increase of the filler content.

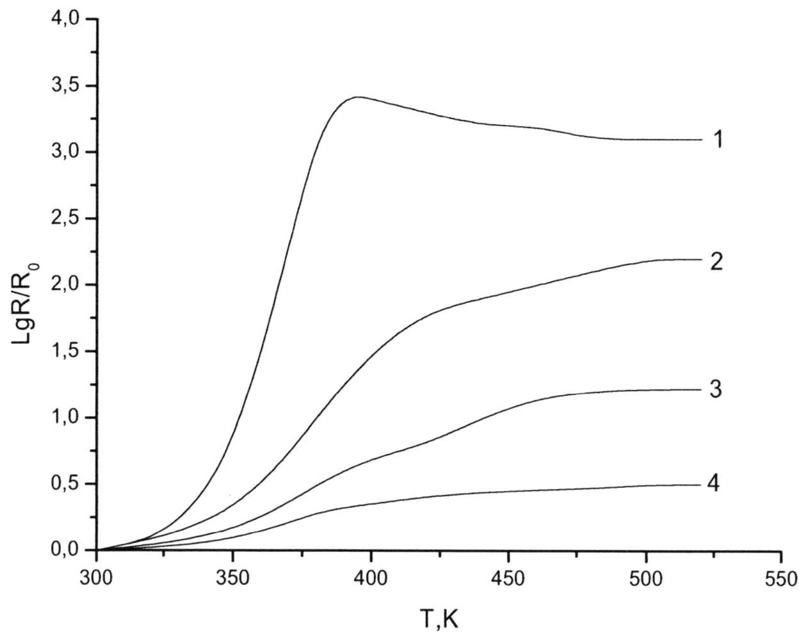

Figure 1. Temperature dependence of ρ_v for composites based on polymerization filled PDMVS with 100 (1), 130 (2), 160 mass part(3) of carbon black P803 and 60 mass part of carbon black P357E(4)

The recepture factor is also displayed at low temperatures, but is very weak in the present case. It was mentioned above that P803 and P357 carbon blacks are mainly different by ratio of three interactions (one of which corresponds to the interphase interaction, and two others – to the intraphase one). Although P803 carbon black is a low – structured one, its interaction with polymer is a little higher than the interphase interaction. Therefore, changes proceeding in the PDMVS elastomer morphology reflect more effectively on ρ_v of the composite containing P803 carbon black, than on those containing highly structured P357 carbon black, which "paralyzes" molecular dynamics of PDMWS. This is well reflected on Figure 1.

Decrease and shift of the temperature maximum of the ρ_v to the side of high temperatures is determined by growth of the materials density and deceleration of segmented mobility. This leads to reduction of the effect of morphological changes in macromolecules on electrically conducting properties of filled rubbers.

It must be noted, that the temperature at which the change takes place depends on pressure, applicable to the samples. Temperature dependences of conductivity for the elastomer with P803 carbon black is significantly changed with mechanical load increasing (Figure 2): it decreases at first time and than the relay effect, inherent for this material, completely disappears. At stretching under 0,4 Mpa the conductivity of the composite becomes independent on temperature. Mechanical effect in the elastomer with P803 carbon black is appeared more clearly. Conformation changes of macromolecules proceed with no obstacles in it because of more week interactions between filler particles and macromolecular segments, than in composites with active P357E carbon black. The effect of mechanical loads on resistance of the first composite is more significant, than in the second one because of strong interactions between composites components. Conducting system based on P803 carbon black also damages (jump like) due to thermo stimulated polymorphous transformations in the polymer matrix. However, orientation of macromolecular segments at stretching of sample avoids this transformations. This leads to disappearance of the relay effect observed on the experiment. Conductivity of the elastomer with P803 carbon black at high tensions is stabilized due to balancing of the processes considered.

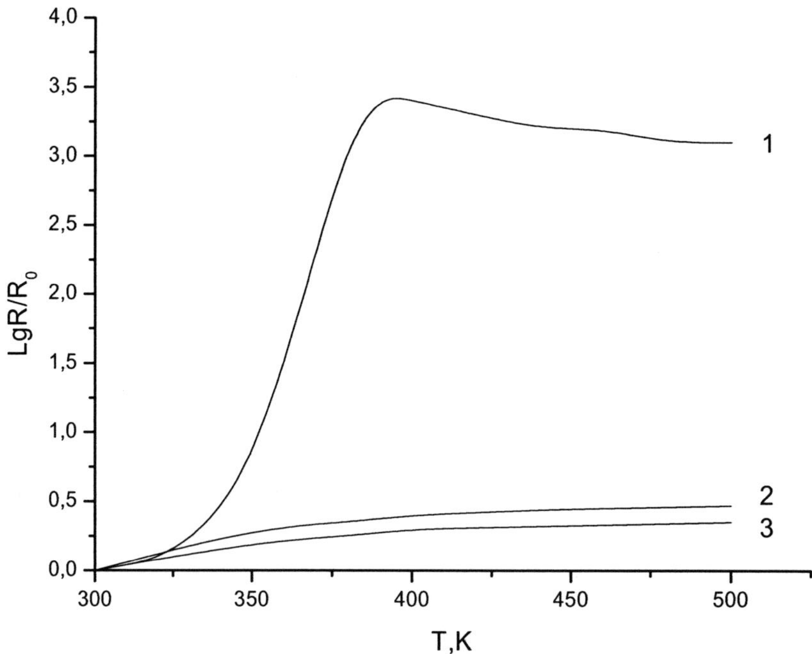

Figure 2. Temperature dependence of electrical resistance for the rubbers based on elastomer polydimethyl-methyl-vinil-siloxane with P803 (1, 3) and P357E (2) carbon blacks under loads 0(1,2) and 0.5 Mpa (3)

The conducting polymer composites considered above display good (to some extent) electromagnetic waves absorption properties. There was performed the experimental determination of the coefficients of reflection, penetration and absorption of electromagnetic waves of centimeter range. The experimental results are presented in Table 1.

According to the data from Table 1, the values of the coefficients sufficiently depend on the concentration and type of the filler, as well as on the technique of the composite obtaining. For example, it extremely depends on PM75 type of carbon black content for absorbed energy of the wave (the maximum occurs near the concentration of 80 mass parts).

However in this case, the materials changes of values of reflection and penetration coefficients are more complicated. The samples produced by additive method with rather low content of the filler (50 mass parts) display an increase of the components of that part of energy, which is spent for reflection and for penetration.

As a consequence, the rest of the energy is absorbed. The processes of scattering and absorption of electromagnetic waves in conducting rubbers are intensified with the increase of the absorbed energy part (at simultaneous decrease of the total part of reflection and penetration). For the samples with rather high content of the filler (90 mass parts) the effect of energy absorption deteriorates by means of the increase of the total reflection surface near the surface of the sample. This happens as response to the increase of the filler particle concentration. In this material electromagnetic energy absorption increases with dispersity and uniformity of the filler distribution in the rubber matrix. For example, according to table 1 the comparison of the sample 4 possessing low homogeneity with the sample 5 possessing high homogeneity shows that the coefficient of the first one is 4-fold lower, than that of the second one at the same content of the filler. Moreover, comparison of the samples 4 and 6 shows that the electrically conducting composite with low content of the filler may possess higher absorption properties, if filler particles are distributed uniformly in the polymer matrix.

Moreover, it was estimated the effect of temperature of current conducting composites on their absorption properties. For this purpose practically all previously used rubber samples were tested. It was found that an increase of the sample temperature leads to an increase of electromagnetic energy absorption (Figure 3).

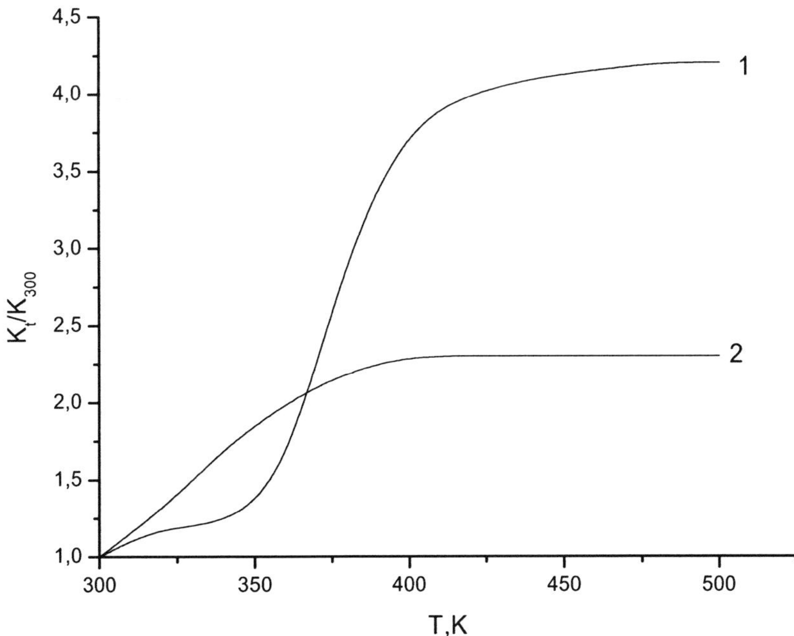

Figure 3. Temperature dependence of absorption coefficient (for cm waves) for CPC based on PDMVS with carbon black P357E (50 mas. part) (1) and 100 mas. part. P803 (2). K_{po} – absorption coefficient at 293 K.)

It is also interesting for the practical use that electrically conducting rubber with the *relay effect* displays a jump-like increase of the wave energy absorption. This is stipulated by the structural phase transition in the polymer matrix of the composite.

Probably the behavior of the curves of the wave energy absorption coefficient with temperature of the sample may have a general explanation. Early it was mentioned that the temperature effect in polymers with dispersed fillers manifests in redistribution of the particles in the polymer matrix with an increase of the average distance between them. It should be taken into account that according to the theoretical bases of electromagnetic wave absorption by solids [8], the deterministic factor for this phenomenon is taken the value of dielectric permeability, which increase induces an increase of conducting material properties. One may suppose that the effective value of dielectric permeability changes due to the interaction of two composite components. Total surface of the interlayer increases at the uniform distribution and increase of the filler dispersity. In other words, the effectiveness of the filler effect on the molecular system increases exhibited by a change of the physical-chemical properties of the polymer.

REFERENCES

[1]   Norman R.N., *Conductive Rubbers and Plastics*, Amsterdam, Elsevier, 1970.

[2]   Gul' V.E., Shenfill L.Z., *Electrically Conducting Polymeric Composites*, Moscow, Khimia, 1984, 240 p. /Rus./.

[3]   *Chemistry and technology of silicoorganic elastomers* /Ed. V.O. Reichsfeld/, Leningrad, Khimia, 1973 /Rus./

[4]   Ivanchev S.S., Dmitrenko A.V. // *Uspekhi Khimii*, 1982, 51 (7), 1178-1800. (Rus.)

[5]   Patent No 1081987 (USSR), 1984 (Rus.)

[6]   Aneli J.N., Khananashvili L.M., Zaikov G.E. //Intern J. *Polymer Materials*, 1994, v.25, p. 235-241.

[7]   Khananashvili L.M., Aneli J.N., Pagava D.G. "Communication of International Conference of Rubber", IRC-90, Paris, 1990, Section 3, p. 228-229.

[8]   Chernov L.A., *Spreading of Waves in a Medium with Random Inhomogeneitis*, Moscow, Ed. of USSR Academy of Sci.1959. /Rus/.

In: Physical Organic Chemistry: Theory and Practice   ISBN 1-59454-275-9
Eds: A. D'Amore and G. E. Zaikov, pp. 133-139   © 2005 Nova Science Publishers, Inc.

Chapter 9

CONDUCTIVITY OF CONDUCTING RUBBERS AT FLOW

J. N. Aneli, *M. M. Bolotashvili and I. G. Ubiria*
Institute of Machine Mechanics of Georgian Academy of Sciences, Tbilisi, Georgia

INTRODUCTION

The conducting rubbers (CR) belong to polymer materials conductivity of which is conditioned by conducting fillers (carbon black, graphite, metal powders). Possibility of alternation of specific volumetric electrical resistance ρ_v in the wide range of values, multi-functionality of application in electrical engineering, durability and simplicity of technology define the increasing interest to CR. At present there are many works dealing with the technology of production and investigation of properties and application of these materials in engineering /1-3/. Irrespective of a wide range of scientific articles about investigation of CR there are still the problems, which are to be solved. For example, the influence of the processes following the loading of materials on the conductivity properties of CR is not yet investigated. The solution of this issue is urgent for designing the elements obtained on the basis of CR sensitive to deformations.

In present work the dependence of value of ρ_v on tensile deformation of materials at flow processes has been investigated.

The rubbers on the basis of poly(dimethyllvinylsiloxan) filled by carbon black of two types P-803 and PME-80V were used as objects for investigation. Composite materials were produced via the polymerization filling method at room temperature (vulcanization agent – diethyl-aminomethyl-trietoxisilane of type ADE-3 /4/ and by the peroxide vulcanization method (vulcanization agent -peroxide of dicumyle)/5/.

The experiments were carried out with using of a simple device allowing to conduct measurements of the values of stretching deformation of samples at constant loading (Figure 1) with simultaneous measurement of rubbers resistance.

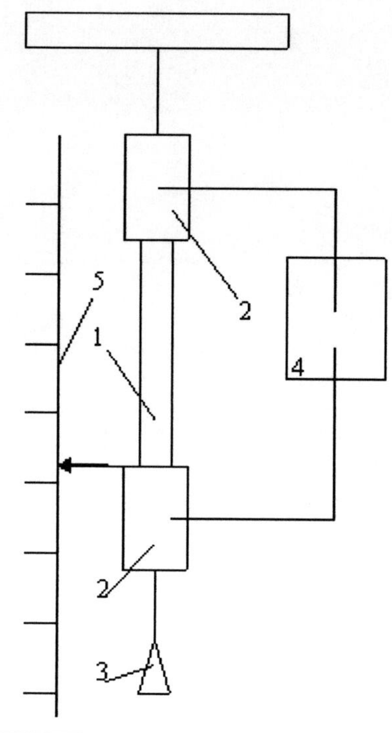

Figure 1. The scheme of measuring of ρ_v relaxation at flow of CR. 1-sample, 2-sample binding and electrical contacts, 3-load, 4-ohmmeter, 5-ruler.

RESULTS AND DISCUSSIONS

The rubbers on the basis of poly(dimethyllvinylsiloxan) filled by carbon black of two types P-803 and PME-80V were used as objects for investigation. Composite materials were produced via the polymerization filling method at room temperature (vulcanization agent – diethyl-aminomethyl-trietoxisilane of type ADE-3 /4/ and by the peroxide vulcanization method (vulcanization agent -peroxide of dicumyle)/5/.

The experiments were carried out with using of a simple device allowing to conduct measurements of the values of stretching deformation of samples at constant loading (Figure 1) with simultaneous measurement of rubbers resistance.

The Figure 2 shows functional dependence of samples extension on time at constant loading at flowing of tested material. The curves of these dependences correspond to the known character of sample extension at flowing qualitatively [6]. Thus, maximal values of relative extension at flowing (stream) for sample 1 containing carbon black of type P803 is higher than those for samples 2 and 3. This difference is explained, in the first place by structural difference between the used fillers. It is known, that carbon black of type P803 is less structured than carbon black of type PME-80V, therefore the amplifier effect of the first filler is less than in case of the second one. Consequently, the extension at flow of sample 1, of a softer rubber, proceeds in a wider range of magnitudes than that of sample 2. These

---

* Main scientific collaborator of the Institute of Machine Mechanics of Georgian Academy of Sciences, doctor of technical sciences, professor. Tel: +(99532) 31-69-67; E-mail: jimaneli@posta.ge

results are explained by differences between magnitudes of elasticity and hardness of samples. In the first case the rubber is obtained by poly-condensation process at participation of the end hydroxyl groups of the macromolecules with inclusion in vulcanization network of vulcanization agent molecules after estrangement from the last triethoxysilane groups and joining to elastomer macromolecules. The polymer system thus obtained is characterized by long and mobile segments, while the sample 3 has more wide vulcanization net obtained as a result of formation of the side bonds through opening of side vinyl groups of elastomer. The mentioned above structural peculiarities define high balance stretching of samples 1 compared to samples 2 and 3.

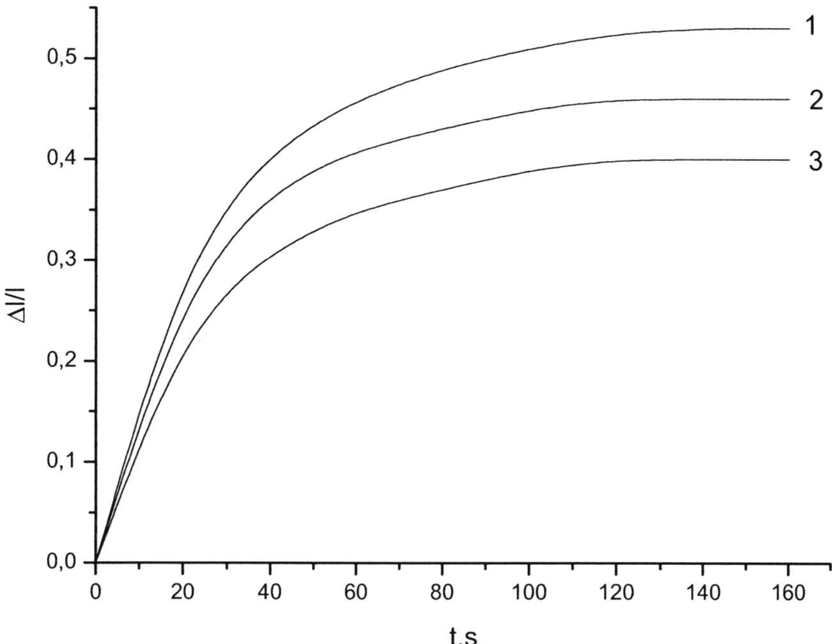

Figure 2. Dependence of lengthening of CR on flow time for samples obtained by additive vulcanization (filler- carbon black of type P803(1) and PME-80V (2)) and peroxide vulcanization (filler - carbon black of type PME-80V) (3) methods.

Curves presented on the Figure 2 reflect the dependence of ρ_v on time beginning from loading start. These dependences were obtained as follows: after loading a sample the change of length and resistance of material during time, which was necessary for achievement of equal magnitudes of measured parameters were fixed. At this time the loading of samples was increased and the dependencies of resistance R and length l of sample were recorded. In such a way the dependence R – T at other loading were recorded. The values of ρ_v were calculated for every separate sample according to the data of R. The curves of dependences ρ_v -T qualitatively have one and the same character for all samples. The quantitative difference between them is explained by a difference in mechanical properties of tested materials. Sharp increase of ρ_v value at the moment of loading of the sample reflects the response of composite conductive system to sharp change of rubbers microstructure because of their momentary stretching.

The analysis of Figure 3 reflects well the connection between mechanical and conducting properties. Thus, the sample characterized by low elastic module (compared with samples 2 and 3) reveals more quick growth in the wide range. This fact is due to velocity of conducting system reconstruction construed from carbon black particles. It must be noted that this velocity is significantly low than that for the process of elastic component of rubber deformation and can be compared with the latter only in the presence of strong interactions between elastomer macromolecules and filler particles. The principal difference between curves presented on Figure 4 consists in a decrease of equal significances of ρ_v for samples 1 and 2 at the increase of strain in them and in the increase of ρ_v for sample 3, although ρ_v for all samples at fixed strain is decreased till definite equilibrated data. According to the model proposed in [7] the stretching of rubber influences on the value of ρ_v ambiguously – for ones it increases but for others –decreases.

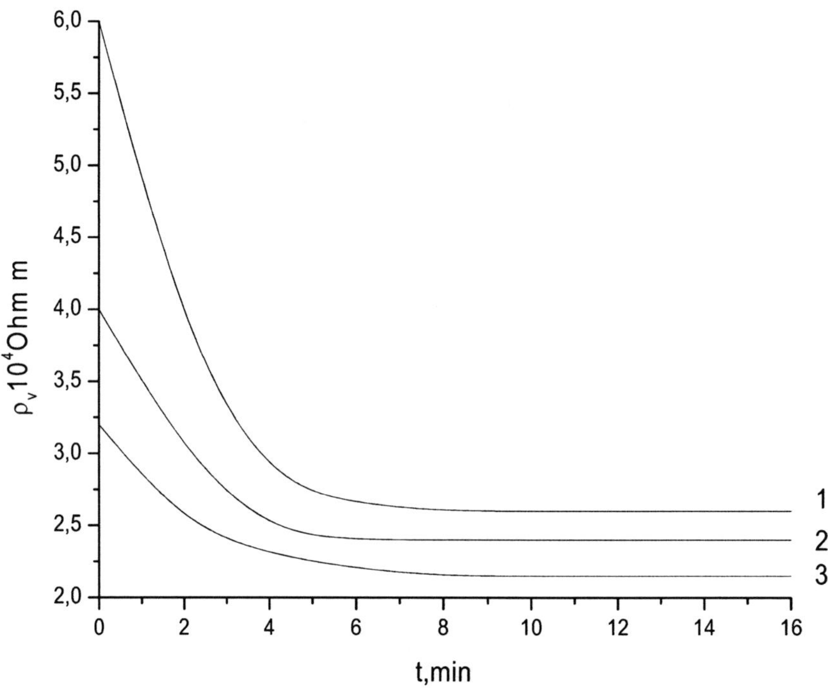

Figure 3. Dependence of ρ_v of CR on flow time for samples 1 - 3 at loading 0,2 Mpa.

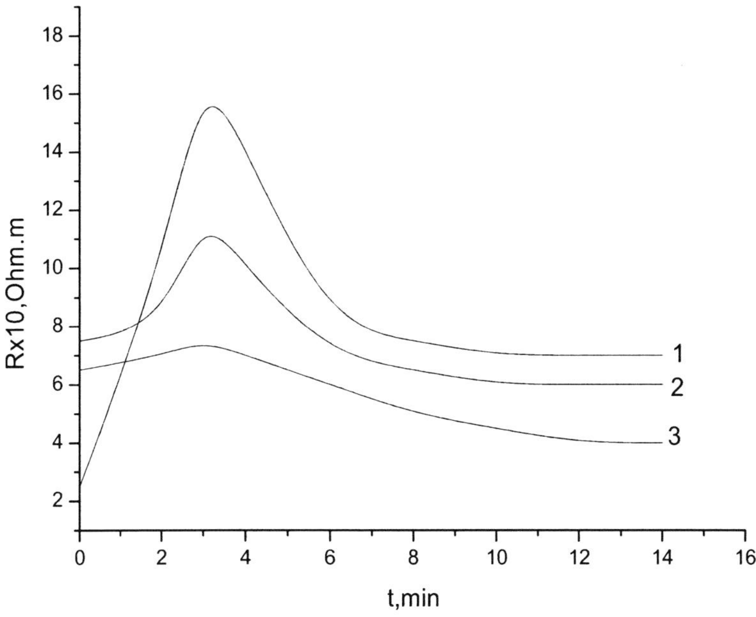

Figure 4. Dependence of ρ_v of CR on flow time for samples 1 - 3 at loading 0.4 Mpa

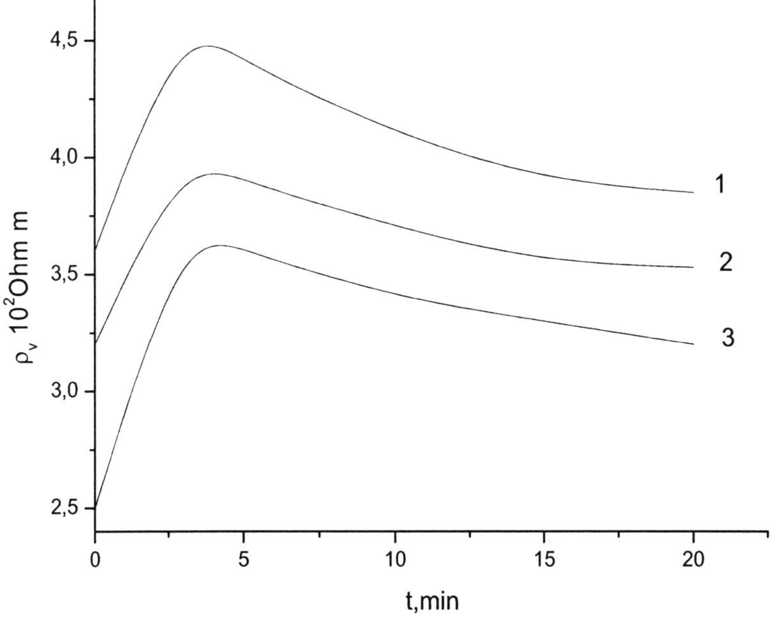

Figure 5. Dependence of ρ_v of CR on flow time for samples 1 - 3 at loading 0.6 Mpa.

It is explained by peculiarities of CR microstructures dynamics. Uniaxial stretching of CR leads to stretching of conducting system too. At this moment two concurrent processes are observed –increase of average distance between filler particles which leads to the increase of resistance and simultaneous narrowing of cross section of rubber with consequent approaching of longitudinal conducting channels in direct of cross section and with partial restoration of ruptured conducting bonds as a result of which ρ_v is decreased. It can be supposed that at stretching of samples 1 and 2 the restoration process of conducting system predominates, while in the case of sample 3 the main process is destruction of this system (it must be noted that the temperature coefficient of resistance of this rubber in contrast to the first two - is negative). The sections of curves presented on Figure 6, which correspond to the flow process can be described by expression $\rho_v = \rho_{vo} \exp(-t/\tau)$, where τ is conducting system relaxation time at flow, ρ_{vo} -magnitude of ρ_v at start of flow.

The curves on Figure 6 show that equal significances of ρ_v achieved after loading at the flow does not depend on the loading regime.

The results described above are repeated if the experiments are conducted under identical conditions at reverse order (that is at the step-by-step or simultaneous decrease of loading of the samples loaded in advance).

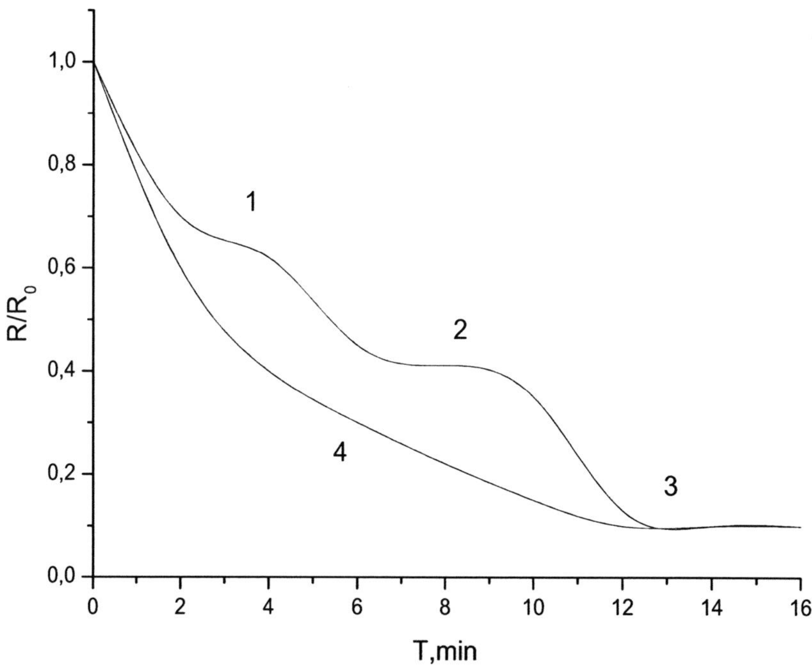

Figure 6. Dependence of electrical resistance for the sample 3 on flow time at step by step loading method (loading: 0,2 Mpa - curve 1, 0.4 Mpa –curve 2 and 0.6 MPa – curve 3) and separately at loading 0.6 Mpa. R_0 – the resistance of rubber before loading

On the basis of the obtained experimental results the following conclusions may be made:

1. The conductive properties of CR significantly depend on the content and mechanical properties of composite materials. The stiff (with high module) CR is characterized by wide range of achievement of equal condition and change of ρv -in less boundaries than "soft" (with low modules) ones. The experiments carried out both at the increase and decrease of loading give practically the similar results.

2. The quasi-stable significances of ρv at flow does not depend on regime of loading, i.e. it does not matter whether the loading (or unloading) of samples is carried out step by step or simultaneously.

REFERENCES

[1] Norman R.N. *Conductive Rubber and Plastics*. Amsterdam. Elsewhere, 1970.

[2] Gul V.E., Shenfil L.Z.. *Conducting Polymer Composites*. M.Khimia, 1984 /Rus./

[3] Aneli J.N., Khananashvili L.M., Zaikov G.E.. *Structuring and Conductivity of Polymer Composites*. New-York, Nova Sci.Publ. 1998.

[4] Aneli J.N., Khananashvili L.M.. *International J. Polymer Materials*,1994,v.25,p.235-241.

[5] Aneli J.N., Koberidze K.G., Mukbaniani O.V., Khananashvili L.M., *Russian Polymer News* 1998, v.3,N4, p.7-9.

[6] Bukhina M.F. *Technical Physics of Elastomers*, M.Khimia,1968 /Rus/.

[7] Aneli J.N., Pagava D.G., Topchiashvili M.I., Khubaeva E.L. Kauchuk I rezina, *Russian journal (Coutchuc and rubber)*, 1984, N11,p.16-18./Rus/.

In: Physical Organic Chemistry: Theory and Practice
Eds: A. D'Amore and G. E. Zaikov, pp. 141-152

ISBN 1-59454-275-9
© 2005 Nova Science Publishers, Inc.

Chapter 10

DEVELOPMENT OF FIRE AND HEAT SHIELD MATERIALS

M. A. Shashkina and A. A. Donskoi
Institute of Aviation Materials, Moscow, Russia
G. E. Zaikov
N.M.Emanuel Institute of Biochemical Physics, Moscow, Russia

INTRODUCTION

Creation of fire and heat shield materials is a complicated analytical and experimental process. In the first stage, data from the literature are analyzed and potential components for materials are selected on the basis of the reference data and thermal tests of separate components. Therewith, heat-physical properties of separate substances, possible physical and chemical transformations under thermal effect, directivity and values of heat effects are taken into account. Information about heat-physical properties can be extracted from numerous reference books. The presence and directivity of heat effects when substances are heated can be estimated by differential thermal analysis.

Fire and heat shield properties of composite materials depend on the component composition. However, it is impossible to describe unambiguously their operation mechanism under high temperature effect on the material by studying transformations of each component separately, because reactions proceeding between components also show up significantly in this case. This determines the second stage of investigations, which includes the study of properties of component mixtures or composite materials. At this stage of doubtless interest are investigations by differential thermal analysis of component mixtures and materials.

The full-scale fire effect on the material is a quite complicated process including many elements of influence on the material. Inflammation and combustion of the material are affected by heat-physical properties and chemical composition of both gas and condensed phases. Therewith, the scale factor with respect to sizes of combustion epicenter and the object present in the fire zone become significant.

Full-scale tests of the article are of the greatest importance as is the accuracy of testing fire and heat shield properties of the material. However, these tests are very expensive and could not be applied during the stage of material development. This method is applied for the final selection of shielding material composition and estimation of its operation in the article.

In this connection, different methods for estimating efficiency of separate components, their mixtures or entire materials are used with regard to the mechanism of one or another suggested component. At the stage of component selection it is desirable to use one of the lowest material consuming studies – the method of differential thermal analysis. This method allows estimation of comparative efficiency of separate components or their mixtures by directivity and values of heat effects proceeding in a definite temperature range. Therewith, the gas phase composition can be specified. The method is efficient in selection of components. However, it does not consider thermophysical properties of either condensed or gas phases, nor the scale factor. Values of heat effects are also estimated comparatively only. This method is applied to the selection of the component structure of the material to be developed. Components possessing the maximum number of sufficient endothermal effects, which consume a large amount of heat, are the most profitable for fire and heat shield materials and the ones with reduced combustibility.

Quantitative estimation of heat effects is performed by the method of scanning differential calorimetry. However, this method requires a complex of research equipment. It can be applied to quantitative estimation of some specific fillers or materials.

Among methods modeling full-scale tests in samples, tests on a solar unit, the method of one-side thermal impact and tests in high-temperature gas or air flow should be noted. These methods allow estimation of material heating-through rate, the rate of material carry-over and temperature of external and the opposite surfaces of the material. The methods are applied during the stage of compounding development of composite material and selection of the most efficient composites. One test method or another from the above-mentioned ones is chosen at the stage of composite material selection based on the operation conditions suggested for the material created. The highest information content is displayed by the method estimating heat amount transmitted through the sample and efficient thermophysical characteristics of degradable materials.

Combustibility of the material and influence of the component composition on it is determined by the oxygen index and temperature profiles in the combustion wave.

Analysis of data from the literature indicates that the following properties are critical for selecting components for the material with reduced combustibility and fire and serving as heat shield:

- susceptibility to physical and chemical transformation with high heat energy consumption;
- optimal heat conductivity and temperature conductivity coefficients;
- the maximum reflection index;
- the minimum transmission index for radiant energy;
- the optimal heat energy absorption coefficient;
- ability of material components to transform the heat energy into other sorts of energy;
- ability to scatter the maximum possible energy into the surroundings.

The study of optical properties of different substances and, in particular, ability to reflect heat energy, shows that different substances display a reflection maximum at different wavelengths. Therewith, the Wien displacement law, which associates wavelength of the maximal irradiation of the black body (Planckian full radiator) with the absolute temperature T as follows, should be taken into account:

$$\lambda_m T = 2,886 \ \mu\text{m·deg}.$$

This law shows that the wavelength, for which the power irradiated by the blackbody is maximal, is inversely proportional to the absolute temperature. To put it differently, multiplication of the maximal wavelength and the absolute temperature is constant and equals 2,886 μm·deg.

The Wien formula can be presented in the form of the following table [1].

Table 1. Wavelength Dependence on Absolute Temperature

Wavelength, μm	9.62	5.77	2.68	1.60	1.37	1.11	0.96
Temperature, K	300	500	1,100	1,800	2,100	2,300	3,000

One may conclude from analysis of the Table that each wavelength corresponds to a definite temperature. As a consequence, to obtain the maximal protection efficiency, components should be selected depending on expected operation conditions reflecting the maximal amount of energy in the given wavelength range.

Infrared rays falling on the substance, like visible light, are partly reflected. Solids give one or several maximums of the reflection index. The highest reflection indices are displayed by such metals as gold, silver, and copper. Aluminum, zinc and their alloys display an anomaly. Minerals and salts possess several reflection maximums that allow selection of the most efficient fillers for composite materials.

Let us present several notes on reflection. In investigations accompanied by observations of infrared radiation reflection by substances, the scientist should remember that:

1. The effect of surface polishing is much less than for visible light. In the far infrared area, unpolished rough surfaces are able to reflect radiation almost as well as polished ones.
2. On the contrary, purity of the surface is of great importance: for example, a dust layer reduces reflection from a metal surface.
3. Water contained in powders reduces their reflection index.
4. Temperature also causes an effect: in some cases, its increase induces shear of reflection maximums, and value of the reflection index expressed as a percentage can be changed.
5. Solid salts possess higher reflection index than their solutions.
6. Variations in angles of incidence of radiant fluxes induce significant, sometimes anomalous changes in reflection indices.

As mentioned above, infrared radiation affecting the substance is partly reflected or dispersed by the surface, but a part of it permeates into the substance thickness. The extent of this part depends on the transmission coefficient of the substance present.

Solids and liquids generally possess high absorption; however, organic substances are more transparent for infrared radiation than mineral ones.

Rubbers are used as the polymeric base for fire and heat shield materials. Natural rubbers differ by good transmittance in thin films with broad absorption bands at 3.43, 6.95 and 12 µm. When aged, rubbers are oxidized, which significantly changes the infrared absorption [1]. Modification of rubbers shows up on optical properties in a manner similar to oxidation. Optical properties are seriously changed by chlorination and sulfochlorination of polymers [2, 3]. A series of investigations has shown that sulfochlorinated and chlorinated polyethylenes and organosilicon rubbers are the most efficient as the polymeric base.

Selection of mineral fillers is closely associated with estimation of their optical properties and susceptibility to physical and chemical transformations with heat absorption under the effect of high temperature. Analysis of the literature shows that different metal compounds are effective with regard to expected operation conditions, temperature, in particular. For example, silicon and aluminum oxides are effective in the temperature range of 900 – 1,200°C. At higher temperatures titanium dioxides are preferable, which is associated with increased reflection indices in the appropriate range of wavelengths.

Organic components cause a negligible effect on optical properties of the entire system. However, their susceptibility to physical and chemical transformations with heat absorption, as well as the ability to be gasified with heat energy elimination induces an enhanced interest to application of similar substances in fire and heat shield materials or those with reduced combustibility. Moreover, application of organic compounds containing halogens and phenyls as substituents can alter the optical properties of the entire system, too.

Studies of developed materials, performed by different methods, have led to conclusions about expected mechanism of fire and heat shield materials and those with reduced combustibility, based on sulfochlorinated polyethylene.

Sulfochlorinated polyethylene is the polymeric base of the material. Differential thermal analysis shows that this polymer degrades in three stages. Mass loss is initiated at 100 – 200°C. Endothermal peaks are observed at 250 and 400°C, the values of which are dependent on chlorine concentration in the polymer. Hence, sulfochlorinated polyethylene provides for heat absorption in the area of 250 and 400°C, heat elimination with gasification products at thermal degradation of the polymer, and reflection of the heat energy at temperatures around 200°C. Injection of mineral fillers increases the yield of solid residue, the content of which in some compoundings exceeds total concentration of mineral fillers. This indicates possible carbonization of the organic matrix in the presence of certain compounds. In the case of sulfochlorinated polyethylene, combination of silicon dioxide with variable valence metal oxides, such as antimony or aluminum, is the most efficient in the presence of magnesium oxide and vulcanizing group.

The study of fire and heat shield properties of samples from sulfochlorinated polyethylene by the method of one-side thermal impact shows that thermal transformations in nonvulcanized polymer proceed very fast with high thermal absorption. However, the polymer operates most effectively only in the case of presence of spatial cross-linking (vulcanized), which allows the spread of heat absorption of the material in time, i.e. provision of durability of the cover from polymeric material commensurable with duration of high-temperature influence. Injection of both mineral and organic fillers into the material compounding allows reduction of the heating-through rate of the protected surface. Combinations of fillers are the most effective in this case.

The method of one-side thermal impact allows general estimation of fire and heat shield properties of samples from composite materials. Unfortunately, these tests give no possibility of separating components of the process and to determine the contribution of every transformation mechanism.

The highest information content is displayed by the estimation method of transmitted flux, which allows estimation of efficient thermophysical properties of degrading materials making possible comparative analysis of material operation with regard to different consumption mechanisms of heat energy delivered to the material. These tests show that materials, incompletely degrading under thermal influence, i.e. the ones containing mineral fillers providing for heat expenditure in the condensed phase or its reflection, are the most efficient. Materials providing for heat release in the gas phase are less effective. This is shown in Figure 1 presenting the change of transmitted flux at temperature effect on the surface of material samples with different fillers.

Test results suggested selection of silicon and antimony oxides, benzguanidine, and chloroparaffin as effective fillers, used in a series of efficient fire and heat shield materials and those with reduced combustibility.

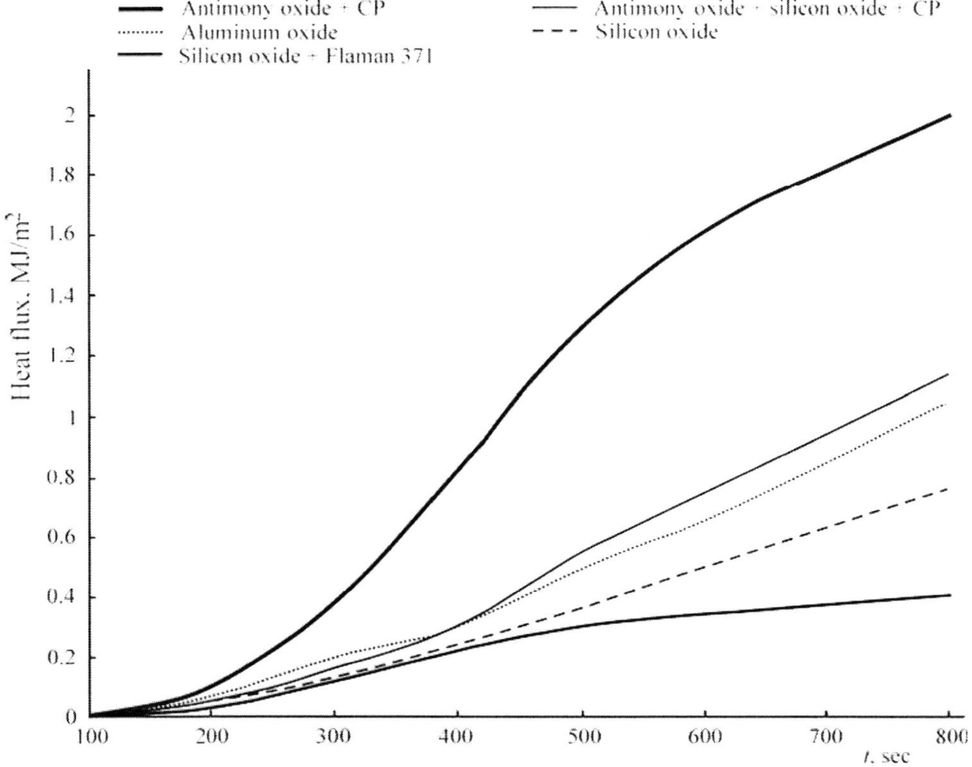

Figure 1. Change of transmitted flux at temperature effect on the sample surface from materials with different fillers.

Developed fire and heat shield materials belonging to the class of materials with reduced combustibility contain mixtures of fillers, which shows up on both thermal properties and strength and elastic ones. This is associated with physical and chemical interactions between material components during the stages of technological processing and under conditions of long-term storage and short-term thermal impact. That is why properties of composite materials containing fillers were studied.

To explain the type of interaction between components in the polymeric matrix, qualitative and quantitative composition of products formed in the condensed phase of vulcanized composites in the temperature range of 200 – 500°C is studied. The temperature range is chosen from results previously obtained of polyethylene temperature field investigation [4], as well as with regard to operation conditions of the material. Thermal tests of samples were carried out by the methodology described in ref. [5]. The phase composition of solid products formed under thermal influence on samples was studied with the help of X-ray phase analysis. Quantitative content of antimony and chlorine was determined with the help of atomic emission analysis [6] and mercurimetry [7]. Thermal behavior of the samples was studied on a derivatograph at temperature increase rate of 5°C/min.

Chemical and atomic emission analyses show that the rate of antimony loss at heating is independent of the polymeric matrix state, and as for chlorine, it somewhat decreases for vulcanized compositions in the temperature range of 350 – 450°C. This is shown in Figure 2 depicting test results of vulcanized and nonvulcanized composites filled with silicon and antimony oxides and containing vulcanizing groups and plasticizer.

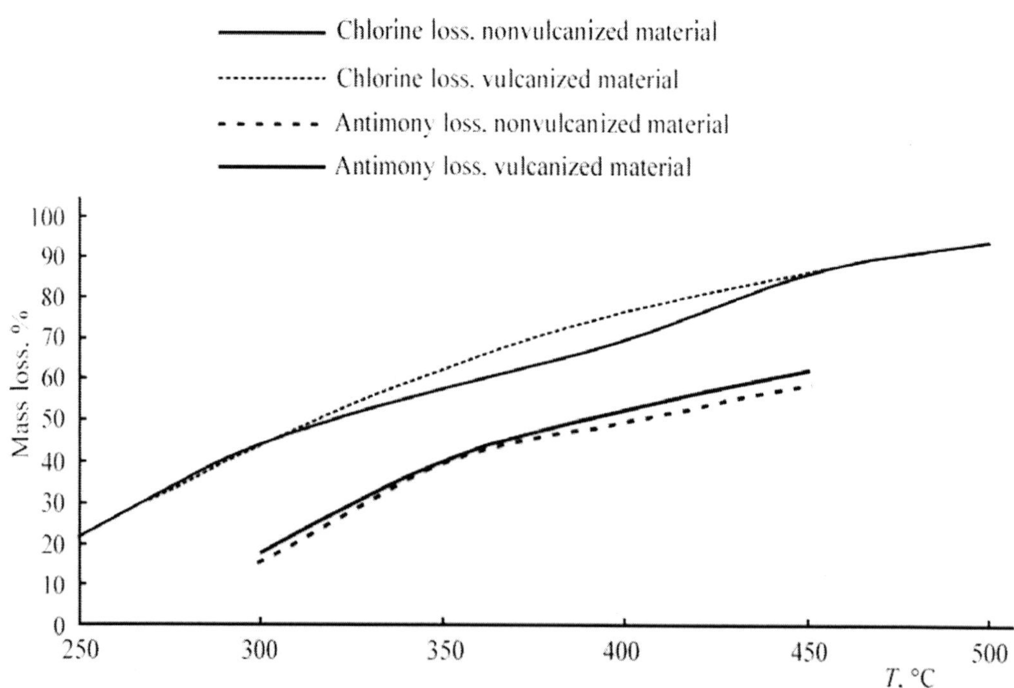

Figure 2. The rate of antimony and chlorine loss at heating samples of vulcanized and nonvulcanized materials

According to the literature [8], mass-spectrometric studies show the antimony trioxide interacts with hydrogen chloride forming antimony chloride and water, but in the temperature range of 200 – 250°C solid antimony oxychloride is formed, which decomposes in the temperature range of 250 – 450°C with antimony trichloride release. In the absence of halogens, solid antimony trioxide does not evaporate at temperatures below 450°C – the point corresponding to thermal degradation of the polymer. As antimony oxychloride is heated up slowly at a rate of 0.2 deg/min, three peaks of antimony trichloride release were observed in the temperature range of 250 – 450°C.

X-ray structural analysis detects antimony oxychloride starting from 250°C during thermal influence on composites containing antimony trioxide. Future temperature increase reduces intensity of antimony oxychloride bands, and diffraction maximums of metal antimony occur. Note that both temperatures and types of products formed in sulfochlorinated polyethylene completely coincide with that of previously studied composite based on polyethylene containing the same synergic couple [5, 6]. The results obtained indicate that the type of interaction between antimony trioxide and chlorine containing compound as synergic couple is independent of the polymer nature (polyethylene, foam polyurethane, sulfochlorinated polyethylene) and the state (vulcanized or nonvulcanized) of the polymeric matrix.

Estimation of oxygen index and the rate of flame spreading by samples based on sulfochlorinated polyethylene with various fillers and their combinations shows values of 42 – 58 OI in accordance with test conditions.

Differential thermal investigations show that thermal degradation of sulfochlorinated polyethylene and composites derived from it under dynamic conditions of heating represents a multistage process, which is affected by all components participating in the composite.

The nature of metal oxide affects the range and rate of volatile release, as well as coke residue yield. The initial polymer transforms into the rubbery state at temperatures above 100°C. This transition is displayed by the endopeak with maximum at 125°C on the sample thermogram. Insignificant mass losses below 200°C (about 4%) are due to release of low-molecular substances present in polymer or formed during technological processing. Intensive degradation of sulfochlorinated polyethylene is observed in the range of 230 – 400°C. The process proceeds by the first order reaction with the following rate constant:

$$k = 4.3 \cdot 10^8 \exp(-94,700/RT) \text{ min}^{-1}.$$

At this stage, substituents are detached from the chain and chlorosulfone groups decay forming sulfur dioxide and hydrogen chloride.

Vulcanized sulfochlorinated polyethylene degrades intensively above 182°C. Mass losses at heating up to this temperature are below 2 wt.%. The first stage of degradation in the temperature range of 182 – 360°C involves little exothermicity, unlike high-temperature exothermal stages. Note that despite reduction of sulfochlorinated polyethylene degradation temperature after vulcanization, its rate is decelerated initially. The activation energy of vulcanizate degradation increases up to 110 kJ/mol, and the reaction becomes of the second order (Table 2).

Table 2. Kinetic Parameters of Thermooxidative Degradation of Composites Based on Sulfochlorinated Polyethylene

Filler composition	Temperature range, °C	Activation energy of degradation, kJ/kg	Pre-exponential multiplicand, min^{-1}	Reaction order
Nonvulcanized, without filler	230 – 400	94.7	$4.3 \cdot 10^8$	1
Vulcanized, without filler	182 – 362	110.0	$1.6 \cdot 10^{11}$	2
Silicon oxide	195 – 380	130.0	$1.2 \cdot 10^{13}$	2
Antimony oxide	220 – 400	167.2	$8.1 \cdot 10^{17}$	2
Antimony oxide + chloroparaffin	110 – 385	145.0	$7.2 \cdot 10^{14}$	1.5
Silicon oxide + antimony oxide + chloroparaffin	120 – 400	140.0	$2.8 \cdot 10^{13}$	2

This is associated with conformation and spatial changes in the network structure of the polymer, and mobility decrease of macromolecules as the result of vulcanization, i.e. spatial cross-linking of macromolecules. Magnesium oxide participating in the vulcanizing system causes an essential influence on the change of macromolecule mobility.

Additional injection of fillers (silicon and antimony oxides) increases stability of vulcanizates. The opposite effect is observed in use of plasticizers (Table 2).

It is common knowledge that the effective value of the activation energy of polymer degradation depends on the initiation reaction. Data presented in the Table show different types of molecular interaction between the polymeric matrix and surface atoms or groups of atoms of metal oxides, which inactivates labile centers of the polymer initiating degradation of cross-linked polymer. The plasticizer, inhibiting interaction with the metal oxide surface promotes polymer degradation. Therewith, it should be taken into account that chloroparaffin and antimony oxide are reactive compounds. As heated under dynamic conditions, chloroparaffin of CP-470 trademark degrades at temperature above 130°C. Its self oxygen index equals 25 [7].

The efficiency of combustion decelerator of this type depends on the ratio of antimony and halogen in the composite, the type of interaction between antimony oxide and halogen-containing compound, and the nature of the polymer nature. All these factors affect the amount of antimony halide released into the gas phase, which is the active inhibitor of reactions in flame.

A typical feature of thermooxidative degradation of composites based on sulfochlorinated polyethylene is reactions with formation of carbonized product and its oxidation at temperatures above 500°C. Polymer transformation into carbonized structure including fragments of condensed aromatic rings proceeds after dehydrochlorination and formation of unsaturated π-conjugated bonds in macromolecules.

Magnesium oxide and silicon dioxide cause practically no effect on coke residue yield. Considering antimony oxide removal from the condensed phase in the form of volatile derivatives, one can conclude that compared with the initial vulcanizate of sulfochlorinated polyethylene, the coke residue yield is increased in this case.

Of interest is searching for the influence of atomic ratio of chlorine and antimony in the composite on combustibility indices of rubbers of this class. Composites containing antimony oxide display variation of the chlorine:antimony ratio only from 7.46 (30 wt. parts of antimony oxide) to 1.6 (70 wt. parts). Additional injection of chlorine with chloroparaffin into highly filled system increases this ratio to 2.16 approaching 2.26 appropriate to the composite with 50 wt. parts content of antimony oxide. It is known that the highest reduction of polymer combustibility is achieved with the help of synergic mixtures of halogen-containing compounds and antimony oxide at chlorine and antimony ratio close to the optimal one for forming antimony trichloride. It is not surprising that composites filled with metal oxides display the highest effect on oxygen index at antimony oxide concentration of 30 wt. parts. Additional injection of chloroparaffin into highly filled composite does not provide for the optimal relation of 3 and causes no increase of the oxygen index. At the same time, as chloroparaffin is injected into composites with 50 wt. parts of antimony oxide, the relation equal to 3.04, approaching the value optimal for forming antimony trioxide, affects significantly the increase of fire shield effect of the material. The highest fire shield effect is observed at chloroparaffin substitution by its brominated analogue containing both chlorine and bromine.

Analysis of derivatograms of these materials shows that differences between them concern only low-temperature stages of degradation. In the case of material with chlorobromoparaffin, slow decrease of the sample mass is observed above 130°C, and up to 220°C, the initiation point of intensive degradation, mass losses become 6% instead of 2% for composites with chloroparaffin.

An additional stage with no exothermal effect and low mass losses is observed in the temperature range of 330 – 430°C. The solid residue yield is equal at 490 and 535°C.

The results obtained indicate that efficiency of fire shield effect is seriously affected by reactions proceeding in the condensed phase, and composition and amount of combustion decelerators transforming into the gas phase. At halogen excess in relation to antimony, hydrogen halide and antimony trihalide are released into the gas phase, and at its deficit the product is mostly antimony trihalide. Unreacted antimony oxide plays the role of inert diluter of the combustible component of sulfochlorinated polyethylene vulcanizates in the gas phase. It is common knowledge that antimony trihalide is a more effective inhibitor of flame reactions, than hydrogen halide. The latter usually acts as diluter in the gas phase at high concentrations.

Antimony halides perform a dual function in flame: they are the source of hydrogen halide and form antimony monooxide participating in catalysis of recombination reactions of active radicals $H^•$, $O^•$, and $OH^•$ in flame by formation of intermediate particles, such as SbOH [9].

Owing to different boiling temperatures of antimony halides (223°C for chloride and 288°C for bromide), the spatial zone is broadened and active particles, affecting combustion, are present in the flame for a longer time.

To define more accurately the mechanism of flame spreading by the material surface, temperature distribution in the combustion wave with oxidant counter-flow was measured. Oxygen concentration in the nitrogen-oxygen medium was close to the flame extinguishing limit (46%). Measurements and calculations are discussed in detail in ref. [9].

Figure 3 shows temperature profiles in the combustion zone of materials based on sulfochlorinated polyethylene filled with 300 wt. parts of antimony trioxide. The following thermophysical characteristics of the studied material were used in processing of experimental results: heat conductivity coefficient, $\lambda = 0.33$ W/m·deg; specific heat capacity, $c_p = 1.3$ kJ/kg; temperature conductivity coefficient, $\alpha = 1.6 \cdot 10^{-7}$ m^2/s; density, $\rho = 1,550$ kg/m^3.

In the flame rim at 3 mm from the material's surface, the temperature reaches 830°C. Maximal temperature outside the rim is 1,230°C. Temperature on the surface beneath the flame rim is 620°C. However, far from the flame rim, the temperature on the sample surface increases up to 960°C due to coke formation and oxidation. The flame rim is determined by location of the maximal temperature gradient in the gas phase.

As mentioned above, the heat flux delivered to the material sample surface consists of three main components: convective component (q_g) transmitted from flame to the surface by convection in the gas phase, heat conductivity in the condensed phase (q_{cs}), and radiation (q_{fr}). Taking into account heat losses (q_l), possible due to irradiation and reflection of the heat energy by the material surface, the total heat flux can be expressed by the following relation:

$$q_s = q_g + q_{cs} + q_{fr} - q_l.$$

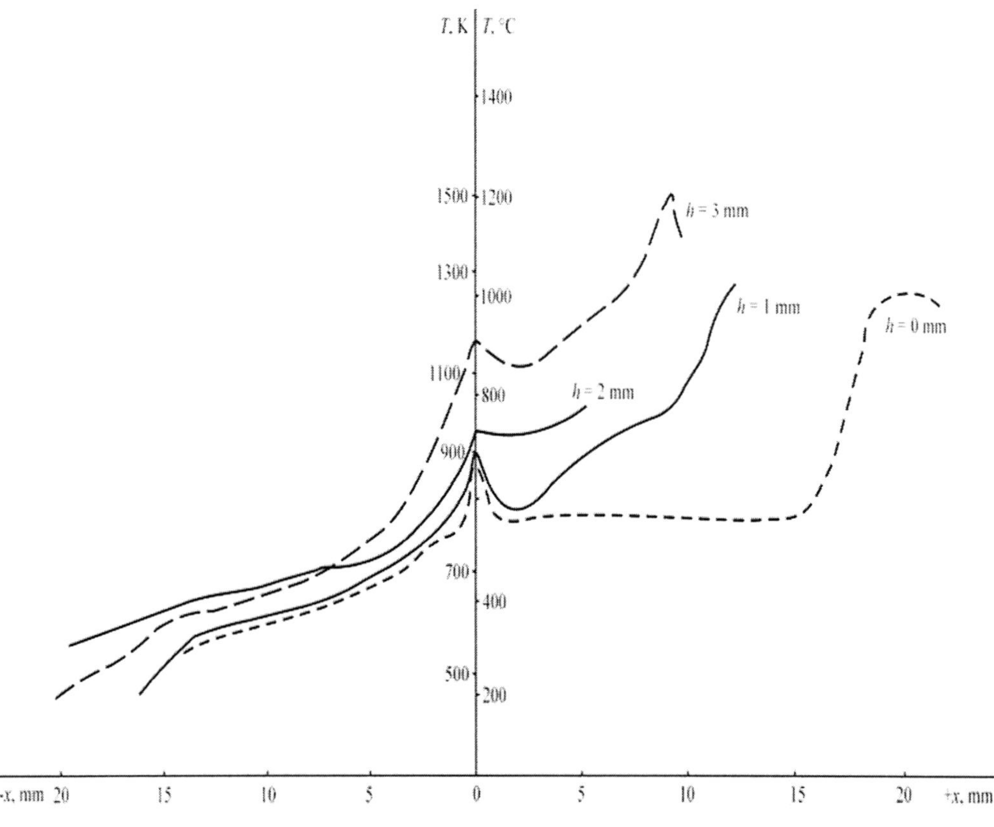

Figure 3. Temperature profiles in the combustion zone of materials based on sulfochlorinated polyethylene. Thermocouple junction from the surface, m: 1 – 0; 2 – 1; 3 – 2; 4 – 3.

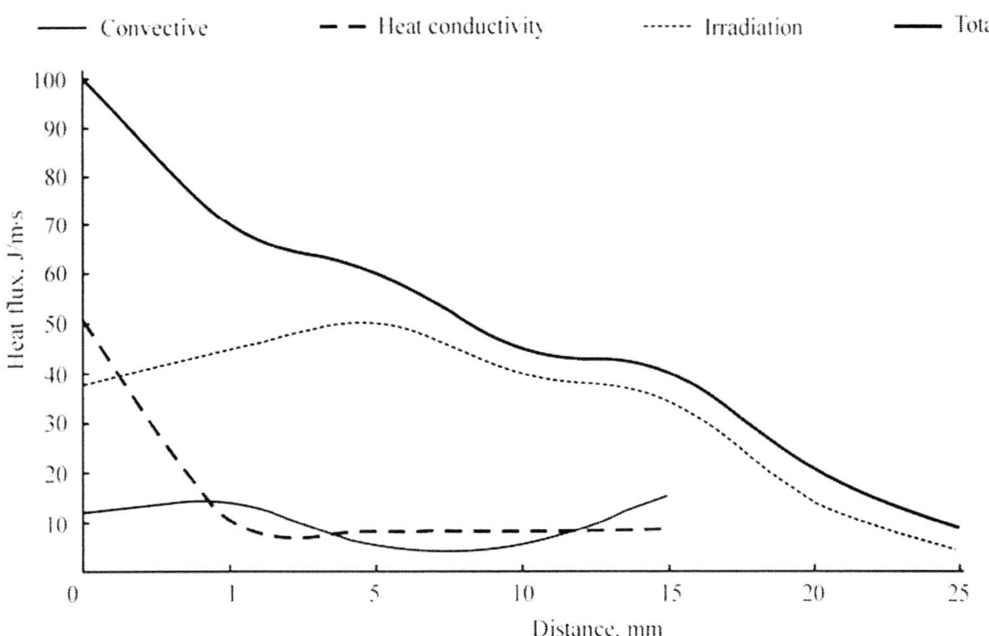

Figure 4. Heat flux distribution during combustion of material based sulfochlorinated polyethylene.

Distribution of heat fluxes by the material surface is shown in Figure 4.

Before the pyrolysis front the flux is consumed for enthalpy change in the surface layer of the material, q_h. The radiation energy absorbed by the material is estimated by the difference between q_h and $(q_g + q_{cs})$. Figure 4 shows that near the flame rim the main contribution into enthalpy change of the surface layer is made by heat transmission by irradiation from flame and heat conductivity by the condensed phase, and far from the rim, mostly by irradiation from flame. The contribution of convective heat transmission by the gas phase near the flame rim gives 9.5% of total heat flux. Contributions of the radiant energy and heat transmission by the condensed phase are 49.1% and 41.4%, respectively.

An essential contribution of heat transmission by irradiation from flame into the total heat transfer is apparently associated with sufficient carbon black formation during material combustion.

The character of radiant heat flux change in the pre-flame zone at the distance of 5 mm before the flame rim indicates a noticeable contribution of heat losses in this zone, caused by reflection and energy irradiation by the material surface.

Inclusion of metal oxides with high index of radiant energy reflection in the infrared range of wavelengths into the composition of polymeric material [1] increases reflective ability of the material.

The data shown indicate a complex mechanism for combustion of composite materials based on sulfochlorinated polyethylene. Therewith, both physical and chemical factors are of great importance. Processes proceeding in condensed and gas phases, and on the interphase show up on the type of heat and mass transfer during combustion of the composite material and on its fire shield properties.

REFERENCES

[1] Deribere M., *Practical Application of Infrared Rays*, Gosenergoizdat, Moscow-Leningrad, 1959, 440 p. (Rus)

[2] Ronkin G.M., Khromenkov L.G., and Iliin B.A., Thermal Analysis of Carbochain Polymers, *Thes. Proc. 18th All-Union Conference on High-molecular Compounds*, Kazan, 1973, p. 125. (Rus)

[3] Ronkin G.M., Mazanko A.F., and Romashin O.P., New Chlorinated Polyolefines and Thermo- and Fire Resistant Materials Based on Them, *Proc. Scientific and Research Institute "Sintez"*, Moscow, 1996, pp. 449 – 508. (Rus)

[4] Bogdanov V.V., Donskoi A.A., Shashkina M.A., and Klimovtsova I.A., Kinetic Stability of Composites Based on SCPE, *Doklady AN Belorussii*, 1994, No. 2, p. 34. (Rus)

[5] Fedeev S.S., Mayorova M.Z., Lesnikovich A.I., Bogdanova V.V., and Rumyantsev V.D., *Vysokomol. Soedin.*, 1983, vol. *25B*(3), p. 150. (Rus)

[6] Bogdanova V.V., Klimovtsova I.A., Filonov B.O., Fedeev S.S., Surteev A.F., and Lesnikovich A.I., *Vysokomol. Soedin.*, 1985, vol. *25B*(1), p. 42. (Rus)

[7] Fedeev S.S., Bogdanova V.V., Surteev A.F., Lesnikovich A.I., Rumyantseva V.D., and Sviridov V.V., *Doklady AN BSSR*, 1983, vol. *27*(1), p. 56. (Rus)

[8] Klimova V.A., *Grounds of Micromethods of Organic Compounds*, Moscow, Khimia, 1967, 101 p. (Rus)

[9] *Polymeric Materials of Reduced Combustibility*, Ed. A.N. Pravednikov, Moscow, Khimia, 1968, pp. 51 – 53. (Rus)

In: Physical Organic Chemistry: Theory and Practice
Eds: A. D'Amore and G. E. Zaikov, pp. 153-158

ISBN 1-59454-275-9
© 2005 Nova Science Publishers, Inc.

Chapter 11

CLIMATIC NATURAL AND ARTIFICIAL AGING OF MATERIALS

A. A. Donskoi and M. A. Shashkina

Institute of Aviation Materials, 17 Radio st., Moscow 119105, Russia

G. E. Zaikov

N.M.Emanuel Institute of Biochemical Physics, Moscow, Russia

INTRODUCTION

Covers from materials with reduced combustibility operate in different climatic conditions, and their durability exceeds 10 years. In this connection, the problem of estimating material behavior under natural climatic aging appears, as well as effectiveness of applying methods of artificial accelerated aging to materials based on sulfochlorinated polyethylene.

Various factors affect polymeric materials during aging: temperature ranged from –60 to +60°C, ultraviolet radiation, humidity, and other factors. Materials with reduced combustibility are multicomponent composites. That is why aging is affected by the material compounding, i.e. interaction of the compound's components during storage and operation. Degradation of sulfochlorinated polyethylene is initiated by peroxides and metal chlorides (zinc, iron, and aluminum). Oxygen initiates dehydrochlorination, but does not affect desulfonation. Degradation and cross-linking reactions proceed in the polymer under the effect of high temperatures. Similar changes may also proceed at room temperature during long storage. Cross-linking is associated with recombination of polymeric radicals of neighboring chains formed during dehydrochlorination and oxidation. Elimination of hydrogen chloride from the system, which is dehydrochlorination catalyst, decreases rates of oxidation and cross-linking of polymeric chains. That is why, to increase stability of materials based on sulfochlorinated polyethylene, hydrogen chloride acceptors are included into the compounding, for example, metal oxides or epoxy resin. For this purpose, magnesium oxide is used in materials created by the authors.

Ultraviolet radiation initiates photo-oxidative processes expressed in dehydrochlorination and oxidation reactions developing by the radical-chain mechanism.

Hence, aging of materials based on sulfochlorinated polyethylene can be represented by the general scheme for transformation of polymers:

1. Change of crystallinity degree and supermolecular structure of polymer.
2. Degradation:
    –   detachment of hydrogen, its substituents, and side chains;
    –   backbone rupture.
3. Structuring:
    –   formation of hydrogen bonds between backbones and oxygen-containing groups, formed during oxidative degradation;
    –   recombination of radicals in neighboring chains, formed during dehydrochlorination, oxidation, and chain rupture.

All the processes enumerated proceed simultaneously during aging, but have different induction periods, reaction rates and changes of reaction rates. That is why changes in the material, mostly typical of one of the processes, accumulate at every particular moment of time and, consequently, one mechanism of aging dominates. Influence of climatic aging on properties of materials based on sulfochlorinated polyethylene was studied under the use of vulcanizing groups of two types (metal-oxide and alcoholic) with different filler concentrations.

In accordance with conditions of long-term storage, full-scale climatic tests were performed in subtropical and middle Russia by 5-year sample exposure on store benches and under direct atmospheric exposure. Stability of materials was controlled by estimation of changes in physicomechanical properties with time. Figure 1 shows results of testing materials with metal-oxide vulcanizing group, filled with antimony oxide in the presence of plasticizer.

Analysis of the Figure shows that aging under different climatic and bench conditions possesses the same general tendency. Exposure during 0.5 – 1 year induces increase of the strength limit at rupture and some decrease of elongation, which is, apparently, associated with prevalence of polymer structure change processes (modification) at this stage. In the initial stage of aging, degree of crystallinity is increased, which is accompanied by rise of strength indices [1]. Therewith, two processes proceed: crystallization and degradation. Chain breaking or weakening of bonds between atoms makes structure reconstruction much easier in the first stage. Moreover, technology of preparing polymeric composites is associated with obtruding a thermodynamically unstable structure onto the polymer, formed at high mechanical stresses during processing in the rubbery and viscous flow states with future fixing of imperfect structure by vulcanization and rapid pressurized cooling. Apparently, conformation transitions proceed during aging, which give both more equilibrium stable structures and unstable mesoform conformations transforming under favorable conditions into thermodynamically more stable molecular structures [1]. Further on, spherulitic structures are enlarged and interspherulitic spaces become clearer. Formation of hydrogen bonds limits mobility of segments. Modification of supermolecular structure of the polymer induces change in the packing density of macromolecules, its inhomogeneity with future development of microdefects and microcracks, and, as a consequence, reduction of strength and elongation. Supermolecular structure change influences conditions of polymer oxidation affecting the rate of oxidant diffusion and defining accessibility of macromolecules for it. Due to simplified diffusion, oxidation is localized on the borders between spherulites and in defect areas of spherulites [2]. Thereafter, the process mechanism is changed, which is indicated by decrease of the strength limit at rupture and bends on relative elongation curves. This means that a year after exposure begins, degradation processes become predominant in the samples. One may suggest that thereafter processes of polymer chain structuring and limitation of their mobility (strength increase with elongation decrease) take place.

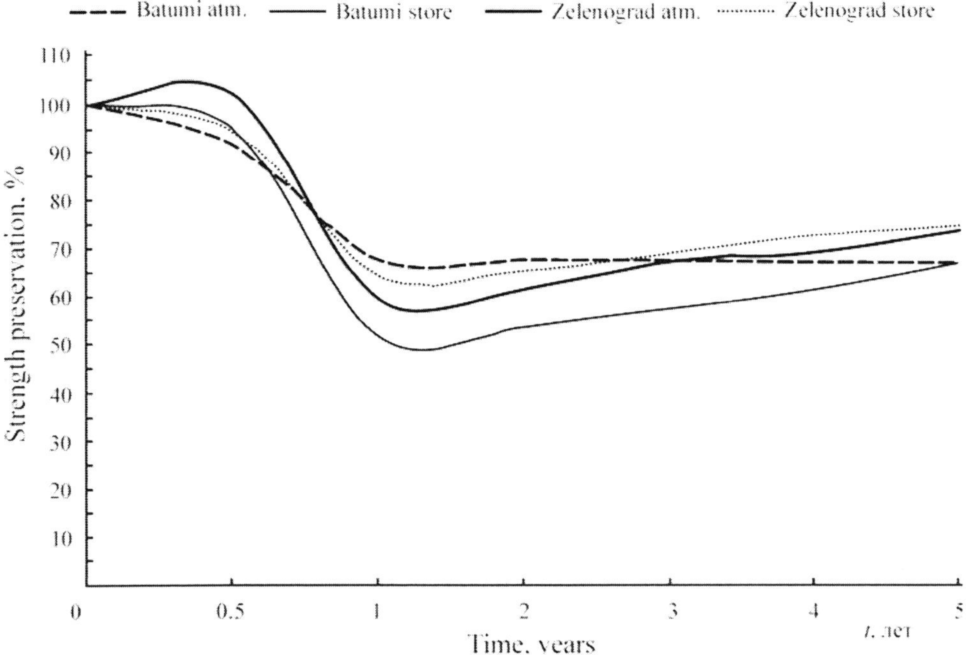

Figure 1. The level of strength preservation during aging materials of metal-oxide vulcanization

Figure 2 shows results of studying material containing silicon dioxide as the filler. Injection of this filler leads to some decrease of strength indices, which is probably associated with weakening intermolecular bonds reducing the strength limit at rupture in the initial period.

The mechanism of transformation processes in the material changes during long-term aging. Therewith, no strength increase was observed at short-term atmospheric exposure and lower total change of strength indices at further sample exposure. Total change of the strength limit under all test conditions for materials of the current compounding did not exceed 20%, whereas in the absence of silicon dioxide it reached 40%. Studies of relative elongation change show that these properties of the material are higher dependent on climate conditions during material aging. In conditions of humid tropical climate, elongation increases, and in the less humid middle Russia climate it reduces. This can be explained by weakening of intermolecular interaction by increase in the distance between polymer chains, which makes intermolecular interaction comparable with the dissociation energy in water absorbed by the material during storage.

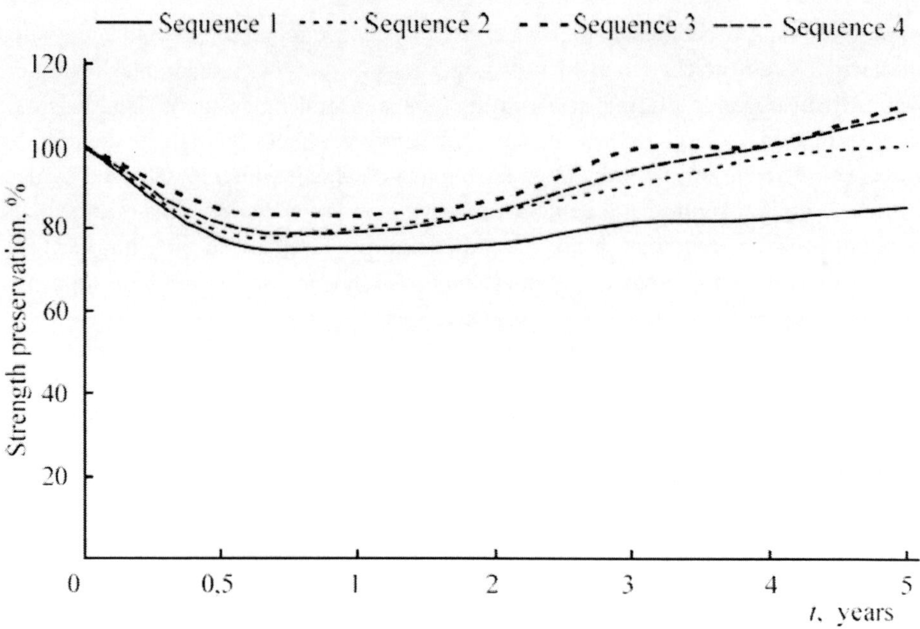

Figure 2. The level of strength preservation during aging of materials filled with silicon and antimony oxides

Injection of some structure formers into polymeric composites gives materials with a structure resistant to various physical and chemical effects of the environment. For example metal oxide substitution by a multiatomic alcohol reduces the tendency of change in strength properties under different climatic conditions to a general form and makes it independent of external conditions (Figures 3 and 4).

Interesting results were obtained at accelerated aging imitating different conditions, durability and storage time of materials, and effects on variations of their physicomechanical properties and operation characteristics. Accelerated tests reproduce effects of heat, humidity, negative temperatures, and seasonal and daily temperature differentials. The mode of artificial aging was chosen with regard to effective activation energy of aging of every particular material, calculated by the change of relative tensile elongation at break at different temperatures. Tests modeled aging over 3, 5, 10 and 15 years. Test results are shown in Figure 3.

In the case of metal-oxide vulcanization, tensile strength limit increases and elongation decreases over the initial five years. Processes of intermolecular interaction and structuring dominate at this stage. Further on, degradation processes lead to dissociation or scission of intermolecular bonds with strength reduction at elongation increase. Up to ten years of aging imitation, the strength somewhat increases again with elongation decrease, which is, apparently, associated with restructuring processes.

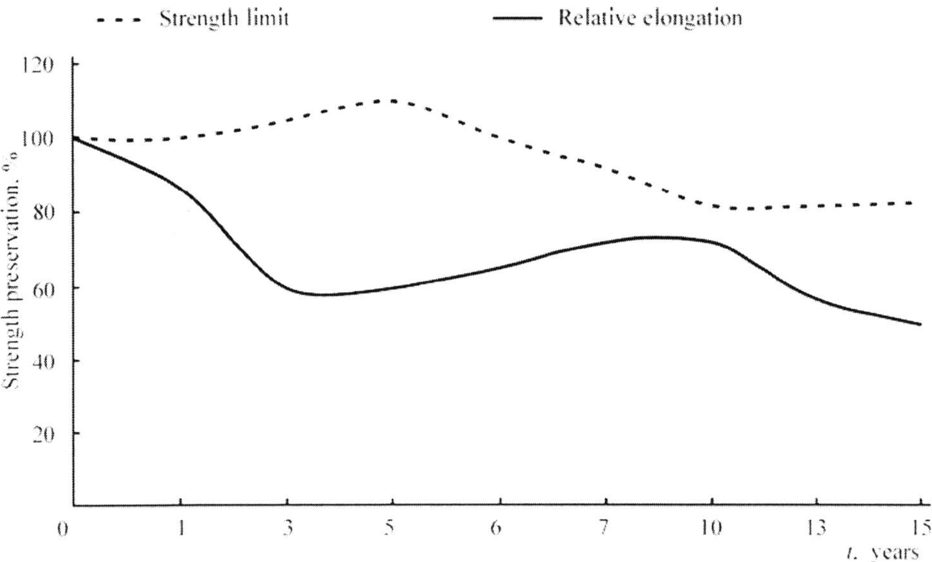

Figure 3. The level of strength preservation during artificial aging of metal-oxide vulcanized materials

Vulcanization by multiatomic alcohols provides properties more stable in time during aging (Figure 4).

Further climatic tests of materials over a period of 15 years show coincidence of results of accelerated and natural aging of materials, i.e. show the effectiveness and reliability of using accelerated aging methods with regard to the activation energy of the process for predicting material properties.

The ability to protect from the effect of fire is the determining one for fire shield materials. In this connection, changes in fire shield materials were estimated during aging. Figure 5 shows test results of change in the level of fire shield properties of a series of polymers, based on sulfochlorinated polyethylene during accelerated aging. Fire and heat shield properties were estimated by the one-side thermal impact method.

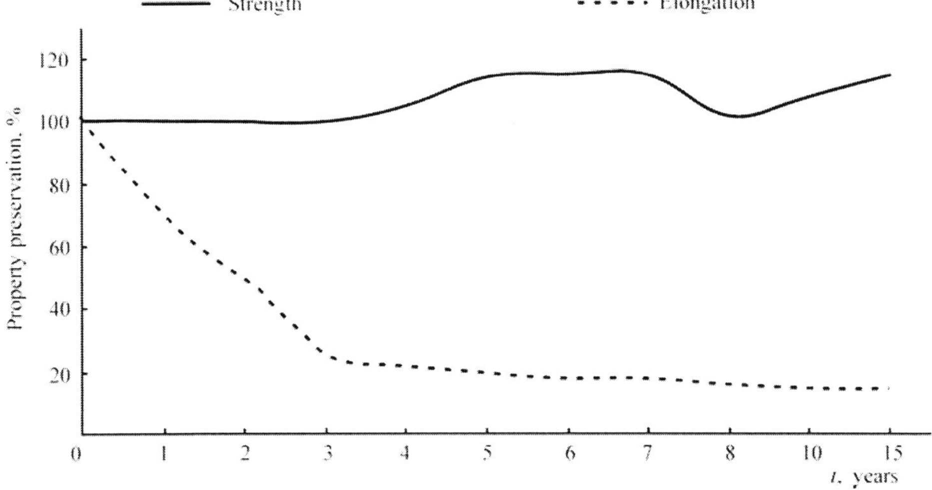

Figure 4. The level of strength preservation during artificial aging of alcohol vulcanized materials

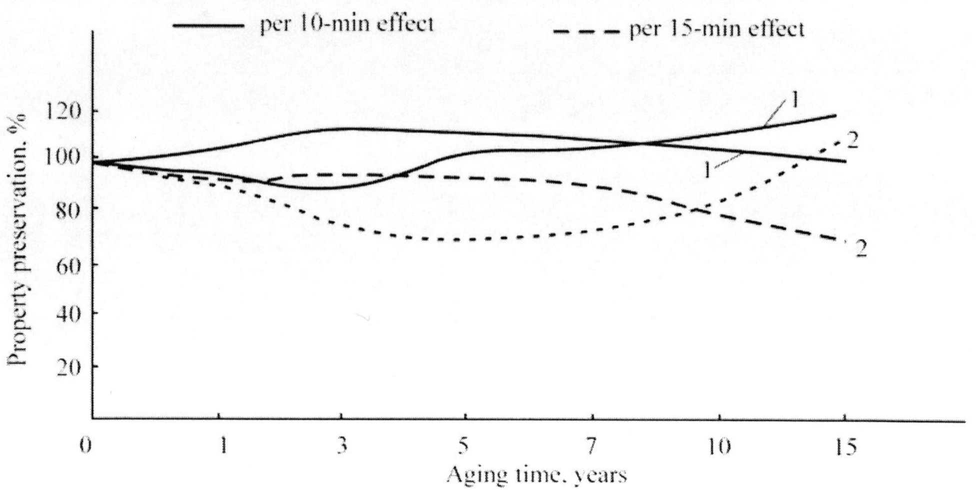

Figure 5. The level of preservation of fire shield properties of materials during artificial aging

Analysis of the Figure shows the high stability of fire and heat shield properties in time of materials based on sulfochlorinated polyethylene of different compositions.

Full-scale climatic tests performed on metal-oxide vulcanized material with a group of fillers under different climatic conditions also display high stability of properties over time [3 - 5].

REFERENCES

[1]   Donskoi A.A. and Shashkina M.A., Elastomeric Fire and Heat Shield Materials, *Scientific-Technical Collection "Aviation Materials at the Border of XX and XXI Centuries"*, Moscow, VIAM, 1994. (Rus)

[2]   Dontsov A.A., *Structuring Processes in Elastomers*, Moscow, Khimia, 1978, 288 p. (Rus)

[3]   Khalturinski N.A. and Berlin Al.Al., Physical aspects of polymer combustion and mechanism of combustion inhibitors, In Collection: Polymeric Materials of Reduced Combustibility, *Proc. IV Intern. Conf.*, Ed. I.A. Novikov, Volgograd, October 17 – 19, 2000, Izd. Volgogradskii Tekhnicheskii Universitet, pp. 9 – 12. (Rus)

[4]   Shashkina M.A., Aseeva R.M., Donskoi A.A., and Zaikov G.E., Fire and Heat Shield Materials Based on Sulfochlorinated Polyethylene, *J. Oxidation Communications*, 1995, vol. *18*(4), pp. 333 – 357.

[5]   Khalturinski N.A., Popova T.A., and Berlin Al.Al., *Uspekhi Khimii*, 1984, vol. *53*(2), pp. 326 – 346. (Rus)

In: Physical Organic Chemistry: Theory and Practice
Eds: A. D'Amore and G. E. Zaikov, pp. 159-166

ISBN 1-59454-275-9
© 2005 Nova Science Publishers, Inc.

Chapter 12

PROPERTIES OF POLYMERIC WAVEGUIDES AS INFORMATION TRANSMISSION CHANNELS

G. M. Rubinstein and N. G. Lekishvili
H.Javakhishvili Tbilisi State University, Tbilisi, Georgia
G. E. Zaikov
N.M.Emanuel Institute of Biochemical Physics, Moscow, Russia

TRANSMISSION BANDWIDTH

As mentioned above, according to structure and principle of the light transmission optical fibers, including PG (Polymeric waveGuides), are subdivided into two types: the ones operating by the principle of complete internal reflection and possessing a step profile of the refractive index (SPRI), and light focusing ones possessing a gradient profile of the refractive index (GPRI). PG of the SPRI type consist of the core with high refractive index and covers with lower ones with clear borders of these layers. For PG of the GPRI type, value of the refractive index gradually changes from the maximum on the axis of the core to the minimum at the surface. For the first time, PGs of the GPRI type have been suggested in 1968 independently in the USSR and Japan [1, 2]. If the refractive index changes along the radius according to the law $n(r) = n_0 \mathrm{sech}\,\alpha r$, then a fiber (or a bar) possesses the property of regular focusing of meridional beams [3, 4].

Information transmitted by optical fibers usually represents a series of short light pulses. Each pulse (in approximation of geometrical optics) may be represented as a great number of beams, each beam entering the fiber at its own angle φ. The range of angles φ depends upon the numerical aperture (1) of fiber of the SPRI type. The path length of every beam is different. That is why, the pulse broadens as passes along the fiber [5 - 8]. At a definite distance in the fiber, the pulse may broaden so high that overlaps with the next one and, consequently, the information transmitted will be distorted or even lost.

The broadening degree of the pulse is usually estimated by retention time (τ) which represents a difference of passing times of the axis and the border meridional ray of the pulse [9]. (It should be noted that this estimation neglects the contribution of non-meridional rays multiple reflected on the core–cover interface, when it is impossible yet to neglect the effect of distorted full internal refraction, i.e. light entering the cover material and passing a part of the way in a medium with a lower refractive index, taking into account of which may change the real value of τ [8, 9], in relation to the above-mentioned definition of this value.) In a rough approximation, τ is expressed as follows:

$$\tau \cong \frac{(NA)^2}{2n_1 c}, \tag{1}$$

where n_1 is the refractive index of the fiber core; c is the light velocity in vacuum. Suggesting that overlapping of pulses does not take place, the maximal cadence rate of pulse transmission in the fiber $\left(V_{T_{max}}\right)$ is estimated as

$$V_{T_{max}} \approx \frac{1}{2\tau}, \tag{2}$$

where $V_{T_{max}}$ equals to the bandwidth of the fiber expressed in Hz.

Because the bandwidth is inversely proportional to distance, more suitable parameter – the bandwidth multiplied by length (Hz·km) – is usually used for estimating the information capacity of the fiber.

Application of estimating expression (1) and (2) to the case of PG of the "ESKA" type ($NA = 0.5$, $n_1 = 1.49$) gives the value of 1.8 MHz·km. Assumption of a small overlapping of margins of pulses enables information capacity to be estimated as 2.5 MHz·km as a maximum [10]. However, the same values of NA and n_1 in the work [11] indicate estimation of the bandwidth as 10 MHz·km, and the work [12] informs about the transmission bandwidth of 2 MHz for PG 24 m long possessing $NA = 0.47$ and $n_1 = 1.492$. It should be noted that analog video signals require the bandwidth of 6 MHz·km, graphic ones with high resolution – 10 MHz·km, and digital video signals – 40 MHz·km [13].

Expressions (1) and (2) indicate that the bandwidth may be increased by decreasing the value of NA. However, such increase will be accompanied by a decrease of input power of the guide.

The bandwidth of optical fibers may be increased by application of light guides with GPRI, in which axis beams are spread in the medium with a high refractive index and, consequently, slower than meridional beams which, in this case, pass by a curvilinear trajectory of the spiral type in a lens-like structure, stipulated by the refractive index gradient. As a result, the difference in times of the beams run, which formed the pulse, decreases and, consequently, the pulse widening decreases which, in its turn, determines width of the transmission band of the fiber. Numerical aperture of the guide with GPRI is determined as follows [14]:

$$NA(r) = \begin{cases} K\sqrt{1-(r/a)^{\alpha}} & (r \leq a) \\ 0 & (r > a) \end{cases}, \tag{3}$$

where $K = \sqrt{n_1^2 - n_2^2}$, n_1 is the refractive index on the guide axis; n_2 is the refractive index on the guide generatrix; r is the distance from the axis to the point of radiation input on the guide cross-section; d is the guide radius; α determines the steepness of the refractive index profile. As expression (3) indicates, NA for a guide with GPRI is always smaller than for a guide with SPRI at the same difference in refractive indices on the axis and on the generatrix.

Power injected into guides with GPRI significantly depends on conditions of input, but may be optimized [15].

Inorganic light guides with GPRI are widely applied. As for PG with GPRI, attenuation in them is not less than 1,000 dB/km [16], and until now we have met no information on production of PG of this type possessing low attenuation.

PG AND LIGHT-EMITTING DIODES (LEDs)

The overwhelming majority of industrial PGs possess the core from PMMA and the cover from fluorine-containing polymers [12 – 15, 17 - 19]. As Figure 1 indicates, the area of these fibers displaying the minimal attenuation lies in the range of 560–570 nm. It would be logic to use as light sources for information transmission in these PD LED the ones irradiating in the close part of the spectrum. However, as indicated in Table 1, LEDs in the green area of the spectrum possess low intensity and long time of triggering [20].

LED in the IR-spectrum display excellent characteristics, and wavelengths of them correspond to the areas of the lowest attenuation in PG with core of PMMA-D_8. However, high price of PG from deuterated PMMA compared with the price of inorganic light guides with significantly lower attenuation significantly limits the application sphere of deuterated PG, the more so as absorption by PMMA-D_8 (see Figure 2) somewhat above 700 nm sharply increases at humidity absorption [21], and PMMA-D_8 is quite moisture-absorbing. That is why, a compromise variant was applied: at the present time, the diapason of wavelengths from 650 to 670 nm is used everywhere for data transmission in PG. This diapason corresponds to the transparency window of PG with PMMA core (shown in Figure 1). LEDs of this sphere possess better characteristics than LEDs of green part of the spectrum. However, wavelength and intensity of irradiation of the latter depend on temperature: when temperature increases up to 50°C, the wavelength of irradiated light transposes by 10 nm to longer waves, whereas transparency window of PG is narrow – from 660 to 700 nm, and irradiating intensity at 60°C gives 70% of the intensity level at 25°C [22]. This thermal sensitivity of LEDs requires additional measures for keeping temperature constant.

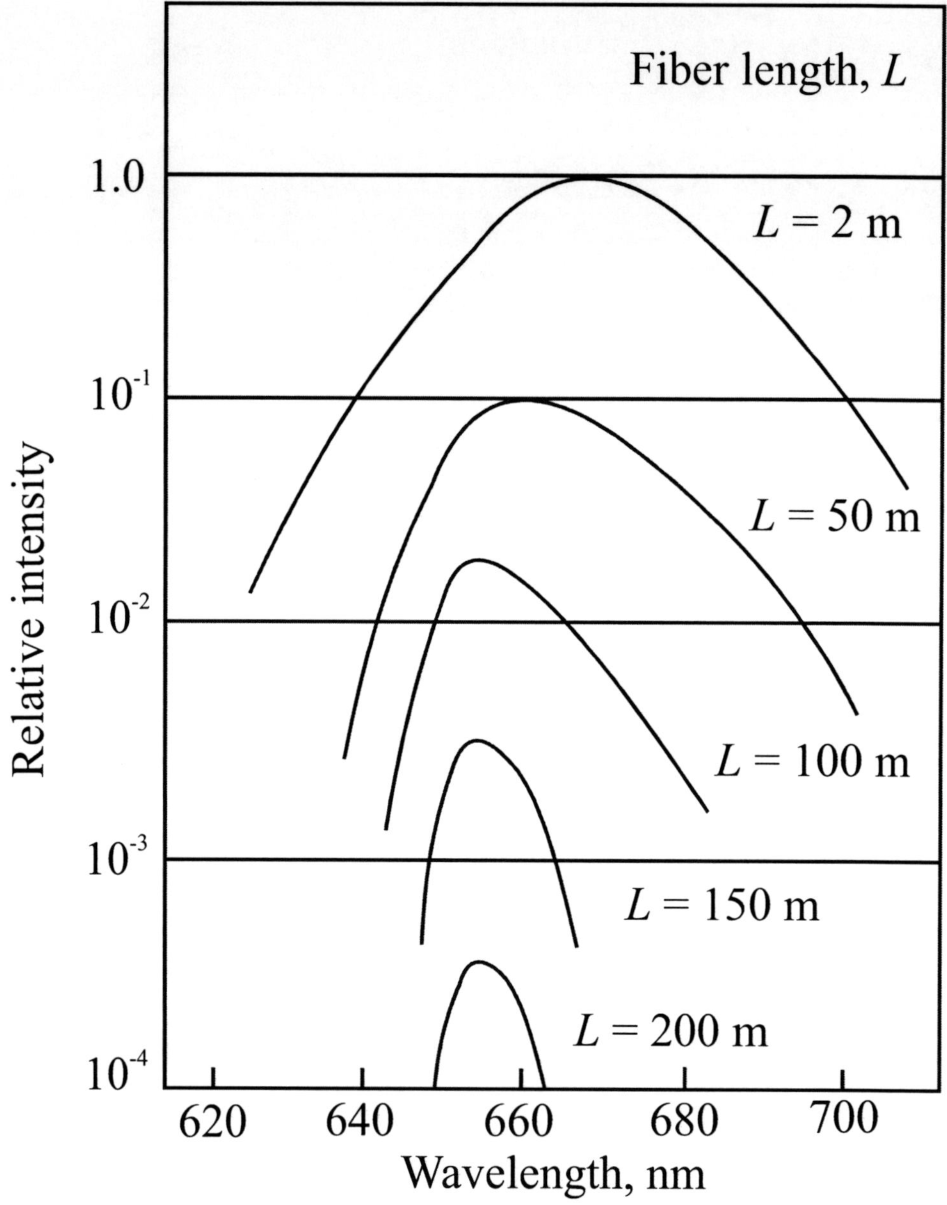

Figure 1. Dependence of spectral distribution of LED-A irradiation on PG length.

Table 1. Characteristics of Light-emitting Diodes [11]

Index	LED in visible spectrum		LED in IR-spectrum	
The main wavelength (nm)	555	660	780	850
Semiconductor	GaP	GaAlAs	GaAlAs	GaAlAs
Intensity (mW)	0.05	1	40	60
Time of triggering (ns)	500	50	15	15

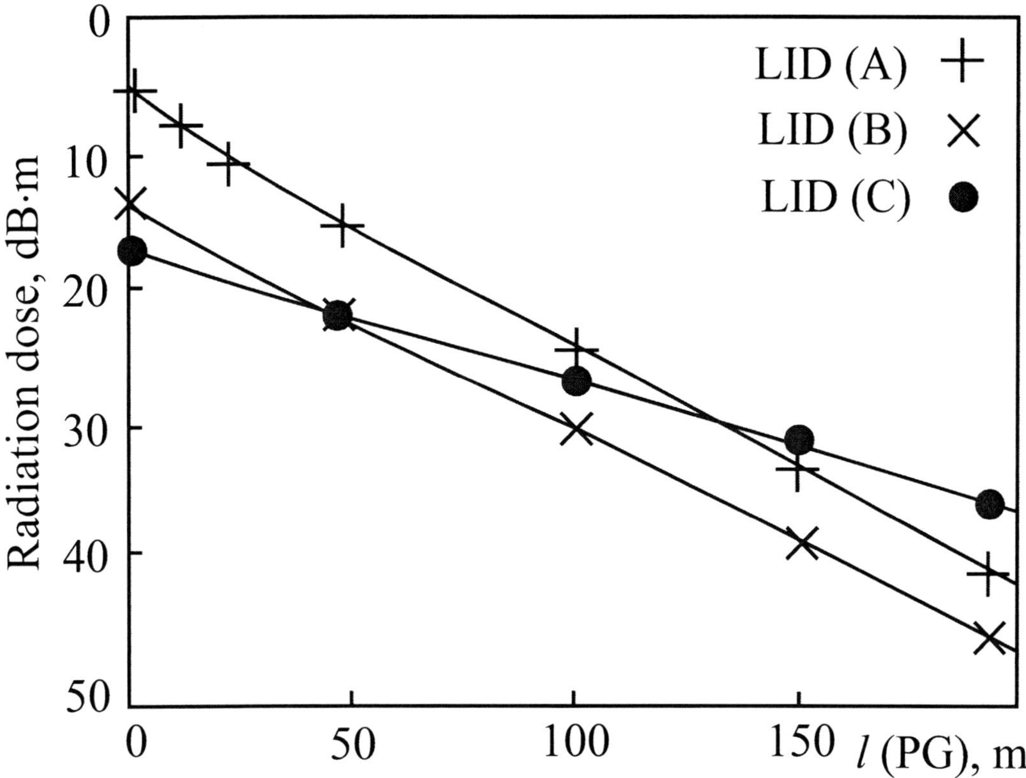

Figure 2. dependence of radiation intensity transmitted by fiber on PG length.

LEDs irradiate with a broad diagram of directivity. That is why, light injected into PG possesses high probability to lose highest modes on inhomogeneities of the core–cover interface, microflexures and other irregularities of structure during running along PG. That is why, attenuation of light from LED is more significant in PG than attenuation of light from a source of parallel beam. This effect is the most noticeable on short distances. As highest modes are lost, the attenuation decreases [5] and becomes constant on a definite length of PG.

LEDs irradiation is not of the monochromatic type, that is why, wavelengths displaced in relation to the center of narrow transparency window of PG, attenuate faster, and spectral distribution of input radiation is distorted. Attenuation of the margin of spectral distribution of LED irradiation is most significant on short lengths of PGs. With regard to increase of fiber length, this attenuation decreases and becomes constant at some length of PG. Hence, wide diagram of directivity and wide spectrum of LED irradiation similarly affect the character of attenuation dependence on the fiber length.

Quite wide-spread LEDs irradiate with the peak at wavelengths of 670 nm (LED-A), 657 nm (LED-B), and 570 nm (LED-C). Figure 2 [18] represents dependence of the spectral distribution of LED-A irradiation on PG length. The maximum on the curve of spectral distribution is displaced to the center of transparency window of PG, from 669 nm after passing PG 2 m long to 657 nm (PG 50 m long) and to 653 nm (PG 100 m long). As PG length increases over 100 m, the maximum of the curve remains in the range of 652–653 nm, and no displacement of the maximum be the wavelength is observed, but the half width of the spectral distribution curve decreases still. In analogous experiments with LED-B irradiation, maximum of the spectral distribution curve displaces from 656 nm (2 m) to 652–653 nm (100 m and longer). The reason of smaller displacement of LED-A irradiation maximum is that the spectrum of LED-B irradiation is better inserted into the transparency window of PG.

Figure 2 [18] represents dependence of radiation intensity transmitted by fiber on PG length.

Intensity of LED-A irradiation decreases abruptly on the initial meters of the fiber, but then attenuation decreases gradually and becomes practically constant (165 dB/km) between 100 and 200 m. At the distance of 200 m, radiation intensity is 41.3 dB/m. In contrast with the case of LED-A, intensity of LED-B irradiation on initial meters decreases not so abruptly and becomes practically constant (162 dB/km) between 50 and 200 m. Radiation intensity at 200 m is 47.2 dB/m. Green light from LED-C with a peak at 570 nm correlates well with the transparency window of PG possessing the lowest attenuation. In this case, displacement of the spectral distribution maximum is absent, however, half width of the curve decreases as the fiber length increases. A small change in attenuation on short lengths is bound to attenuation of the highest modes and the change of half width of the spectrum curve. The level of green LED-C light intensity is lower, than for red LED-B and LED-A on short lengths of PGs. However, after 150 m it reaches their level and keeps intensity of 36.6 dB/m up to 200 m and longer. As it is indicated by the above-said, transmission of a signal along the fiber by distances longer 200 m is possible, if green LED-A is applied. However, long time of its triggering will decrease already short width of the transparency band of PGs.

It is obvious that appearance of LEDs of green spectrum with short time of triggering and higher irradiated intensity will sharply broaden the sphere of application of PGs and will allow full consumption of advantages of these waveguides.

Unfortunately, optical properties of perhalogenated polymers accessible now are not suitable, and they cannot be applied in PGs. Not only polymers themselves, such as polytetrafluoroethylene (PTFE) and polychlorotrifluoroethylene (PCTFE), display high crystallinity degree [23, 24], but also their monomers are not block polymerized. Consequently, true losses will increase due to natural and technological dispersion. Moreover, if fluorine-containing polymers possess amorphous structures (fluoroalkylmethacrylates), then the higher concentration of fluorine is, the lower is the glass transition temperature [25] and, as a consequence, heat resistance.

Therewith, potential optical abilities of even crystalline halogenated polymers are noted in work [26]. Problems associated with crystallinity were by-passed by cooling down the polymer melt at the rate of ~100°C/s. In this case, the amorphous state of the melt was "frozen" before crystalline spheres were formed [27]. However, this method is not quite suitable for obtaining PG, because PG obtained in this way will be able to recrystallize as a result of temperature variation under exploitation conditions.

Ability to crystallize decreases in the presence of large atoms or molecules in side groups of polymeric chain. Such side groups distort ability to formation of the ordering: for example, the presence of a volumetric chlorine atom in PCTFE is the reason that the crystallinity degree of this material is lower than of PTFE. Similar mechanism of crystallinity decrease is natural and irreversible. That is why, it is preferable before fast cooling.

If efforts of chemists-synthetists are successful and an amorphous perhalogenated polymer is synthesized, difficulties in obtaining a material possessing optical transparency associated with impossibility to application of the method will possess a tremendous potential abilities as the material for PG with low losses.

The best compromise is searched for yet – synthesis of partially halogenated materials – which enables amorphous nature of polymer to be combined with ability to block polymerization promoting a decrease of IR-absorption due to decreased concentration of hydrogen. Good results are displayed by a combination of fluorination and deuteration [28, 29]. Of interest is the information by Asahi Garasi Co. about synthesis of a new fluorine-containing polymer, named by the Company as "Sytop", which possesses fiber-forming properties [30, 31].

REFERENCES

[1]   *Patent No. 308,111*, 1976 (USSR). (Rus)
[2]   *Application No. 47-28,056* (Japan).
[3]   Mikaelyan A.L., *Doklady AN SSSR*, 1951, vol. *81*, p. 569. (Rus)
[4]   Marcus E.D., *Appl. Opt.*, 1979, vol. *18*, p. 2073.
[5]   Blagidze Yu.M., Jibladze M.I. *et al.*, *Quantum Electronics*, 1973, vol. *4*(16), p. 97. (Rus)
[6]   Rubinstein G.M. and Perel'man N.E., *Quantum Electronics*, 1974, vol. *1*, p. 983. (Rus)
[7]   Jibladze M.I., Perel'man M.E., Rubinstein G.M. *et al.*, *Proc. Int. Conf. on Laser*, December 15 – 19, 1980, New Orleans, 1981, vol. *80*, p. 464.
[8]   Jibladze M.I., Perel'man M.E., Rubinstein G.M. *et al.*, *Izv. AN SSSR, Ser. Fiz.*, 1979, vol. *43*, p. 292. (Rus)
[9]   Senior J.M., *Optical Fiber Communications*, Prentice-Hall, 1985.
[10]   Emslie Ch., *J. Mater. Sci.*, 1988, vol. *23*, p. 2281.
[11]   Nara S., *Keiso*, 1982, vol. *25*, p. 17.
[12]   Maier J., Lieber W., Heiniein W., Croh W., and Herbrechtsmeier P., *J. Theis. Electr. Lett.*, 1987, vol. *23*, p. 1208.
[13]   Glen R.M., *Chemitronics*, 1988, vol. *1*, p. 96.
[14]   Gloge D. and Marcatili E., *Bell. Syst. Techn. J.*, 1973, vol. *52*, p. 1563.
[15]   Bastawros A., *Optik*, 1986, vol. *74*, p. 57.
[16]   Ide F. and Yamamoto T., *Sen'i Gakkaisi*, 1984, vol. *40*(4-5), pp. 248 – 250.
[17]   *Patent No. 4,593,974* (USA).
[18]   Fujimoto S., *Proc. Soc. Photo-Opt. Instrum. Eng.*, 1987, vol. *799*, p. 139.
[19]   Kitazawa M., *Papers Presented at the Fourth European Fiber Optic Communication and Local Area Networks Exposition*, June 23 – 27, 1986, Amsterdam, the Netherlands, 1986, p. 172.
[20]   Weisner W., *Electronikschau*, 1988, vol. *63*, p. 60.
[21]   Kaino T., *Appl. Opt.*, 1985, vol. *24*, p. 4291.
[22]   Yamaguchi M., Yamamoto K. *et al.*, *National Tech. Rep.*, 1988, vol. *29*, p. 13.
[23]   Kaino T., *J. Polym. Sci.*, 1987, vol. *25*, Part A, p. 37.
[24]   *Plastic Optical Fibers*, Proc. SPIE, Sept. 1993, vol. *1799*, Boston.
[25]   *Patent No. 4,687,295* (USA).
[26]   *Patents Nos. 1,431,157; 1,449,150; 2,161,954* (Great Britain).
[27]   *Patent No. 3,524,369* (FRG).
[28]   *Application No. 60-26,014* (Japan).

[29]   *Patent No. 4,576,438* (USA).

[30]   Ohtsuka Y., *Oputoronikyusu (Optronics)*, 1986, vol. 5, p. 67.

[31]   Chagulov V.S., Lekishvili N.G., Rubinstein G.M., and Sanadze N.S., *Patent No. 1,671,032*, 1991 (USSR). (Rus)

In: Physical Organic Chemistry: Theory and Practice
Eds: A. D'Amore and G. E. Zaikov, pp. 167-209

ISBN 1-59454-275-9
© 2005 Nova Science Publishers, Inc.

Chapter 13

DEGRADATION OF ALIPHATIC-AROMATIC POLYIMIDES – POLYALKANIMIDES

E. V. Kalugina

Polyplastic Co., 14A, General Dorokhov st., Moscow, Russia

K. Z. Gumargalieva

N.N.Semenov Institute of Chemical Physics, Moscow, Russia

G. E. Zaikov

N.M.Emanuel Institute of Biochemical Physics, Moscow, Russia

INTRODUCTION

Polyalkanimides (PAI) are fatty-aromatic polymers derived from aliphatic, usually linear diamines and tetracarboxylic aromatic acids.

For the first time PAI were synthesized in 1955 by polycondensation in melt of pyromellitic acid and C_7 and C_9 diamines [1]. In the former USSR, Academician V.V. Korshak *et al.* were the first who synthesized PAI by the one-stage high-temperature polycondensation in solution [2]. The studies of PAI properties, derived from 3,3'4,4'-diphenyl tetracarboxylic, 3,3'4,4'-diphenyloxide tetracarboxylic, and 3,3'4,4'-diphenylsulfone tetracarboxylic acids, as well as on pyromellitic acid and C_6 and C_8 diamines indicated the prospect of PAI application as heat-resistant construction fiber polymers and the binder for abrasion tools.

Industrial development of PAI was determined by the increase of industrial demand in heat-resistant wire insulation. Experts of Raychem Company tested many polymers, including polycarbonate and polysulfone. The results of these investigations gave raise to design of technology and organization of manufacturing PAI derived from pyromellitic anhydride and dodecamethylene diamines or tridecamethylene diamines. These PAI possess high hardness number at room temperature, are strong at temperatures above 150°C, and may be effectively processed by extrusion. In 1970's, Raychem Company already produced a great variety of wires with insulation from PAI [38 – 40], PAI-based films and fibers, and construction materials of *Polyimidal* and *Poly-X* trademarks for the automobile industry. Rohm und Haas

GmbH produces PAI derived from dodecamethylene diamine under *Kamax 201* and *Kamax 301* trademarks [3]. In Russia, pilot production of PAI and composite glass-filled materials derived from them is also realized.

Chemical structure of PAI with variable fatty chain length is the following:

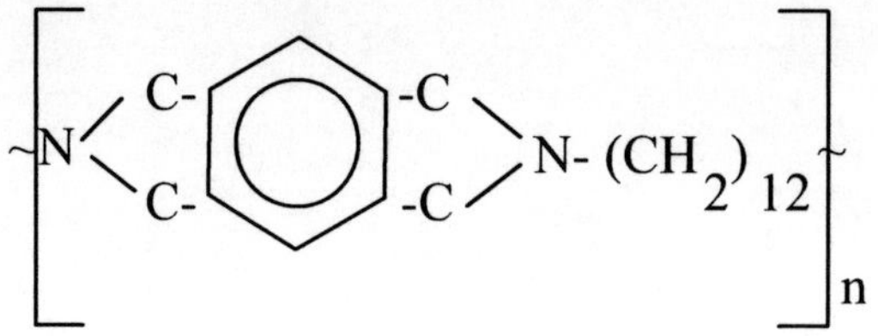

It provides the variety of macromolecule packing in polymeric body; therefore, homological sequences of PAI became the object of many X-ray diffraction studies which, besides works on synthesis, gave a large volume of fundamental publications on PAI also concerning degradation transformations [1 - 8].

From applied positions, in the PAI family polydodecamethylene pyromellitimide (PAI-12) attracts attention due to its excellent physical and mechanical, heat physical and dielectric properties [9 - 11]. In the sequence of engineering thermoplasts, PAI-12 is considered as material for electrotechnical purposes, having the working temperature range of 150 - 200°C. As is indicated by "UL – temperature indices", PAI are present in the same sequence with polysulfones.

PAI-12 (hereinafter, PAI, if not compared with different polyalkanimides) is synthesized in sequence by polycondensation of dodecamethylene diamine and pyromellitic dianhydride in N-methylpyrrolidone solutions at 40 - 50°C, further polyamidoacid cyclization at 150°C, separation of precipitated polymer with the cyclization degree over 90% (IR-spectroscopy data on the ratio of absorption bands at 1780 and 1720 cm^{-1}), and then powder washing and drying.

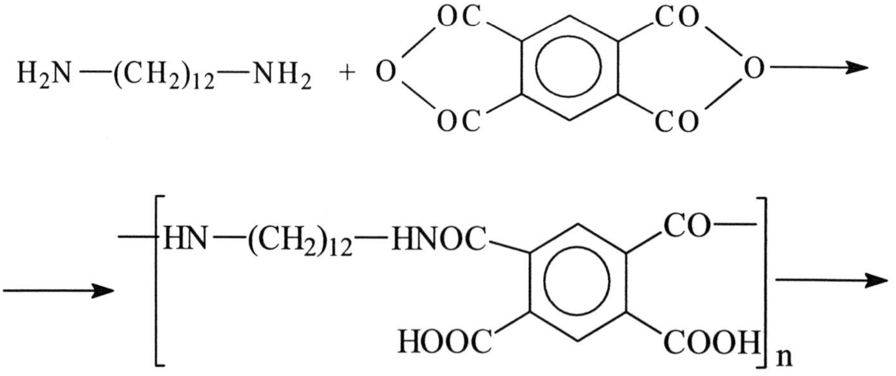

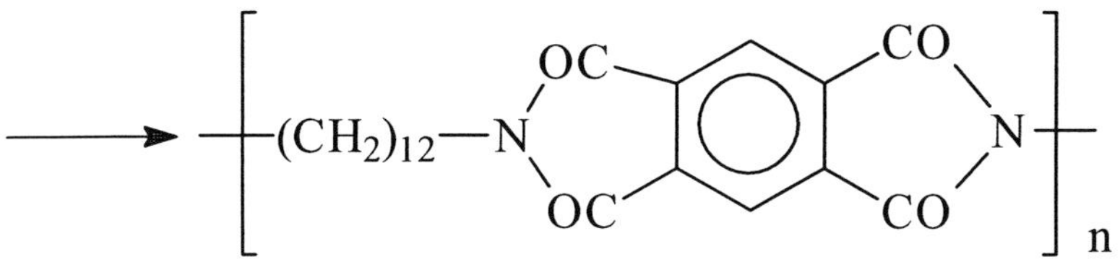

Polycondensation in amide solvent proceeds through formation of intermediate salts of the following structure:

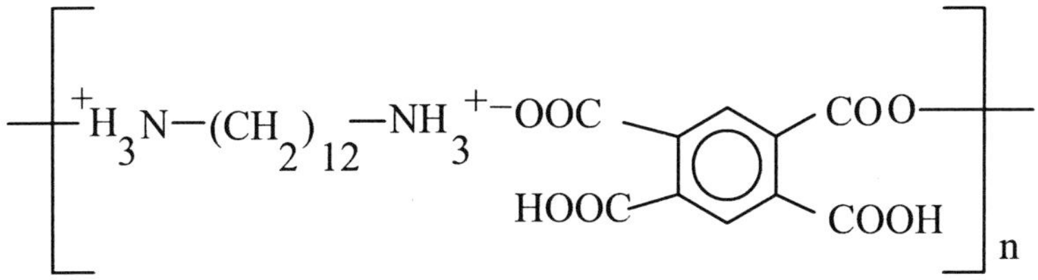

which was determined with the help of IR-spectra and potentiometric titration technique [54]. Studies of condensation in model systems, for example, phthalic anhydride – lauryl amine or pyromellitic acid – lauryl amine ones, also indicated possible formation of di-, tri- and tetrasalts at a single aromatic ring which, in turn, indicated quite high reactivity of all carboxylic groups at the aromatic ring. At polycondensation appropriate group reactions must lead to branching of macrochains, which is the apparent reason for low gel-fraction concentration (below 5 wt.%) in the marketable end PAI. Low defectiveness of the structure is also indicated by comparison of calculated ($[C] = 69.1\%$, $[H] = 6.8\%$, $[N] = 7.3\%$, $[O] = 16.8\%$) and experimental ($[C] = 69.0\%$, $[H] = 7.5\%$, $[N] = 7.1\%$, $[O] = 15.4\%$) of PAI elemental composition.

End groups significantly affect thermal oxidative stability of PAI. In PAI synthesis pyromellitic dianhydride is taken in some excess that defines predominance of carboxylic end groups in macromolecules compared with amine ones or blockade of amino groups by acetic anhydride. Marketable end PAI are have the melt viscosity equal 10^3 - 5×10^5 P or $\eta_{sp} = 0.8$ – 1.5 (0.5% solution of m-cresol – tetrachloroethane mixture). According to X-ray diffraction data [6, 7] PAI has 40 - 60% crystallinity degree, and the chain conformation of the polymer extracted from the reaction mixture is coiled in the amine component. At temperature about 270°C, resulting conformation transformation, the chain is straightened and becomes bladed; PAI melting point is 285°C with the maximum of appropriate endothermic DSC peak at 298°C.

The following production phase – extrusion, is used either for powder pelletization or obtaining composite materials (glass-, mineral-, etc. filled), derived from PAI. The temperature mode by zones is 270 - 330°C. The mass was press molded at 330 ± 10°C to a mold, heated up to 150 - 180°C. These stages are the most temperature aggressive for the polymer. That is why the study of degradation transformations in the melt is the urgent problem.

POLYALKANIMIDE DEGRADATION IN MELT

Thermal Degradation

Chemical instability of PAI macromolecules in melt becomes noticeable already at 300°C due to the increase of gel-fraction content and decrease of specific viscosity in solution (Figure 1). The processes are speeded up with temperature, and the effective activation energy of PAI gel formation in the temperature range of 300 - 450°C equals (50 ± 10) kJ/mol.

Judging by kinetics of mass losses and $C_1 - C_3$ hydrocarbon extraction (Figures 2 and 3), pyrolytic reactions in PAI are detected already at 300°C, though after tens of hours of heating are required for detection. At the initial stage, the seeming zero order of kinetics of mass losses and hydrocarbon extraction is determined by transformations degrees, negligible even at 400°C. The effective activation energy of mass losses, equal (232 ± 20) kJ/mol, correlates with E_a values for release of methane, ethylene, ethane and propane equal (192 ± 15), (160 ± 15), (238 ± 15) and (170 ± 15) kJ/mol, respectively.

PAI pyrolysis at 300°C proceeds with release of light (volatile) hydrocarbons only; at 350°C or higher carbon oxides and low-volatile products are also released from PAI. The latter precipitate near the hot zone, shaped as yellow-brown oil-like blushes and light crystals.

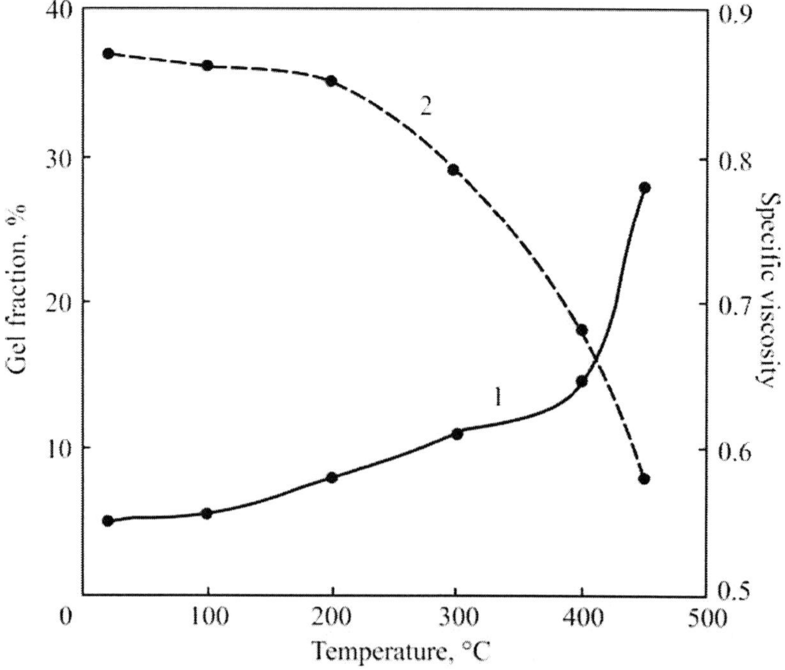

Figure 1. Kinetics of gel-fraction accumulation (1) and sol fraction viscosity variation (2) at PAI degradation in vacuum

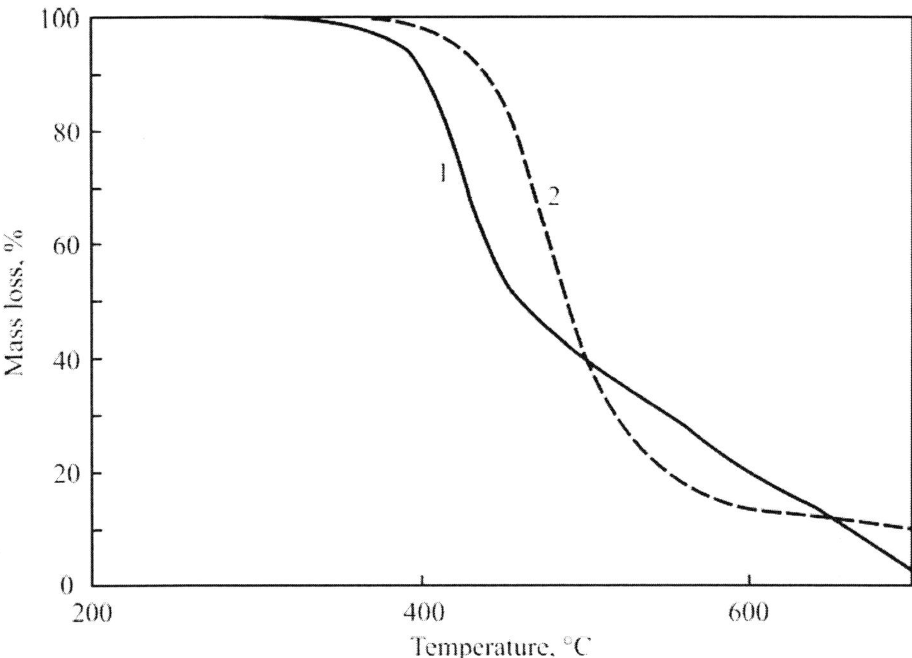

Figure 2. Thermogravimetric analysis (TGA) of PAI at the heating rate of 5 deg/min in air (1) and argon flow (2)

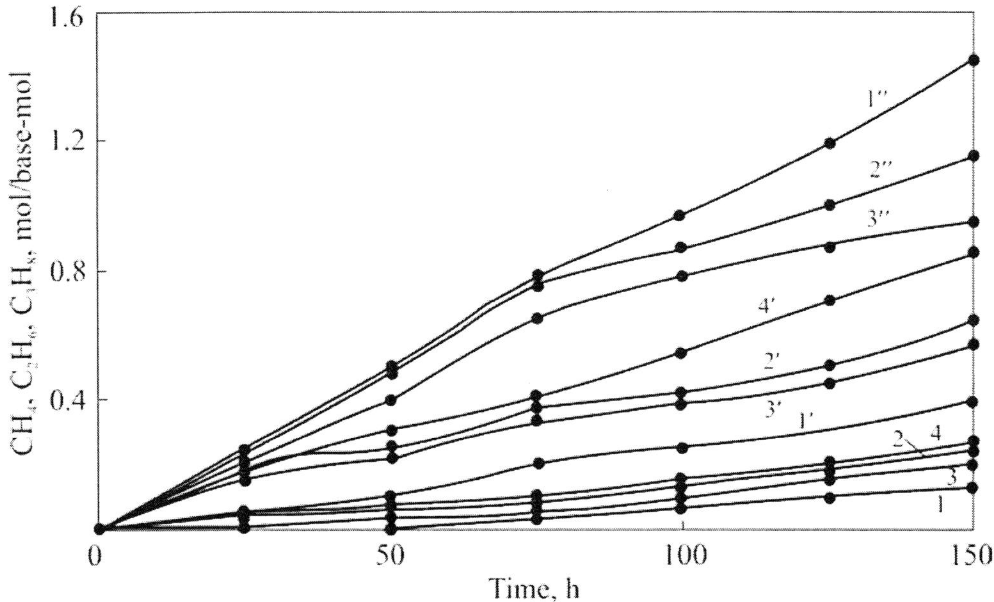

Figure 3. Extraction kinetics of CH_4 (1, 1',1"), C_2H_4 (2, 2', 2") and C_2H_6 (3, 3', 3") at 300, 350 and 450°C, respectively, and C_3H_8 (4, 4') at 350 and 400°C at PAI degradation in vacuum

According to ^{13}C NMR-spectroscopic and mass-spectroscopic data, this oil-like blush is formed by the following homologues of phthalamide, nitrile phthalamide and pyromellite diimide:

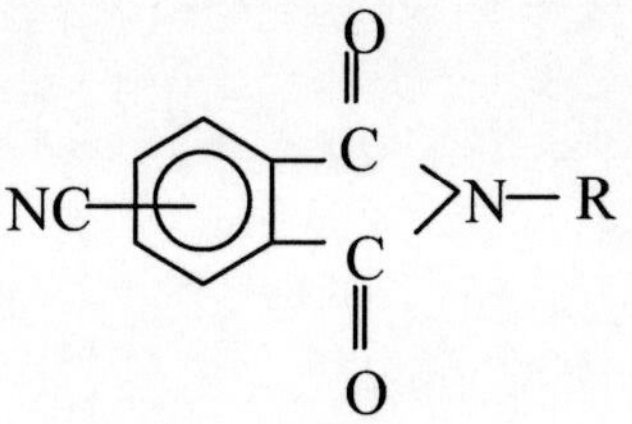

$$R= CH_3, \ C_{11}H_{23}, \ C_{12}H_{25}$$

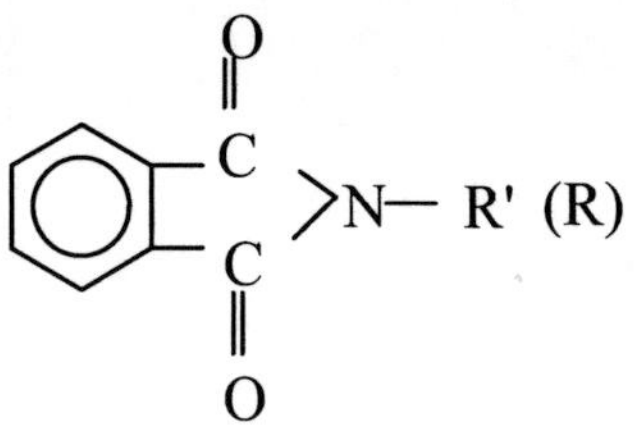

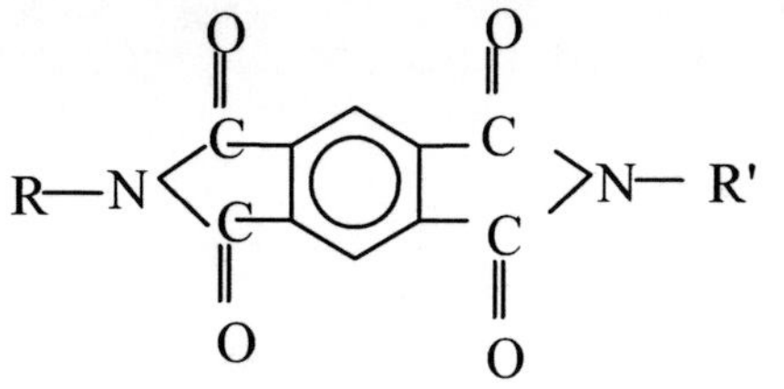

$$R= CH_3\text{-}(CH_2)_n, \quad R'= (CH_2)_n\text{-}CH=CH_2$$

Homological sequences are traced well by mass-spectra of degradation products measured at low ionization potential with sequences of molecular ions, differed by 14 wt. units:

$(CH_2) - 217 - 231 - 245 - 259 - 273 - 267 - 301 - 315;$

$201 - 215 - 229 - 243 - 257 - 271 - 285 - 299 - 313 - 327 - 341;$

$230 - 244 - 258 - 272 - 286 - 300 - 314 - 328 - 342 - 356 - 370 - 384 - 398 - 412 - 426 - 440 - 454 - 468 - 482 - 496 - 510 - 524 - 538.$

Natural fractionation of the products with respect to distance from the hot zone and determination of mass-spectra at programmed heating allow detection of the yield distribution in homological sequences. This distribution indicates preferable breaks of methylene chains at their ends (a great amount of methyl phthalimide is formed) or almost at the middle of them.

A low volatile product precipitating as crystals (judging by IR-spectra of crystals and ethanol [12]) represents ammonium bicarbonate, the formation of which during PAI pyrolysis is possible at the interaction of primary products (ammonia and CO_2) in the hot zone, i.e. NH_4HCO_3 is the secondary product.

As the initial chemical structure and chemical structures of the products are compared, it is indicated that pyrolytic transformations of PAI proceed in two directions: by methylene chains and heterocycles. Being the most thermally labile fragments of PAI macrostructure, ethylene chains begin degrading at a noticeable rate when polymer melts, i.e. at 300°C, approximately, though even at this temperature the pyrolysis rate is also negligibly low (judging by total yield of hydrocarbons): about 0.1% of methylene chains per hour are damaged.

At 350°C, products of heterocycle decomposition occur: carbon oxides, phthalimide and nitrile phthalimidestructures, but the yield of these products is very low yet. For example, the release rate of CO_2 – the main product (by yield) of heterostructure degradation - is by one and half orders of magnitude lower than hydrocarbon release rate, not even taking into account the release of oligomeric fractures with residues of broken methylene chains. However, heterocycle pyrolysis becomes clearer with temperature increase. At 400 – 450°C IR-spectra display the progressive degradation of heterocycles, first, by a decrease and then full disappearance of absorption bands at 1720 and 1780 cm^{-1}, related to v_s(C=O) and v_{as}(C=O) of the imide cycle [13], respectively. One more sign of this degradation is occurrence and intensification, and then full dominance (at 450°C or higher) of v(C=N) absorption band of nitrile groups [14]. Aliphatic structure degradation leads to the conjugation system increase in pyrolyzate, detected by intensive coloring of the coke residue and paramagnetic manifestations (ESR signal represents a singlet with g-factor equal to free electron). The analysis of polymer degradation always implies two points of view on this process. From practical positions, purely thermal (pyrolytic) reactions must not make obstacles at normal cycle of PAI processing lasting several minutes. However, thermally, PAI is not absolutely inert in the temperature range of processing. Though the pyrolysis rate is still negligibly low, at overheatings and long stay of the material in dead volumes pyrolytic reactions will cause serious damages of chemical structure of macromolecules and, correspondingly, a decrease of polymeric material quality.

Moreover, thermal transformations in PAI are of definite scientific interest. The structure of PAI which represents (very simply) a hybrid of polyethylene and polyimide has not ever been considered in terms of reactivity in thermal reactions. Basing on the previous experience in the polymer science, it might be instinctively forecasted that methylene chains are thermally stable compared with cycloarylene structure, but the experiment is the only think that may determine the features of flexible and rigid fragment reaction in the chemical structure and, apparently, their interrelations.

Let us start from low temperatures (300°C), when methylene chains break and heterocycle is stable still. Typical feature of this process is rather narrow selection of C_1-C_3 hydrocarbons – the degradation products. Random breaks would cause occurrence of wide selection of products from methane to dodecane and release of alkylpyromellitimide oligomers. In the sensitivity range of modern instrumental techniques of analysis (GLC, mass-spectroscopy) the attempts to detect these substances failed. This means that either these substances do not occur at all or, which is most probable, their synthesis rate is negligibly low versus light hydrocarbons. As a consequence, methylene chain breaks in PAI have some specificity.

It is common knowledge that similar to cracking of alkanes, pyrolysis of PE and other hydrocarbon polymers is the radical-chain process [15]. The signs of radical process are observed for pyrolysis of methylene chains in PAI, which are high activation energies and synthesis of not single but several unitypical products (hydrocarbons up to C_3) with approximately similar yields. Finally, of great importance is the similarity principle concluded in equal structure (methylene chains) and equal chemistry. The chain type of the process is determined directly by chain initiation and propagation rate changes. This was made at the study of hydrocarbon cracking, when the origin of one product or another in one elementary reaction or another may be simply determined. At polymer pyrolysis direct proofs of the chain process proceeding are not obtained. Indirect proofs are used: correlation of effective activation energies with C-C bond strength and identification of products, which synthesis might be explained as the consequence of the chain transfer reaction.

These signs are also displayed by pyrolysis of PAI methylene chains: effective activation energies of C_1-C_3 hydrocarbon release fall within the range of 160 - 240 kJ/mol, which is much lower than any estimations of C-C-bond energy both in purely carbon surrounding (330 - 360 kJ/mol) and nearby the heteroatom (290 – 330 kJ/mol) [16]. Homolytical break of α-C-C-bond happens simpler compared with more distant bonds. Single O and N atoms in the chain reduce α-C-C-bond strength by 12 – 20 kJ/mol [17]. High delocalizing ability is displayed by aromatic ring: delocalization energy of benzene radical equals 30 - 40 kJ/mol [18]. Following the additivity concept in chemistry, one would expect an analogous effect for pyromellitimide fragment. The authors of the current monograph have calculated molecular diagrams for methyl phthalimide and dimethyl pyromellite diimide – compounds modeling the boundary fragment of the elementary unit of PAI – by CNDO/2 technique using the software developed in L.Ya. Karpov Research Institute of Physical Chemistry. The geometry of molecules is shown in [19]. Diagrams in Figure 4 show charge distribution on atoms of the models.

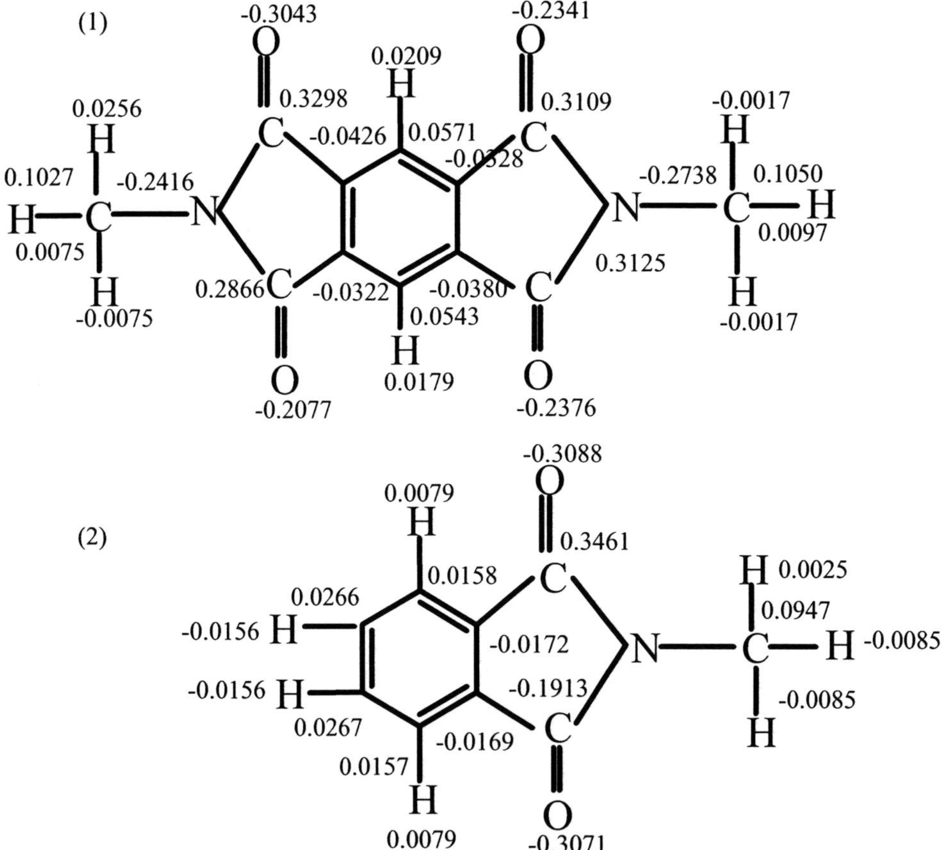

Figure 4. Molecular diagrams of PAI model compounds: dimethyl pyromellite diimide (1) and methylphthalimide (2)

The calculations show development of π-electron system in the model, which also affects alkyl substituting groups, to higher extent at α-C atom rather than at a distant one. Valent π-

electron released from α-C atom at homolytical break of C-C and C-H bond is engaged in the π-system of arimide cycle, e.g. delocalization energy is released, by the value of which the strength of α-C atom bonding in alkyl substituting group is reduced.

Therefore, since strength of aliphatic C-H-bonds is much higher than that of C-C-bonds [16], the most probable initiation act is the weakest α-C-C bond break in the methylene chain:

Though the end alkyl macroradical I is stabilized by the arimide structure, it dies eventually, most likely, by detachment of H-atom and sequential formation of methyl pyromellitimide end group. End macroradical II of quite long methylene chain is quite reactive, especially in the polymer melt. Actually, macroradical II is identical to corresponded macroradical of methylene chain in PE.

The situation becomes more similar due to close viscosities of the melts of marketable end PAI and common PE trademarks (up to 10^5 P). This definitely indicates equal molecular mobility in melts. The probability of two subsequent breaks in the same methylene chain of limited length is low. Therefore, long hydrocarbons are not detected in products of PAI pyrolysis at 300°C. More probable is depolymerization of macroradical II with ethylene detachment, which is the main product by output at 300 - 350°C.

$$\longrightarrow C_2H_4 + II$$

Ethylene polymerization heat known from the literature [20] equals 121 kJ/mol, which correlates well with experimentally determined activation energy of ethylene release at PAI pyrolysis (158 ± 12 kJ/mol). End macroradicals of type II are capable of realizing the chain transfer by detaching H atom from CH_2-group in intermolecular or intramolecular reactions. Intramolecular detachment of H atom as a monomolecular reaction is kinetically more profitable than the interchain exchange [21], if nonstressed transition complexes are synthesized. This possibility of macroradical II isomerization through six-term transition complex formation leads to detachment of C_3H_7 radical then formation of C_3H_8 and end allyl groups:

Compared to other hydrocarbons, besides ethylene, relatively low effective activation energy of propane release which equals (170 ± 14) kJ/mol correlates with suggested mechanism of its formation in specific reaction of radical isomerization. As temperature increases, differences in the reaction energy becomes smoother, and already at 350°C products of PAI pyrolysis possess compounds of oligomeric type with full selection of lengths of methylene chains.

Thermal instability of the imide cycle is observed already at 350°C, when products of its decomposition – carbon oxides, phthalimide and nitrile phthalimide structures – are primarily detected (formed). Schemes of pyrolytic reactions of polymellitimide cycle are widely discussed in the literature, devoted to degradation problems of classical aromatic polyimides [22]. Generally, these schemes are unambiguous, but some reactions - imide-isoimide regrouping, in particular, which leads to formation of a "crude" for CO_2 (one of the main products) – are unique:

This reaction explains the occurrence of both CO_2 and nitrile phthalimide homologues in PAI pyrolysis products with some molecular ions in the following mass-spectrum: $201 - 215 - 229 - 257 - 271 - 285$.

The only way for CO formation is direct break of isoimide cycle and its decomposition. Besides molecular hydrogen release, some products may be hydrogenated at high temperature in hydrogen-fertile systems (which are PAI) with active transfer of H atoms from methylene chains. This is confirmed experimentally by detection of ammonium bicarbonate (the "witness" of ammonia presence in the system) in the pyrolysis products:

The set of heterocycle pyrolysis products in PAI conforms to that of polypyromellitimide derived from 4,4'-diaminodiphenyl ester of the Kapton type - the representative of aromatic polyimide family, most well known and studied.

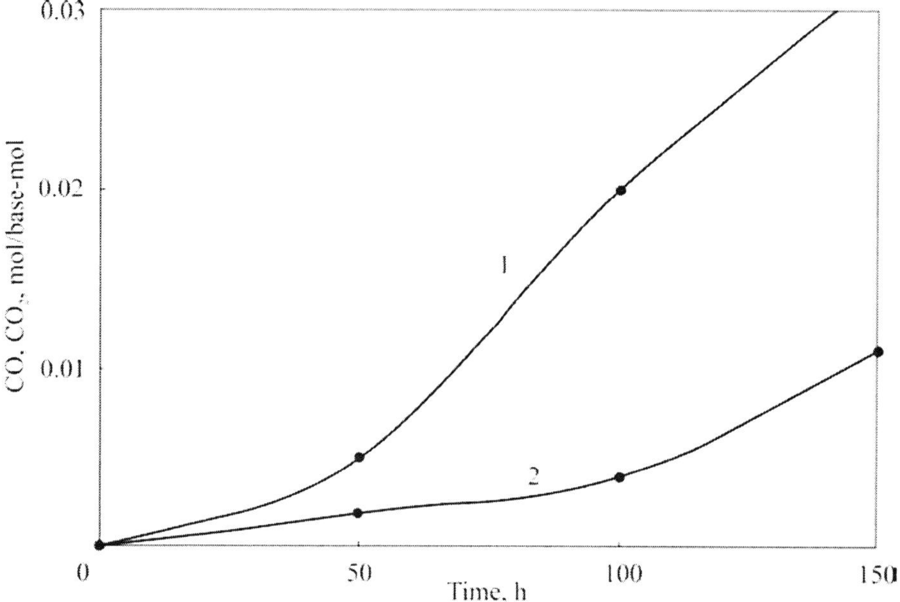

Figure 5. Kinetic curves of CO (1) and CO_2 (2) release at PAI degradation in vacuum at 350°C

Of interest is another thing: if the radical process of methylene chain degradation in PAI affects stability of heterocycles in its structure compared with classical polyimides, in which chain type of the process is rather ambiguous at low temperatures, when direct detachment of H atom from the ring is thermally hampered still. Among organic compounds, the strength of $C_{aromatic}$-H-bonds is maximal [16]. Figure 5 shows kinetic curves of carbon oxides release during PAI thermal degradation at 350°C. There are no data in the literature on transformations of aromatic polyimides under current conditions. Therefore, these products obtained at pyrolysis of fatty-aromatic and fully aromatic polyimides may not be accurately compared. It has been shown [23] that thermal processing of aromatic polyimide PM proceeding during 2 hours at 485°C in the oxygen-free environment releases 0.15 and 0.3 mol/base-mol of CO_2 and CO, respectively. Even a rough comparison of total yields of these products shows much lower stability of the heterocycle in fatty-aromatic polyimide. Apparently, chemical structure of PAI represents one more example of interrelation and agreement of structural components, absolutely different in their structure. Elementary unit of the macromolecule is presented by an entire construction. Some elements of the structure are displayed under some conditions, and the rest elements - under other conditions, but signs of nonadditivity will always be found in the behavior, indicating the interaction of elements. The concept of additivity is a suitable tool of investigation, which gives an opportunity to assess the system response. Therefore, real deviations from additivity create a broad palette of reactivity and properties observed for polymers, which macromolecules are "built up" from the same "bricks".

Thus, in the temperature range of PAI processing, its pyrolytic transformations, generally associated with degradation of methylene chains, is possible. Since high-temperature inhibition of alkyl radicals, for instance, such radical "traps" as conjugation systems or stable radicals, is low effective, the processing mode must be strictly observed and overheatings avoided.

Heating in air is of greater danger for PAI, because its degradation is sharply intensified due to oxidation.

Thermal Oxidative Degradation

The formal assessment by TGA technique in air (Figure 6) shows that among other polymers with one or another PAI fragments in the structure (namely, aliphatic polyamide derived from dodecalactam (PA-12) and polypyromellitimide derived from 4,4'-diaminodiphenyl ester (polyimide PM-1 or *Kapton*), thermal oxidative stability of PAI is medium. Figure 6 also shows TGA curves for high density polyethylene (HDPE), which structure models very long methylene chain (3 – 6 branches and 0.6 – 0.8 unsaturated bonds per 1000 C atoms in the methylene chain) [24], and polycaproamide (PA-6) – the polymer with relatively short methylene chain.

It is clear why thermal stability of PAI in air is lower compared with aromatic polyimide. The whole experimental experience in questions of oxidation of hydrocarbons, hydrocarbon and heterochain polymers [21, 25] testifies about higher reactivity of aliphatic structures in thermal oxidation reactions, which confirmed by well-developed theory of gas- and liquid-phase oxidation [25, 26]. This theory associates kinetics of oxidative reactions with C-H bond strength, chain process specificity, and the presence of intermediate products which branch the kinetic chain. *Apriori* thermal oxidative degradation of PAI is developed in the aliphatic chain. PAI possesses higher thermal oxidative stability compared with other polymers of similar structure, specifically with the closest analogue – PA-12. Judging by kinetics of H atom detachment by nitroxyl radical in heptane, pentane and linear PE [27], the strength of C-C and C-H-bonds changes weakly along the methylene chain. The energy of middle C-C-bond breaks in $C_{12}H_{26}$ and C_6H_{14} equal 334 kJ/mol. approximately [16]. The bond strength in alkanes increases by 10% approaching the chain end [16]. On the contrary, C-C- and C-H-bond strengths in nitrogen-containing aliphatic compounds are reduced by 8 - 20 kJ/mol compared with methylene groups more remote from the heteroatom [17]. Similar effect of another heteroatom (oxygen) on thermal oxidation kinetics of polyethers and strengths of appropriate bonds was shown before [17]. Calculations and experimental determination of C-H-bond strengths in individual compounds are confirmed by studies of thermal oxidative degradation of aliphatic polyamides, for which the primary attack of oxygen on α-methylene group was shown [28]. The above-shown results of quantum-chemical calculations of the models indicate delocalizing ability of the arimide fragment even in relation to valent σ-electrons of methylene C atom. Therefore, from positions of pure chemistry, one would hardly expect any advantage in thermal oxidative stability, which really exists in PAI compared with the analogues – aliphatic polyamides.

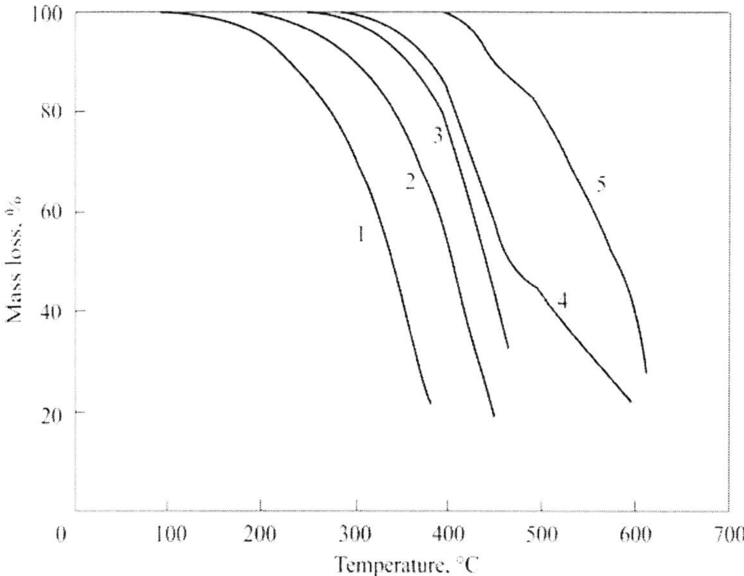

Figure 6. TGA curves for PE (1), PA-12 (2), PA-6 (3), PAI (4), and PI (5), in air at 5 deg/min heating rate.

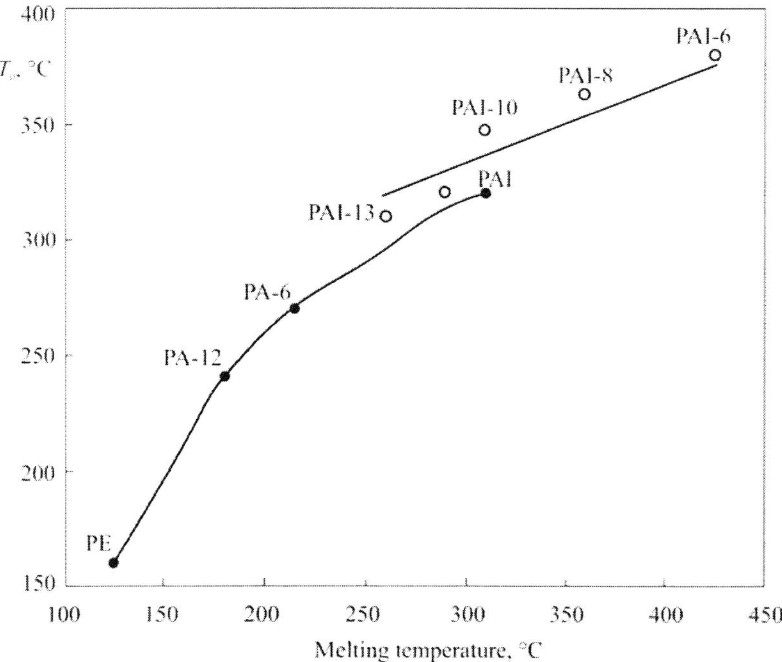

Figure 7. Dependencies of melting point and initial degradation temperature (T_0) according to TGA data on PE, PA-12, PA-6, PAI (1) and polyalkanimides (2) at the heating rate 5°/min in air

The modern concept of solid-phase degradation reactions [29] is illustrated by the dependence of almost all elementary reactions of chain oxidation, O_2 and degradation product transport due to molecular movements in the polymeric body. Even the initial mass loss temperature (T_0) – the gross parameter of thermal oxidative stability – in the sequence of polymers having similar chemical structure of the elementary unit is bound to macrochain mobility, estimated by melting points of the polymers. The polymers in the following sequence: PE, PA-12, PA-6 and PAI, possess melting points equal 125, 181, 215 and 298°C and T_0 equal 60, 240, 270 and 320°C, respectively. As is obvious, both parameters display almost linear dependence (Figure 7). Similar dependence was also observed for some polyalkanimides with methylene chain from 6 to 13 carbon atoms long: PAI-6, PAI-8, ..., PAI-13, correspondingly (Figure 8).

The thermogram of PAI in air (Figure 8) possesses high intensity exothermal DTA peak with the maximum at 290°C and two less intensive exothermal peaks with maximums at 350 and 450°C. The first peak corresponds to temperatures approaching the initial mass loss temperature, whereas other peaks correspond to advanced PAI degradation. A thermogram, recorded in the absence of O_2, possesses a split up endothermic peak, related [6] to the phase transition of PAI (279°C), at the place of the first exothermal DTA peak location.

The exothermal effect associated with O2 presence fully masks endothermic effects of both phase transitions.

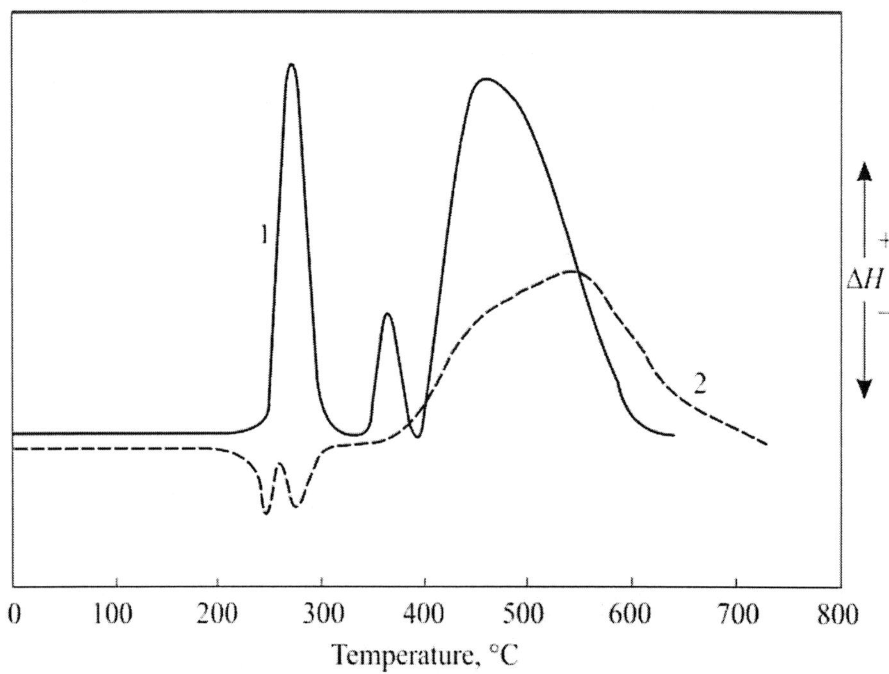

Figure 8. DTA curves for PAI in air (1) and argon flow (2). The heating rate is 5°/min

Similar situation is observed for other polyalkane pyromellitimides - PAI-6, PAI-8, PAI-9, PAI-10, PAI-11, and PAI-13 – for which the intensive exothermal peak masks endothermic peaks of phase transitions and melting. The dependence of PAI chain conformation features on the evenness (the number of C atoms) of the methylene chain is considered [6]. It is proved that contrary to PAI with even number of methylene groups (except for PAI-6), a spiral twofold axis is typical of uneven aliphatic chains of PAI and PAI-6. In the case of massive framing (heterocycle), this axis prevents chain packing e.g. crystallization. PAI with even chains are highly crystalline polymers, and uneven ones are liquid-crystal with the long-range order along the chain and short-range order in the chain location plane [6]. The phase transition of primarily coiled structure to more stretched one is absent in uneven PAI and PAI-6, whereas in even PAI it is prior to melting.

Therefore, the connection of exothermal DTA peak in the presence of O_2 in the system and its location near the initial mass loss (intensive PAI degradation) in the temperature field indicates its oxidative origin. Putting the mechanism of PAI interaction with O_2 aside, note that the process in even PAI is characterized by almost twice higher hear release rather than in uneven PAI and PAI-6 (Table 1). This shows obvious correlation between thermal oxidation kinetics and morphological features of PAI structure.

Since the rate of heat release correlates with thermal oxidation rate, at molecular mobility defrost (1st phase transition) and further melting the oxidation rate of even, highly crystalline PAI is twice higher compared with uneven, liquid-crystal PAI. NMR wide bands of uneven PAI also display a narrow component indicating the presence of non-crystalline domains. For more details on the liquid-crystal structure formation in uneven PAI, refer to X-ray diffraction analysis results [6].

Table 1. Temperature Transitions and Heat Effects in PAI

Polymer	Phase transition temperature, °C [47]	Melting point, °C (by DTA peak maximum) [47]	DTA effect,[*] kJ/base-mol		
			Phase transitions (total)	Initial oxidation (1st DTA exopeak)	Exoeffect correction with respect to phase transition heats
PAI-6	-	430	10.3	139.1	149.4
PAI-9	-	303	23.3	76.1	99.4
PAI-11	-	292	23.4	77.0	100.4
PAI-13	-	263	24.6	107.9	132.5
PAI-8	349	362	18.5	190.3	208.8
PAI-10	308	318	26.9	191.9	218.8
PAI-12	279	298	31.0	203.5	234.5

[*] According to known data from the literature, thermoanalyzer is calibrated by $NaNO_3$ and $AgNO_3$ standard substances, which melting temperatures fall within the range of PAI phase transition temperatures. Calibration check of PET melting heat gave (40.6 ± 4) kJ/mol, which correlated well with literary data [51] - 47.3 kJ/mol.

At dynamic heating, a rigid, low mobile structure of even PAI somewhat stabilizes the polymer. Methylene chains in PAI, thermodynamically labile at high temperatures (in PE, these methylene chains are intensively oxidized and break at these temperatures), are kinetically unable to overcome the activation barrier of unfavorable (possibly for propagation and kinetic chain branching reactions) morphological structures. The polymer is somewhat "overheated", which promotes similarity with the idea of "explosive" depolymerization, put forward [30] in the analysis of TGA data on depolymerizing polymers at extremely high heating rates (over 200°/s). The authors [30] suggested characterize degradation by critical temperature, at which intermolecular links preventing emission of products – backbone fractures somewhat break at ones, and the polymer "explosion-like" degrades.

In PAI hydrogen bonds are absent [6]. Therefore, thermal oxidation kinetics is limited by dormancy of molecular motions in macrochains. We are not to discuss the mechanism of morphology effect on kinetics of PAI thermal oxidation in the radical-chain process. One may refer to the famous monograph written by Buchachenko and Emanuel [29], where this question is thoroughly discussed on the example of solid polyolefins. The oxidation chemistry of these substances and their low-molecular analogues – liquid hydrocarbons, is studied quite well. The following different thing is of importance in the context of the current monograph. The intense oxidation during a short time (within the 1st exothermal peak at the thermogram – 10 min or less) leads to noticeable degradation-induced changes, increasing in the polymer symbate with oxidation proceeding (traced by the DTA curve run, Figure 9). The gel-fraction content in PAI within exothermal peak of oxidation is auto-excited accelerated up to 15%, and specific viscosity of the sol-fraction decreases from 0.88 to 0.52 dl/g.

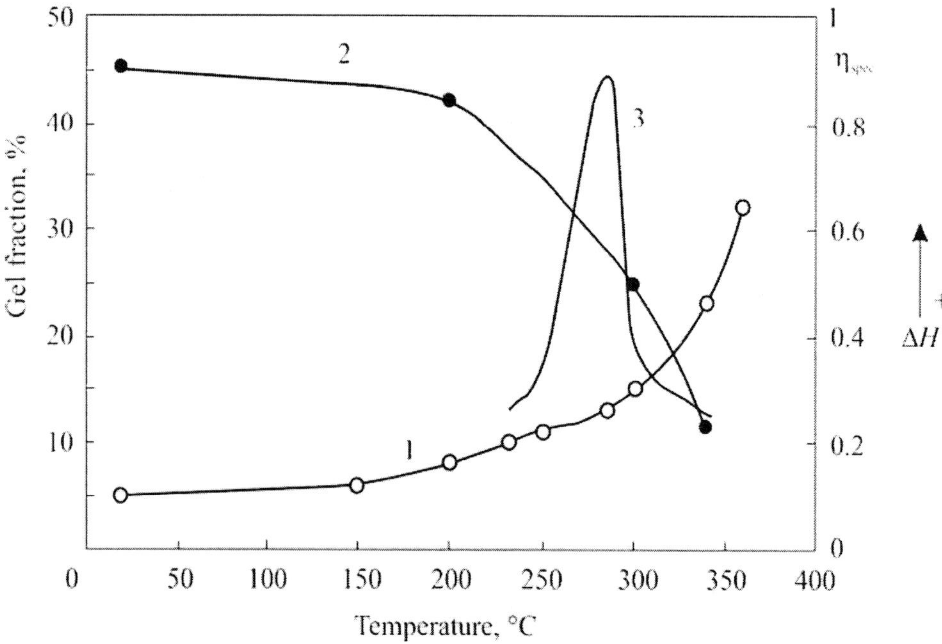

Figure 9. Gel-fraction content (1) and sol-fraction viscosity (2) in exothermal DTA peak for PAI (3) in air (the heating rate 5°/min)

The first stage of PAI-12 powder processing (extrusion, compounding) is similar to thermal analysis, and the sample is subject to dynamic heating in both cases. At initial stages of vacuum extruder the residual pressure equals 30 – 50 mm, e.g. oxygen is always present in the system, even not taking into account oxygen dissolved in the polymer. The gross heat release corresponded to the 1st exothermal DTA peak increases from 203.5 to 277.8 kJ/mol with the heating rate of 5°/min. As a consequence, in contrast with the extremely short range of dynamic heating (320 - 350°C) at vacuum extrusion of the powder, danger of PAI-12 thermal oxidation is significant at polymer melting. This is confirmed by difficulties in its processing into articles and problems with granular material quality.

Hindered PAI oxidation at low temperatures (below 250°C) does not show its full absence. For example, PAI heated in air at 200°C during 100 hours absorbs more than 1 O_2 mol/base-mol. The 1st exothermal peak is absent on the thermogram. It is replaced by an endothermic peak associated with melting of the polymer. On the thermogram of oxidized sample the first exothermal peak is absent, and an endothermic peak associated with the polymer melting occurs at its place. The first exothermal peak is a suitable index of PAI-12 oxidation degree, if O_2 diffusion into the sample is free. For example, if a bulky sample melted in air is grinded, the powder thermogram possesses oxidation exotherm at the place of endotherm of melting.

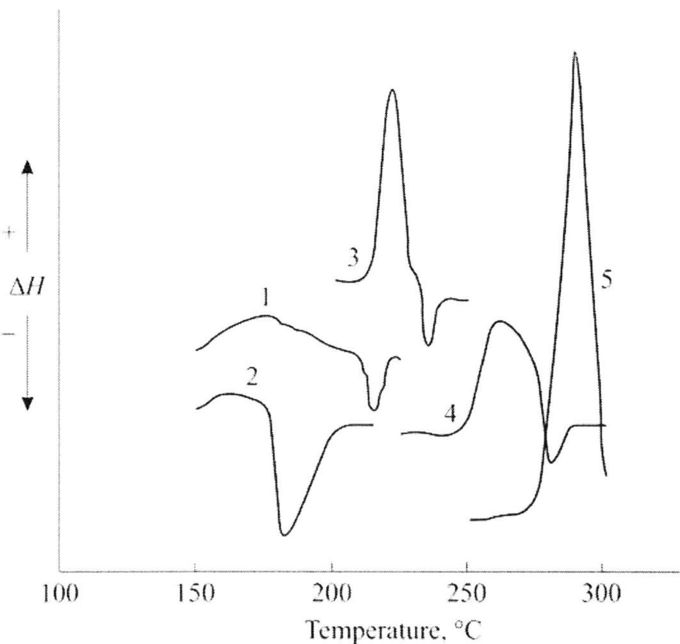

Figure 10. DTA curves for PA-6 (1), PA-12 (2), PBT (3), PET (4), and PAI-12 (5) heated in air (heating rate 5°/min)

The phase transition effect on exothermal DTA peak, similar to PAI, is observed for PET and specifically for poly(butylene terephthalate) (PBT). These are crystalline polymers with chemical structure affined to PAI, namely, by combined methylene chains and aromatic rings (Figure 10). Judging by significant intensity of exothermal peaks preceding endothermic melting peaks on DTA curves, oxidation processes, primarily hindered due to the structure rigidity, are sharply intensified at pre-melting of these polyolefins. On the contrary, oxidation of aliphatic PA is described by a fuzzy exothermal peak. As shown in the thermogram, it begins at temperatures much below PA melting point (Figure 10). Exothermal PA peaks are symbate to weight increment due to atmospheric oxygen binding. No weight increments at PAI, PET and PBT heating are observed. DTA curves illustrate the difference in thermal oxidation kinetics for crystalline polymers having methylene chains in their structure. Purely aliphatic polymers with highly developed molecular motion in the solid phase are simply oxidized at dynamic heating at temperatures much below their melting points. Solid oxidation of fatty aromatic polymers is hindered, but becomes intensive at phase transition temperatures. DTA data is sufficient for thermal stabilization of engineering polymers. For example, for aliphatic PA, these data currently indicate the necessity in protection of polymers both at processing and operation, whereas PAI and aromatic polyester should be stabilized at processing only. However, the final decision about thermal stabilization at particular stages of polymer "life" may be made on basis of comprehensive study of polymer features concerning its thermal behavior in temperature intervals of material processing and operation.

When heated, PAI absorb O_2 and release carbon oxides, water and formaldehyde (Figure 11). Thermal oxidation kinetics (with eliminated diffusion hindrances) is at first glance of the breakthrough type typical of high temperature oxidation of hydrocarbon polymers [31]. At 250 - 300°C during initial 5 - 20 hours, PAI absorb up to 8 O_2 mol/base-mol. Further on, the process is abruptly slowed down, but the polymer manages to absorb more 2 O_2 mol/base-mol during next 100 - 200 hours. The two-stage type of thermal oxidation process is clearly observed at 350°C, when the first fast stage is finished by absorption of 8 O_2 mol/base-mol during last 2 hours, and then the second, slow stage with almost linear dependence of O_2 absorption rate lasts for hours. At low temperature oxidation of PAI (250°C or lower) the second stage is not observed. As transferred to semi-logarithmic coordinates, corresponded kinetic curves of O_2 absorption and those describing the first stage of high-temperature oxidation are transferred to lines that formally indicate the first kinetic order. The Arrhenius temperature dependence of O_2 absorption initial rate constants does not fit a straight line, but displays a break in the temperature range of phase transitions in PAI (Figure 12).

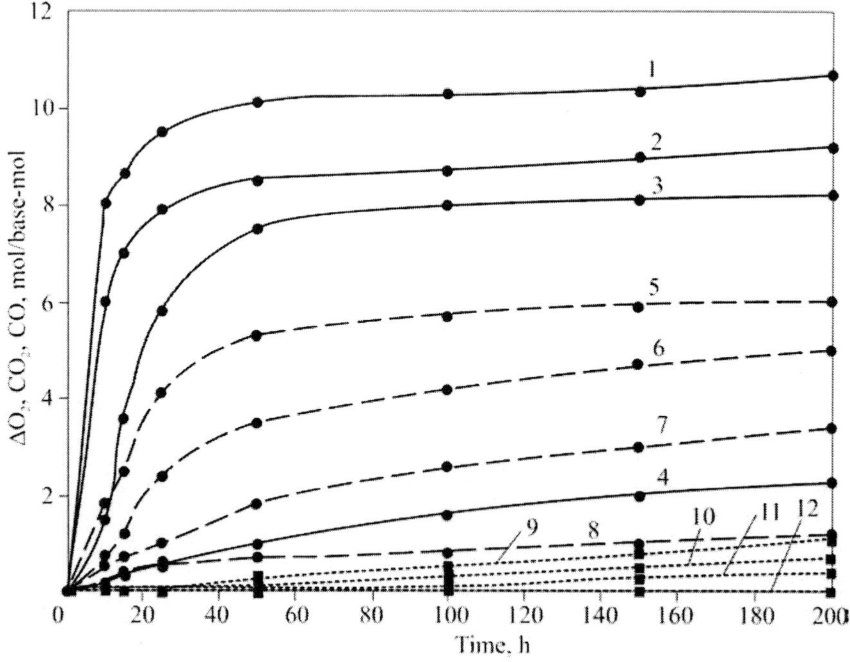

Figure 11. Kinetics of O_2 absorption (1 – 4), CO_2 (5 – 8) and CO (9 – 12) release at thermal oxidation of PAI in air at 350, 300, 250 and 200°C, respectively

Kinetics of carbon oxide release is identical to O_2 absorption kinetics. Effective activation energies of CO and O_2 release at PAI thermal oxidation in melt equal (163 ± 10) and (135 ± 10) kJ/mol.

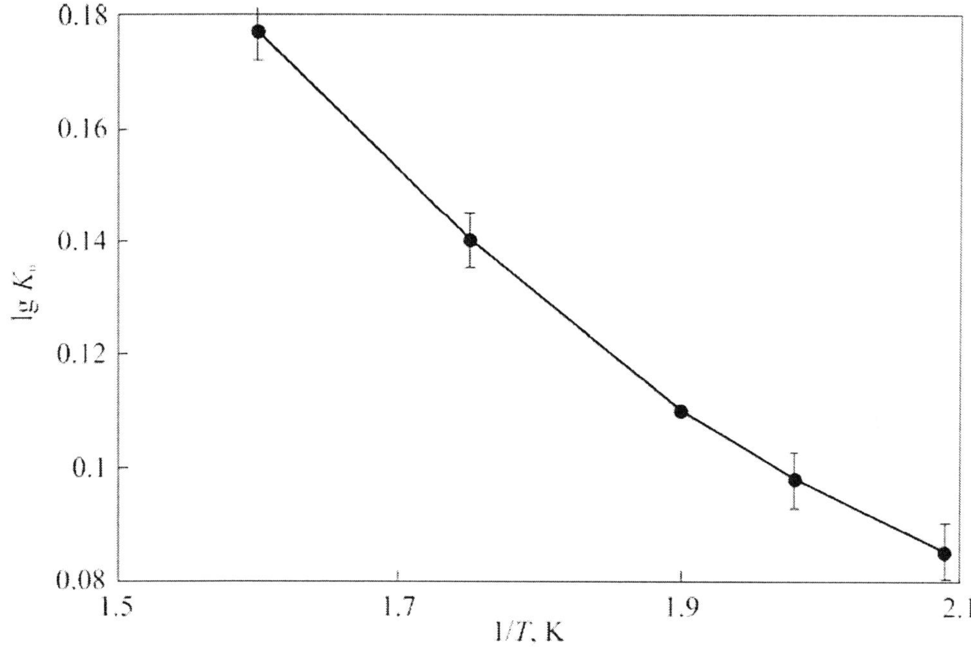

Figure 12. The Arrhenius dependence of initial rate constants K_0 of O_2 absorption on temperature of PAI thermal oxidation in air

The two stage kinetics of high temperature oxidation of aromatic polybenzoxazole containing the methylene bridge in the diamine component is described in the literature [32]. The first, fast stage is associated with aliphatic fragment oxidation, and the second, slow one – with oxidation of the aromatic structure. The case is similar for PAI. The below IR-spectroscopy and mass-spectrometry data indicate that methylene chains are primarily oxidized both at low temperature of PAI oxidation (the operation temperature range and close temperatures) and high temperature oxidation (temperature of processing)with exposures correlating with the processing cycle duration. Therefore, aromatic structure is not practically touched. At processing temperature the rate of aromatic structure oxidative degradation is, at least, two orders of magnitude lower than the methylene chain oxidation rate, i.e. choosing the way of PAI thermal stabilization one may neglect oxidation of the aromatic structure.

The quantity of oxygen released from PAI at thermal oxidation (at the 1st stage), totally with carbon oxides, water and formaldehyde equals a half (molar) of O_2 absorbed by the polymer from the gas phase (Table 2).

Though the release of carbon oxides possesses quite high temperature coefficient: $E_a = (134 - 163) \pm 10$ kJ/mol, temperature increase to 350°C leads to a noticeable reduction of the part of these products in the oxygen balance. Moreover, the release ratio of CO_2 and CO changes with temperature: it is 10:1 (molar) at temperature below 300°C, and 6:1 at 350°C. The increase of CO release is associated with heterocycle decay. As shown above, heterocycles participate in PAI pyrolysis at 350°C. The rate of these decays is low compared with methylene chain breaks. Moreover, as observed from mass losses, the gross rate of PAI pyrolysis at 350°C is by an order of magnitude lower than thermal oxidation rate. Therefore, a noticeable increase of CO output is explained by oxidative pyrolysis of heterocycles. Previously, using an isotope label in O_2, it is shown that at temperatures of full thermal stability of imide cycles oxygen activates purely thermal reactions of heterocycles in aromatic polyimide.

Table 2. Oxygen balance at PAI Thermal Oxidation

Exposure conditions[*]	O_2 absorption, mol/base-mol	Total oxygen amount released with CO_2, CO, H_2O, CH_2O, mol/base-mol
350°C, 1 h	2.4	1.1
300°C, 2 h	3.3	1.5
250°C, 5 h	1.4	0.7
200°C, 50 h	0.9	0.5

[*] Different exposure times are caused by the process kinetics at the mentioned temperatures. The data relate to the 1st oxidation stage.

As PAI oxidizes, relatively low amount of H_2O is released (2 – 5% of CO_2 release), whereas H_2O is the main product released at hydrocarbon polymer thermal oxidation, but in another temperature range (up to 200 - 220°C) [22, 33]. The variety of oxygen-containing organic oxidation products (aldehydes, ketones, alcohols, ethers, acids) is typical of thermal oxidation of polyolefins generally and methylene chain in PE specifically. Total output of these substances exceeds the release of carbon oxides [34]. The variety and output of these products at thermal oxidation of aliphatic PA is lower [28]. At temperatures up to 300°C, thermal oxidation of methylene chain in PAI leads to formaldehyde release and trace amounts of other aldehydes. At high temperatures, oxidative degradation of methylene chain in PAI in melt is clearly displayed by sharp increase of formaldehyde output which, besides CO_2, becomes the basic volatile degradation product. According to ^{1}H NMR-spectra, PAI oxidation in melt gives various degradation products. The main contribution to the spectrum is belonged to aliphatic chain protons: signals with chemical shifts of 0.85 – 86 ppm correspond to end CH_3-groups; γ- and more remote CH_2-groups – 1.23-1.28 ppm; α- and β-CH_2-groups – 3.6 and 1.6-1.7 ppm, respectively; N-CH_3-groups – 3.2 ppm. Signal intensity integrals show the length of methylene chains in thermal oxidation products equal C_5-C_6 and, basing on the contributions from protons in N-CH_2-groups, about 20% of methylene chains are linked to imide cycles or other nitrogen-containing groups, amino groups, for example. The rest methylene chains (R) are distributed between the following structures: R-$CH_2OC(O)$ – 4.0 – 4.1 ppm, and R-$CH_2C(O)$ – 2.2 – 2.3 ppm. Signals with chemical shifts of 7.69 – 7.82, 7.93 and 8.24 ppm relate to phthalimide, nitrile phthalimide and pyromellitimide structures, respectively. In C NMR-spectrum, chemical shifts ranged within 132.3 – 133.8, 123.9 – 126.6 and 117.2 ppm correspond to the mentioned structures. Here, methylene chains are observed at 38.7 – 14.08 ppm at corresponded distance from nitrogen atom: for example, chemical shift of 38.7 - 38.1 ppm corresponds to α-CH_2 group, 27.0 – 26.8 ppm to β-CH_2 group, etc.

A lower amount of absorbed oxygen is released with volatile products. The rest oxygen is accumulated in the polymer (refer to Figure 13 showing temperature dynamics of changes in

PAI elemental composition, induced by thermal oxidation at different temperatures). At 200°C, a noticeable oxygen accumulation in PAI takes about 1000 hours, whereas at 350°C change of the initial elemental composition C - 69.0%, H - 7.5%, N - 6.7%, O - 16.8% to 66.4, 6.0, 5.9 and 21.7%, respectively, takes 1 hour only.

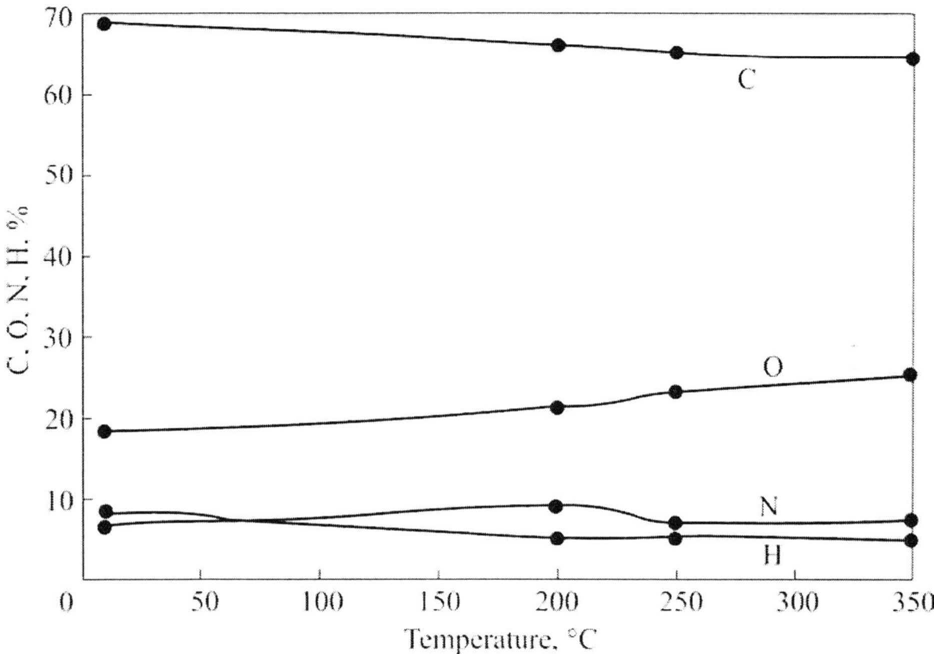

Figure 13. The change of PAI elemental composition at thermal oxidation in air: at 200, 250, 300 and 350°C and 500, 200, 10 and 1 hour exposure, respectively

Degradation changes in PAI are reflected by IR-spectra. The Fourier distribution IR-spectrum of initial PAI has the following absorption bands: a doublet at $v_{(C-H)arom}$ = 3038 and 3101 cm^{-1}; a doublet at $v_{(C-H)aliph}$ = 2848 and 2921 cm^{-1}; a doublet at v_s = 1720 cm^{-1} and $v_{as(C=O)}$ = 1780 cm^{-1} of imide cycle, respectively [13]; and deformation oscillations of imide cycle at 728 cm^{-1} [14]. Intensive bands at 1122, 1157, 1367, 1397 and 1563 cm^{-1} are related to valence and deformation oscillations of bonds which form skeleton (backbone) of the macromolecule.

Compared with IR-spectrum of the initial polymer, spectra of PAI oxidized at high temperatures, but at low exposure at first glance display no significant differences, except for consecutive increase of the absorption band at 3470 cm^{-1}, which relates to associated $v_{(OH)}$ [14], as well as broadening of the long-wave wing at 1720 cm^{-1} band base and occurrence of a bending at 1620 cm^{-1}. Computer treatment of degraded PAI spectrum and its comparison with the library of Fourier distribution IR-spectra, measured on Perkin-Elmer 1710 FTIR spectrophotometer has given a formal evidence of possible presence of the following structures: aromatic imide, for example, phthalimide, maleimide or glutimide, ureal and carbamate, and various linear or cyclic carbonyl groups. Obviously the first evidence of aromatic imides, currently pyromellitimide group, is unambiguous. Therefore, since computer searches by the most intensive bands in decodable spectrum, the rest structures suggested should be taken into consideration, and new evidences of their occurrence must be searched for.

Quantitatively, changes in optical densities of basic spectrum bands were estimated using the band of 728 cm^{-1} (Table 3).

Table 3. Relative Optical Densities $\dfrac{D_x}{D_{728}}$ of Absorption Bands in IR-spectra

of Oxidized PAI (at 1 h Exposure)

Temperature, °C	$\dfrac{D_x}{D_{728}}$ at x equal:			
	2921 cm^{-1}	2848 cm^{-1}	1720 cm^{-1}	1780 cm^{-1}
Before oxidation	1.2	0.9	2.5	1.1
275	1.1	0.8	2.6	1.1
300	1.0	0.8	3.8	1.1
350	1.3	0.9	4.7	1.3

The band intensity at 1720 cm^{-1} regularly increases with thermal oxidation which is illustrated by differential IR-spectra. No regularity is observed in changes of other bands. These IR-spectroscopy data contradict to the hypothesis of primary oxidation of methylene chains in PAI. Actually, thermal oxidation of PAI at 300 and 350°C during 1 hour results in absorption of 1.6 and 2.4 mol O_2/base-mol and release of 0.1 and 0.5 mol CO_2/base-mol (dominant output product), respectively. Thus resulting oxidation, the oxygen content in PAI becomes nearly twice higher, whereas methylene chain damages calculated by CO_2 as the degradation index in the carbon balance give 2%, approximately. One may make a different suggestion: in oxidized structures one oxygen atom substitutes two (ketone) or, most frequently, one (aldehydes, alcohol, ether) hydrogen atom. As a consequence, the mentioned level of O_2 absorption induces the loss of hydrogen from methylene groups equal 10 – 20%. The lower level falls within the error of IR-spectrum quantitative assessments, where 20% difference would be noticeable. IR-spectra do not allow methylene chain oxidation assessment, but clearly indicate the increase of carbonyl group content.

The Fourier IR-spectra in the range of 1650 – 1750 cm^{-1} observed at high resolutions (2 cm^{-1}) indicated identity of the band shape at 1720 cm^{-1} with the maximum at 1713 cm^{-1}, detected at sensitivity increase for the initial and oxidized PAI (Figure 14). Propagation of a new carbonyl band somehow from inside (to say "from the same root") of the initial one related to $\nu_{(C=O)}$ of the imide cycle. This may be dually explained. Primarily, thermal oxidation and degradation of methylene chains intensify symmetrical oscillations of the cycle by means of asymmetrical oscillations. This explanation is obeyed by the above-mentioned increase of negligibly small damage of methylene chains, detected at the moment of spectrum record. Moreover, if judged by the area of bands within 1,650 – 1,800 cm^{-1}, at 1,720 and 1,790 cm^{-1} both the relation of optical densities changes and carbonyl absorption significantly increases. Secondly, it may be explained by occurrence of new C=O-groups as the oxidation product.

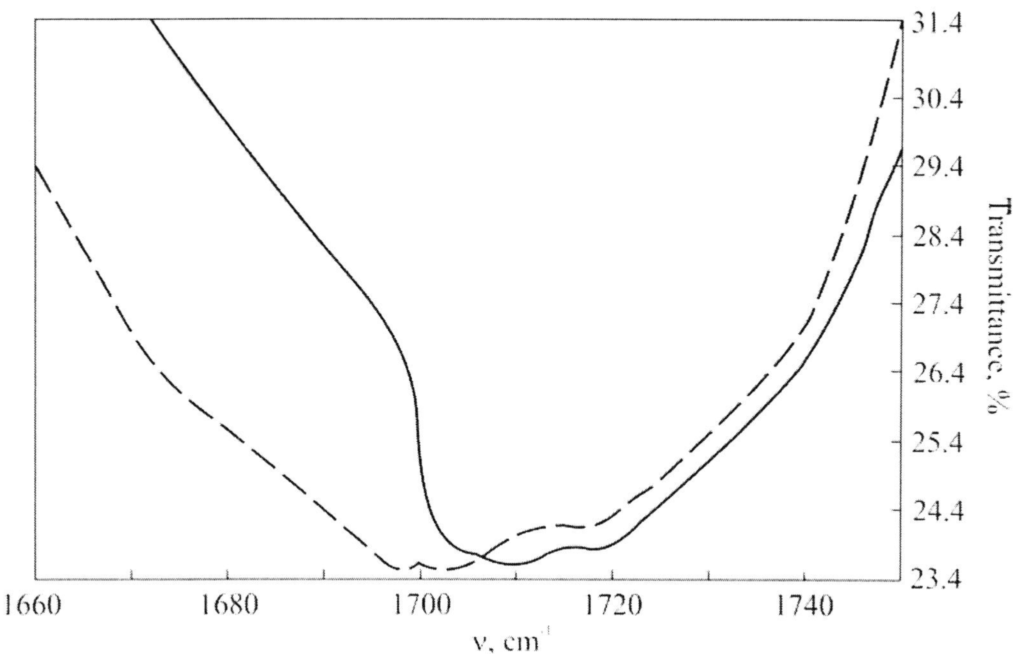

Figure 14. High resolution (2 cm^{-1}) Fourier IR-spectrum for PAI before (1) and after (2) thermal oxidation in air at 350°C during 1 hour

The literature shows investigations of aliphatic PA thermal oxidation by IR-spectroscopy [35]. The occurrence of new C=O-groups, resulted by low-temperature oxidation of PA-6, PA-66, PA-11, PA-12, and the so-called "soluble" N-methoxymethylated PA-6 is displayed in absorption increase in the range of 1700 – 1750 cm^{-1} at the short-wave wing of the initial amide band.

The authors have not performed the band separation, but, anyway, consider that the bending contains two maximums at 1715 - 1719 and 1731 - 1736 cm^{-1}, which they relate to a new imide structure, synthesized by α-CH$_2$-group oxidation.

For N-methoxymethylated PA-6, the most intensive, new carbonyl band in IR-spectrum is observed at 1719 cm^{-1}. As shown in the literature, the feature of oxidized PAI IR-spectra may be explained by occurrence of new α-C=O-group in the methylene chain. The alternative attack of molecular oxygen and preferable oxidation of internal methylene groups would form a situation similar to thermal oxidation of PE backbone, when a broad, multi-structural band occurs in the range of 1700 – 1800 cm^{-1} in IR-spectra of oxidized PE. Analysis of this multi-structural band has detected the contribution of aldehydes< ketone, ether, and carboxylic groups into $\nu_{(C=O)}$ [36].

Besides gaseous and polymeric products, PAI thermal oxidation produces low-volatile substances condensed near the hot zone. Natural fractioning of these products by the reaction chamber and programmed heating in the mass-spectrometric analysis with direct injection to the source of ions make mass-spectra, measured at low ionizing stress (12 – 20 eV) with mostly molecular ions fixed, simpler. These mass-spectra simply trace series of molecular ions with mass step of 14 mass units: 216-230-244-258-272-286-300-314-328-342-356-370-384-398-412 etc. to 552; 231-345-259-273-287-301-315; 243-257-271-285-299-313-327-341-355; 371-385-399-413-427-441-455-469-483-511-525-539; 543-557-571-585-599-613-

641-655-669-683-698. Moreover, mass-spectra possess separate molecular ions either interfering into homological sequences, but being more intensive rather than the neighboring ones and all sequence members (for example, peaks with 160, 230 and 543 m.u.), or singles outside the sequence (for example, 73, 82, 214, 109 m.u.), or inside the sequence, not occupying a sequence member place (for example, 460 m.u.).

These peaks belong to substances with somewhat "economic" structure, e.g. the structure with no breaks of the methylene backbone with mostly arylene and heterocyclic end groups, as well as methylene or ethynyl products of heterocycle degradation. In mass-spectra of thermal oxidation products, homological sequences of molecular ions mainly correspond to analogous sequences in mass-spectra of PAI pyrolysis products. Initial two sequences with uneven values for molecular ions correspond to mass-spectra of pyrolysis products by both distribution and multiplicity (ions-satellites of sequence members with mass numbers $\pm(1 - 3)$). E.g. they respond to methylene backbone fractures having phthalimide end group, alkane phthalimides with end methyl or ethyl group. The rest two uneven sequences include higher order oligomers than these formed in pyrolysis. For example, molecular ion with 543 m.u. starts the following homological sequence:

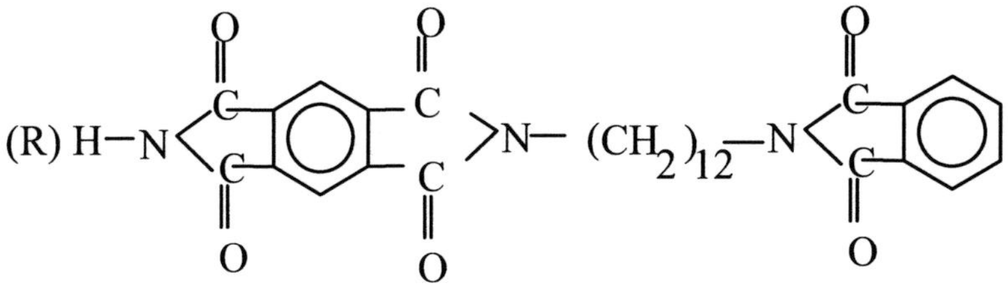

$$(R) \; H-N< \text{[pyromellitimide]} >N-(CH_2)_{12}-N< \text{[phthalimide]}$$

It is traced up to the last member of $R = C_{12}H_{25}$ sequence with the molecular weight 711. Therefore, thermal oxidation gives high yield of products representing the elementary unit fragment or fractures without methylene backbone fragments, or preserving it in full by means of end blockade similar to the above-mentioned product with molecular weight 543. Pyromellitimide and dodecamethylen diphthalimide are among these products.

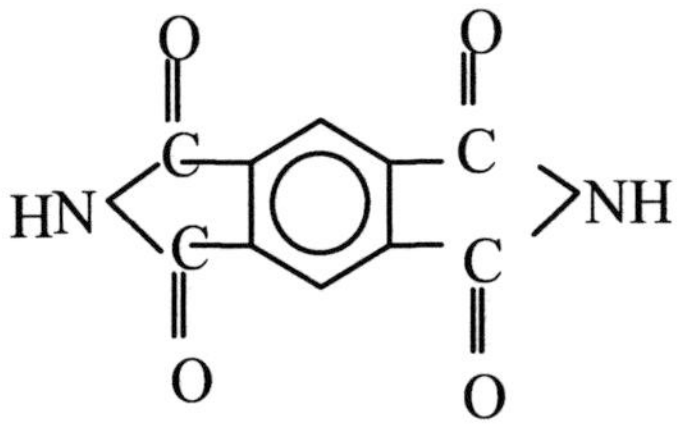

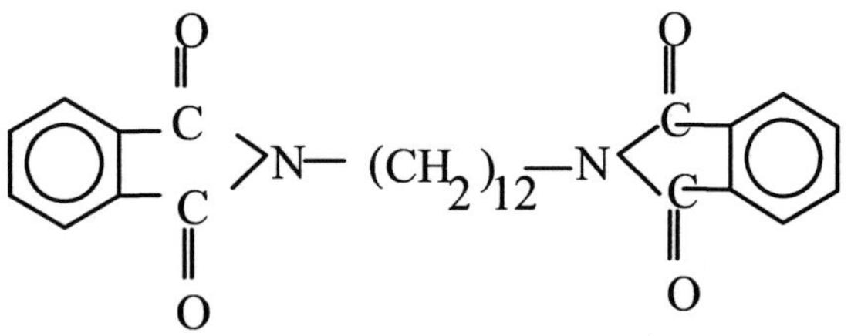

$$HN< \text{[pyromellitimide]} >NH \qquad \text{[phthalimide]} >N-(CH_2)_{12}-N< \text{[phthalimide]}$$

This molecular ion sharply interferes between alkane pyromellitimides. At thermal oxidation, homological sequence with even molecular ions is much longer than at pyrolysis. Tracing the intensity distribution and multiplicity, the second part of the sequence with molecular weights of 370 or higher is found analogous to some molecular ions of alkane pyromellitimide sequence formed at pyrolysis. The beginning of the even sequence of thermal oxidation products from 244 and 216 m.u. is not shown in mass-spectra of pyrolysis products. The intensity distribution in this half of the even sequence possesses the maximum at 230 m.u. Molecular ions in the first and the second half of the even sequence possess different multiplicities. Obviously, the even sequence is not a single homological sequence, but is formed by two independent sequences, one of which is represented by alkane pyromellitimides. Another sequence might be formed by oxidation products, containing C=O groups in the methylene backbone, for example:

This oxygen-containing structure (possibly, an alternative of OH-containing structure) allows composing a sequence of homologues corresponded to experimentally obtained one. Enumeration of other alternatives does not give positive results, for example, nitrilephthalimide alkanes, which also form an even sequence. ^{1}H and ^{13}C NMR-spectroscopy data indicate the presence of carbonyl-containing aliphatic structures in thermal oxidation products.

Single, intensive molecular ions with 57, 73, 97 and 109 m.u. belong to nitrogen-containing products of pyromellitimide structure degradation. Comparison with the mass-spectra library shows that amides and pyrrolidone derivatives are most probable.

Judging by ^{13}C NMR-spectroscopy data, the part of polymelliiimide oligomers decreases with temperature in thermal oxidation products; vice versa, the contribution of phthalimide and substituted phthalimide oligomers, for example, nitrilephthalimide oligomers increases (Figure 15). These data allowed assessment of the mean molecular mass of thermal oxidation oligomeric products: 1,200 – 2,000 at 300 - 350°C and 350 – 700 at 400 - 450°C. Hence, the average length of methylene backbone twice decreases.

Finally, as PAI are subject to high-temperature thermal oxidation, the main degradation changes proceed in both methylene backbone and aromatic fragment, which is the imide cycle.

The unique property of PAI which differs it from the known polymeric materials, is high resistance to short-term (up to 30 s) thermal impact (260-280°C) without meltbacks and deformations that allows flow soldering of metal reinforcement [11].

Thermal aging (during 90 days) at 100 - 200°C [37] shows decrease of blow viscosity, stress at break and relative elongation of PAI at simultaneous change of polymer morphology (X-ray analysis and electron microscopy data). Thermal aging (1,000 hours at 170°C) and cycles of heating from -60 to +170°C do not change the dielectric loss tangent and permeability, but an abrupt (almost by an order of magnitude) decrease of surface and volumetric electrical resistance is observed [11].

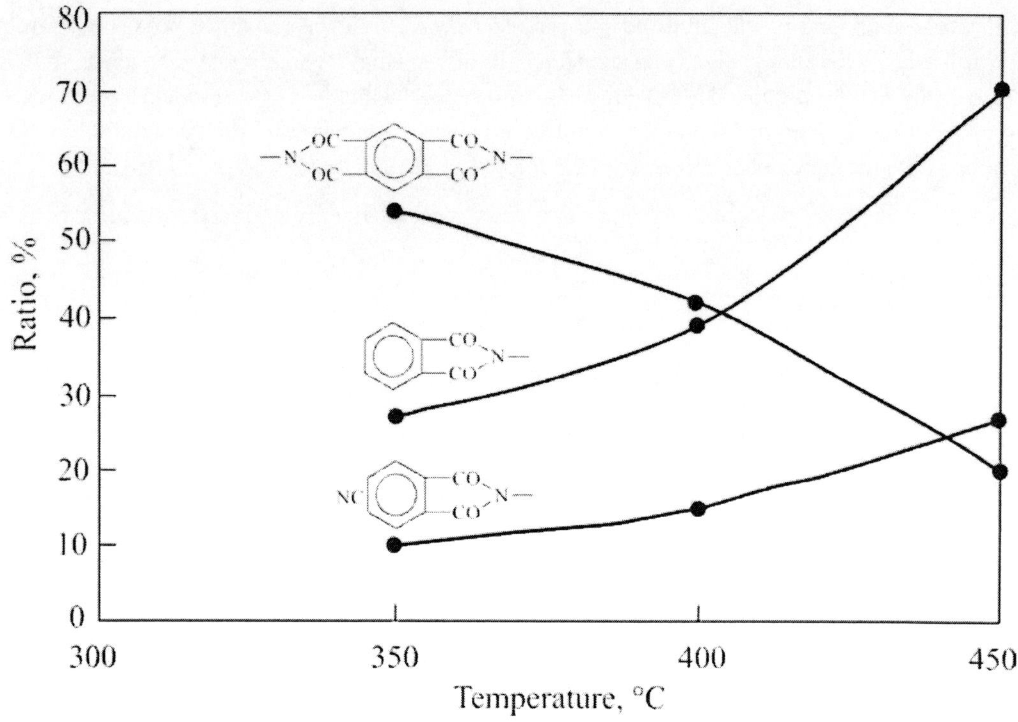

Figure 15. Changes in the ratio of structural fragments in oligomers at thermal oxidation of PAI (according to ^{13}C NMR-spectroscopy data)

Obviously, the mentioned macroscopic changes in PAI processing and operation temperatures are induced by definite changes in chemical structure of the polymer due to heat and oxygen effect.

Solid-phase Oxidation

As obvious from IR-spectra of PAI subject to thermal oxidation at 200 - 250°C during 1,000 hours, the polymer, though slowly, is oxidized in the solid phase in the temperature range of operation: concentration of C=O-groups increases and methylene groups are clearly exhausted ($v_{(C-H)}$ bands at 2,800 – 2,900 cm^{-1}). Thermal oxidation during 1,000 hours at 250°C, i.e. under rigid operation conditions, leads to full burn off of methylene backbone and partial destruction of imide cycles (the occurrence and extension of $v_{(C-H)}$ band at 1,620 cm^{-1}). Similar observations were made [38, 39] at thermal oxidation of aromatic polypyromellitimide PM-1.

It is found that slow oxidation of aromatic nuclei happens already at 250°C. After several thousand hours, this process causes destruction of cycloarylene structure and the loss of macroscopic properties. Kinetics of O_2 absorption at 250°C (Figure 16) shows that thermal oxidation rate of fatty aromatic polyimide PAI is an order of magnitude higher than the oxidation rate of aromatic polyimide. Though in this case the main contribution is made by fatty chains, high reactivity is displayed by cycloaromatic structure in PAI should also be taken into account.

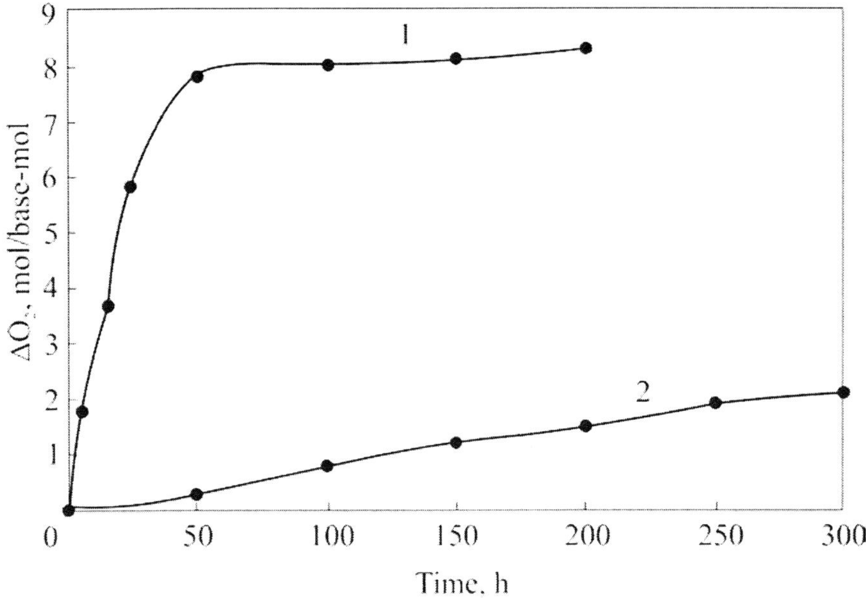

Figure 16. Kinetics of O_2 oxidation at thermal oxidation of PAI (1) and PM-1 (2) in air at 250°C

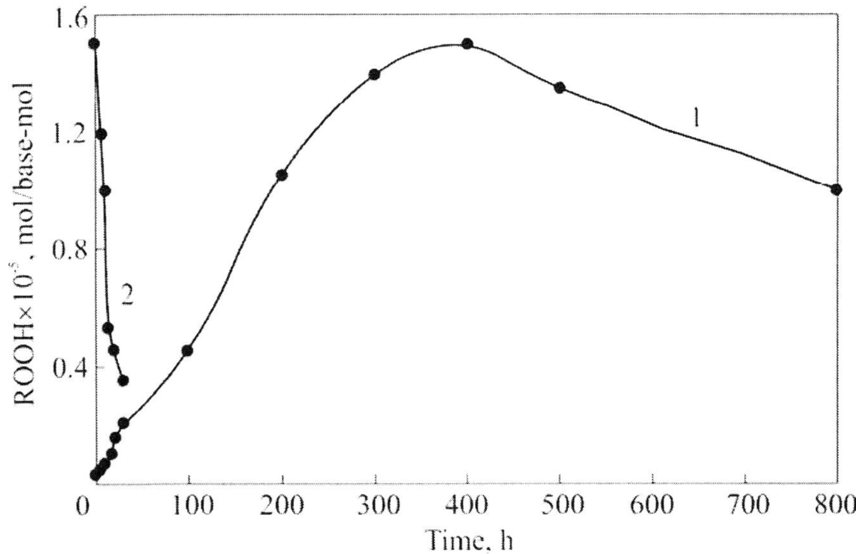

Figure 17. Kinetics of ROOH hydroperoxide accumulation (1) and decay (2) at PAI thermal oxidation at 200°C in air

As known in kinetics of liquid-phase reactions, oxidation of purely aromatic substances is induced by hydrocarbon additives as the source of peroxy radicals [25].

Predominant oxidation and degradation of fatty chains during thermal-induced aging noticeably exhausts the C and H element concentration in PAI (refer to Figure 13). As shown in experiments, 500 hour thermal oxidation at 200°C decreases C and H contents from 69.0 and 7.5 to 65.0 and 6.0%, respectively, with corresponded increase of oxygen concentration. Hydrogen is reliably detected during 800 hours of thermal oxidation. Hydrogen release kinetics correlates with the gel formation kinetics (Figure 18).

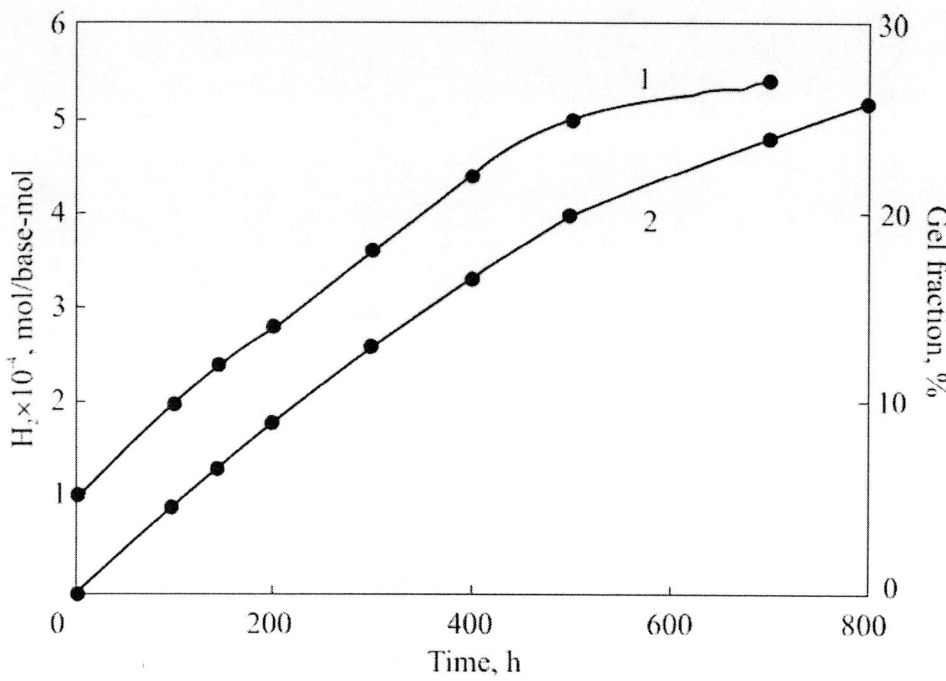

Figure 18. Kinetics of gel-fraction accumulation (1) and H_2 release (2) at PAI thermal oxidation at 200°C

Thermal aging causes formation and accumulation of hydroperoxides (the iodimetric technique) in PAI (Figure 15). The kinetics of hydroperoxide ROOH accumulation is extreme. Accumulated ROOH concentrations ($\sim 1.5 \times 10^2$ mmol/base-mol) are negligibly small (by five orders of magnitude lower) than absorbed amounts of O_2. Nevertheless, these negligible amounts are reliably detected during initial 800 hours of thermal oxidation. Meanwhile, ROOH is not practically accumulated in methylene chains of PE, which oxidizes at extremely high rate of 2×10 mol/kg·s [10] at 200°C.

In this case, high-temperature branching is assumed to happen via degradation of peroxy radicals. As known from the literature [42], ROOH is accumulated in PA-66 at 180°C. Similar to PAI, current kinetics of ROOH accumulation is extremely shaped, where maximal concentration $[ROOH_{max}] \approx 1.5 \times 10^{-2}$ mmol/base-mol is already reached 45 min after initiation of heating. In aliphatic PA, oxidized at 200°C, no hydroperoxides were detected. The absorption rate of O_2 at 200°C decreases in the sequence as follows: PE >> PA-66 > PAI (10^4, 0.1 and 0.01 mol/base-mol·h, respectively), whereas ROOH stability increases with the opposite order. Both sequences (directly and oppositely, respectively) correlate with the molecular mobility of polymers. At 200 °C, PE represents a melt, and PA-66 and rigid-chain PAI are solid. This comparison is valuable from positions of detecting the effect of the phase state and macromolecule mobility on kinetics of the process proceeding in similar chemical structures in methylene backbones. Moreover, it is also important from positions of hydroperoxide exclusion from the process of high-temperature oxidation of polymers. Though thermal oxidation of PAI forms hydroperoxides, this classical direction of the oxidation process is apparently secondary due to the following reasons. Assume that the process is developed by the classical scheme with respect to additional, elementary reactions, suggested for hydrocarbon polymer thermal oxidation [34, 40]. Therefore, following from the balance equation for hydroperoxide in autooxidation reaction:

$$\frac{d[ROOH]}{d\tau} = \alpha \cdot W(O_2) - K_p[ROOH]^n,$$

where α is hydroperoxide output per absorbed O_2; K_p is ROOH decay constant; n is the reaction order of ROOH degradation. At the extreme point, where $\dfrac{d[ROOH]}{d\tau} = 0$:

$$\alpha = \frac{K_p[ROOH]_{max}}{W(O_2)_{max}}.$$

At 200°C, ROOH degradation (Figure 17) obeys the law of kinetic order one ($n = 1$) with $K_p = 1.7 \times 10^{-2}$ min^{-1}. E.g. at 200°C PAI autooxidation constant equals $\sim 10^{-2}$, which is much below values characterizing autooxidation of polyolefins in the solid phase [40]. Autooxidation of PAI at temperatures low for this polymer mostly develops by another way, but not via hydroperoxides, possibly by previously suggested mechanism of high-temperature oxidation [40]. Definite similarity of cases follows from analysis of α for initiated, low-temperature oxidation and autooxidation of polyolefins [40]: α regularly decreases from 0.7 – 1.0 to 0.04 – 0.6 with oxidation temperature increase from 45 – 90 to 130 - 135°C.

The solid-phase oxidation of PAI at operation temperatures or intensified thermal aging (200 - 250°C) proceeds slowly, without releasing any heavy products. However, long exposures (about 1 year) induce formation of light-yellow pyromellite diimide crystals on the surface of molded PAI samples (Figure 19). Other heavy products of degradation, released at thermal oxidation of PAI in melt, are not identified here, though in this case random macrochain breaks occur, too, which is confirmed by permanent decrease of sol-fraction viscosity (refer to Figure 9). However, molecular mass of oligomers is so high that they do not practically diffuse into the solid phase and release from the surface, but remain in the polymer.

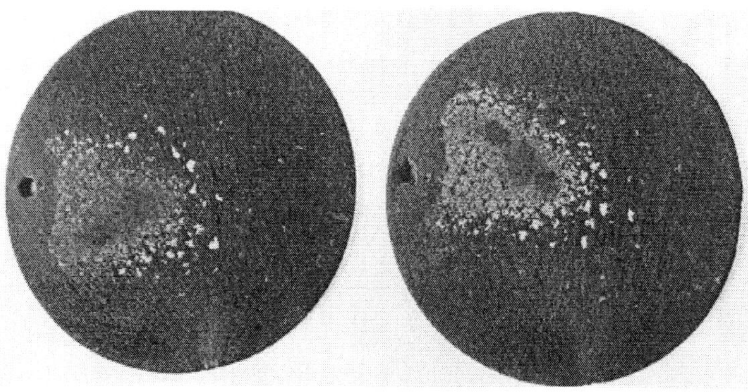

Figure 19. Aged polyalkanimide samples. Light crystals on the sample surface represent pyromellite diimide

At statistical development of thermal oxidation along the macromolecule, formation of pyromellite diimide (attack 1 and 2) and other macrochain fractures, for example, methylpyromellitimide (attack 1 and 3), are of the same probability. Volatility of the latter is similar to that of diimide, i.e. formation of this product would be clearly observed.

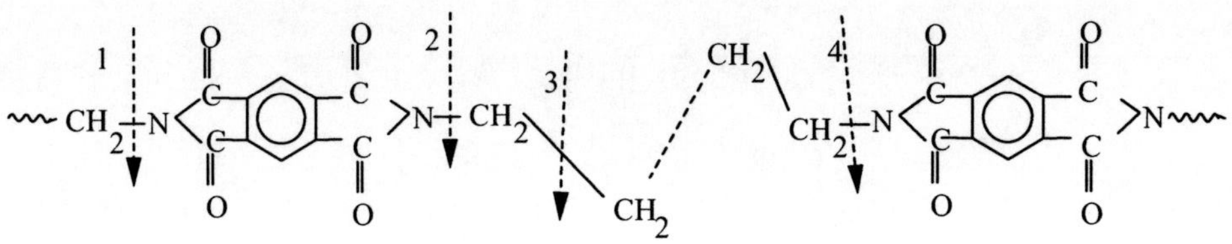

Similar explanation may be given to the alternative of attack 1 – 4 with release of undecane pyromellitimide. However, pyromellite diimide is the unique heavy product of PAI solid-phase oxidation. To put it differently, polymellit diimide formation testifies that, besides the mechanism of random O_2 attacks along the macrochain, the mechanism of α-CH_2-group oxidation activation in neighboring methylene chains, separated by pyromellitimide fragments, does exist or even dominates at relatively low temperatures.

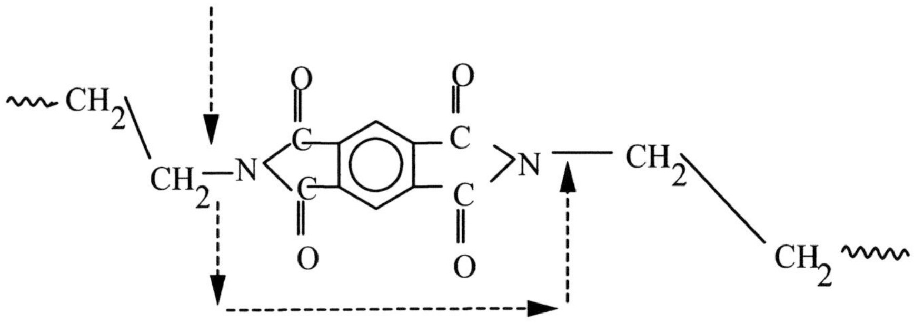

Quantum-chemical calculation of dimethylpyromellitimide model (refer to Figure 4) shows that σ-bonds of methylene groups are affected by π-systems of pyromellitimide fragment. Occurrence of new π-electron as a result of α-C-H-bond break at O_2 attack will affect the π-system of pyromellitimide fragment and, further on, on σ-bonds of α-CH_2-groups in neighboring methylene chain, probably making them weaker or activating this group in any different way. A hypothetical mechanism of oxidation activation in PAI is somehow similar to the known allyl activation of PVC dehydrochlorination, when the structure defect, mainly π-system in the double bond or ketoallyl group, forms instability of neighboring Cl atom or entire elementary unit. In this labile fragment degradation is induced by Cl detachment. Hence, π-system develops originating the polyene sequence, which activates the following vinylchloride units [41]:

$$\text{wwwC-CH=CH-CH www} \xrightarrow{-HCl}$$

$$\longrightarrow \text{wwwC-CH=CH-CH=CH-CH-CH}_2\text{www}$$

In the literature, another mechanism of oxidation center transfer is known, aimed at the immediate environment, but not statistically anywhere. At solid-phase oxidation of polyolefins, besides hydroperoxide groups, statistically disposed along macrochains, regular ROOH blocks are observed and frequently dominate [42]:

Formation of blocks is associated with kinetic disadvantage of intermolecular chain transfer

$$RO_2 + ROOH + R$$

and, vice versa, advantage of peroxy-radical isomerization via a six-term transition complex - the so-called "chemical relay":

Obviously, such transfer of free valence in PAI is impossible due to geometrical hindrances:

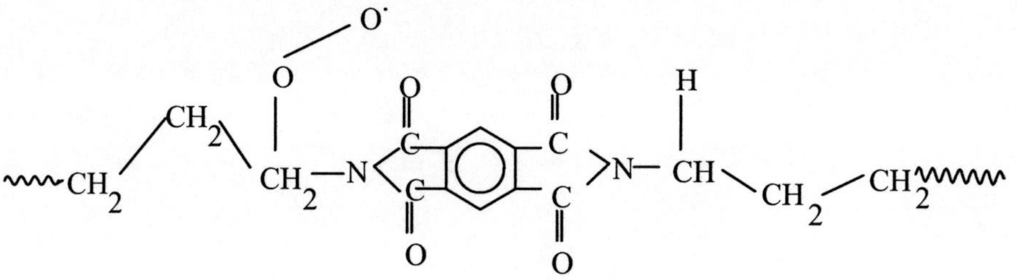

Most probably, RO_2 isomerization along self methylene chain would be expected. However, this does not create conditions for pyromellite diimide synthesis. Apparently, the hypothesis of α-CH_2-group electron activation by π-system of pyromellitimide fragment, linking them, has no alternatives.

Let us consider the probability of pyromellite diimide formation compared with longer macrochain fractures in neighboring methylene chains. Occurrence of the latter would be expected as a result of O_2 attack at all α-CH_2-groups. In the initial state, before the first O_2 attack, all α-CH_2-groups are basically equivalent, i.e. strengths of C-H-bonds in them are equal. Further on, dealing with the experimental fact of exclusive pyromellite diimide synthesis at low temperatures (below 250°C), the authors suggest that any O_2 attack and middle macroradical formation result in transiting excitation from π-system in peromellitic fragment to the electron system of neighboring α-CH_2-group due to occurrence of a free valence. As a result of additional α-C-H-bond weakening, oxidation is mostly realized at that place. Therefore, bonds break, and pyromellitimide "preparation" deteriorates.

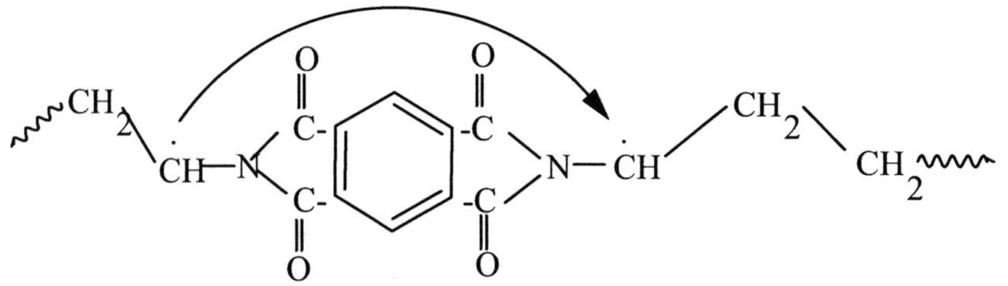

The fact of bond weakening in radicals compared with their saturated analogues is known [16]. The known from the literature maximal weakening of C-H-bond, equal 21 kJ/mol, may be accepted [16]. The decrease of C-H-bond strength in α-CH_2-group, neighbor to previously attacked one, by 21 kJ/mol means that at 200 - 250°C the probability of new attack at this place is 100 – 120 times higher, than in other, not neighboring α-CH_2-groups. This means that precipitation obtained at PAI thermal oxidation would possess 1% or slightly less amount of other oligomers, for example, dodecamethylene dipyromellitimide. However, the practice shows that heavy products, formed at PAI aging up to 9,000 hours, give nothing except for yellowy crystals of pyromellite diimide.

Remind that pyromellite diimide is the unique heavy product of low-temperature oxidation of classical polyimide PM-1, or Kapton [39]. Moreover, the product of similar structure compared with the primary macromolecule – 2,2'-(1,4-phenylene)-bis-(3-

phenylpyrazine) (PPP), is synthesized at aging of polyphenylquinoxaline [43]. Terephthalic acid amid was also detected at thermal oxidation of polyphthalimides [44].

Formation of PDI, TPC amide and PPP indicates a specific feature of thermal oxidation of poyheteroarylenes, which has not been discussed before in the literature. Mechanical description of their formation, i.e. oxidation or "burning off" of one of pyromellitic (or phthalamide) fragment boundary first and subsequently induced destruction of another boundary may not explain the absence of competition from the side of analogous, but larger fractures even in the presence of a noticeable difference in C-H-bond strengths in induced and rest α-CH$_2$-groups. Two explanations may be given:

1. Induced weakening of C-H-bond is much higher – 41 kJ/mol, for example. In this case, the probability of PDI formation is by three orders of magnitude higher, than for other diimides. However, so abrupt strength loss by one bond in the α-CH$_2$-group, induced by a free valence formation at the periphery of conjugated fragment, must be accompanied by simultaneous strength loss of other bonds in this group (for example, C-C-bonds with β-CH$_2$-group). Taking into account lower strength of C-C-bonds compared with C-H-bonds, an additional strength loss would lead to preferable thermal breaks of α-C-C-bonds and noticeable content of methylpyromellitimide in degradation products. However, this was not observed even at high temperatures.

2. The second hypothesis is corresponded to the idea about molecular complexes of O$_2$ with substrates, which is not practically used in thermal oxidative degradation of polymers. The existence of O$_2$ molecular complexes with many individual organic compounds was proven (particularly, using NMR-spectra) in works by Buchachenko *et al.* [45]. As suggested before [43], oxidation of hydrocarbons is initiated by molecular complex formation with O$_2$. Formation of O$_2$ molecular complex with pyromellitic fragment and its consecutive decaying with oxidation initiation simultaneously on two methylene chains principally solves the problem of synchronization required for exclusive pyromellite diimide detachment, but not any other products of this type.

There are no direct proofs of such complex formation with PAI, as well as with other polyheteroarylenes. By analogy with [43], the authors studied a possibility of O_2 molecular complex formation with model PAI compound – didodecyl pyromellite diimide. To remove humidity and probably sorbed O_2, the model was preliminarily heated up at 120°C in vacuum during 1 hour and then dissolved in chloroform. The blend was degasified with argon first in a flask and then in a cuvette. UV-spectra of the model were measured compared with control cuvette, filled with chloroform, which was treated similar to the solution. After measuring the spectrum of argon-degasified sample, cuvettes were blowdown with pure O_2, and spectra were measured again. After O_2 blowdown spectra of the model display increased intensity that may be associated with O_2 absorption (direct influence) by didodecyl pyromellitimide (Figure 20).

Another experiment was performed on PAI. As heated in air, PAI intensively absorbs O_2, specifically during melting, when molecular motions are defrost. This is accompanied by an intense exothermal DTA effect. One may suggest that intensive absorption of O_2 in the melting range is associated with direct interaction between O_2 and imide fragments proceeding by the above-mentioned scheme. However, such interaction may also proceed at room temperature, possibly not only with oxygen, but, for example, with I_2, though oxidizing power of the latter is much lower. Thermograms for PAI, vacuumed at 200°C during 1 hour to remove humidity and sorbed O_2 and then exposed in I_2 vapors during 1 and 24 hours, were measured in air and compared (Figure 21). It was observed that the intensity of exothermal peak decreases symbate with time of exposure in I_2 vapors. Seemingly, at room temperature O_2 (currently, I_2) really forms a complex with PAI fragments located on the surface. Molecular mobility increases with temperature, and zones previously located "deep inside" become capable of complex formation. Unfortunately, there is no I_2 in the system anymore, but free access of O_2 is provided, and it occupies free places and, thus, forms complexes. Some already mentioned features of PAI thermal oxidation and, the more so, its resistance to thermal oxidation, discussed below, may be explained only from these positions.

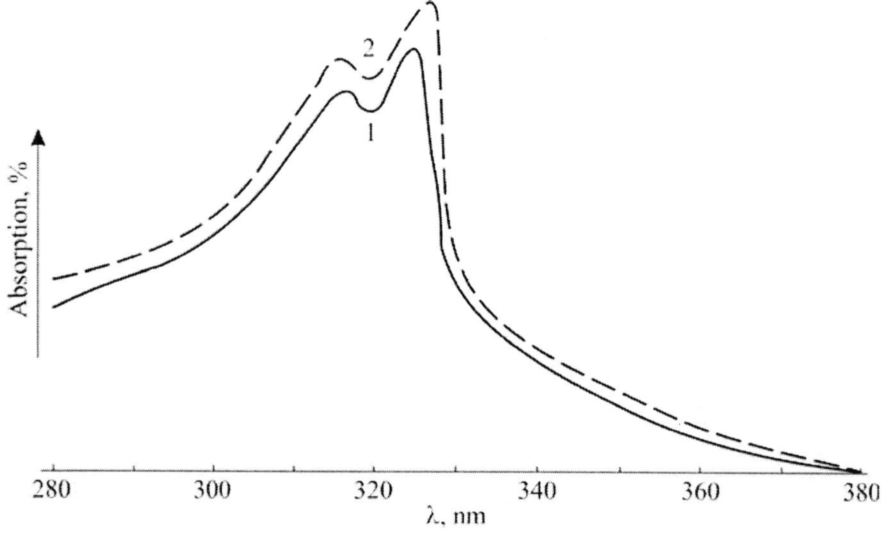

Figure 20. UV-spectra of model PAI compound – didodecyl pyromellite diimide, before (1) and after (2) blowdown by pure O_2

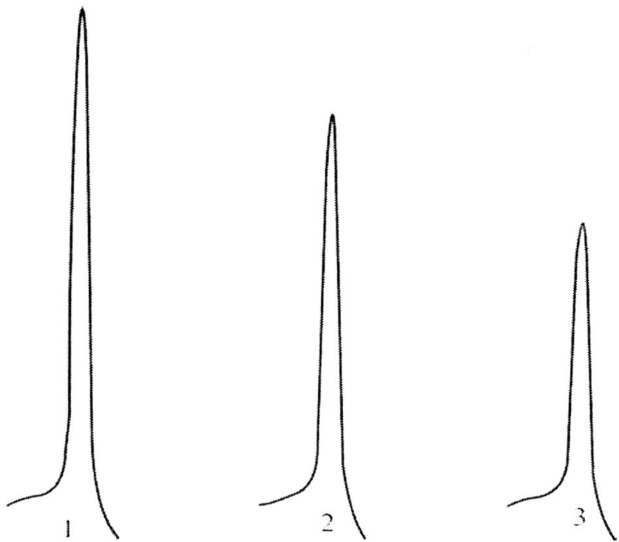

Figure 21. PAI thermograms before (1) and after processing by I_2 during 1 (2) and 24 hours (3)

Let us consider another example – thermal oxidation and thermal stabilization of aliphatic polyamides, which by phenomenology are close to PAI. Thermal oxidation of PA, described in the literature [28], is identical to the classical scheme of hydrocarbon oxidation. Hence, these schemes do not explain some facts, for example, a significant increase of PA thermal stability at N-alkyl substitution. Moreover, high antioxidant activity of Cu^{2+}/I^- system may not be explained in the framework of hydroperoxide mechanism, though it exceeds all classical antioxidants in efficiency.

Deactivation of peroxy-radicals or hydroperoxides by π-system is not confirmed because of specific action of the system in PA only and, at most, the absence of stabilizing effect in polyolefins which are exclusively subject to radical-chain oxidation reactions. Besides Cu, other transition metals are also effective in PA. Moreover, the sequence of their antioxidant activity coincides with the Poling sequence of electronegativity [46]. The latter, in turn, correlates with the ability to complex formation or stability of unitypical complexes. The idea of molecular complexes suggests an alternative to the act of thermal oxidation initiation and a possibility of inhibition. The idea that molecular complexes of organic substances with O_2 participate in oxidation occurred, when they were detected by Evans [43]. Complex formation with hydrocarbon is reversible reaction due to thermal instability, and it has no consequences for the substrate. Therefore, this reaction is not considered in schemes of thermal oxidation of polymers, though properties of the complex (for instance, with heterogroups) may be absolutely different, and the ways of transformation may destroy the polymer. Anyway, addition of the radical-chain scheme with this act allows explanation of thermal oxidation features of PAI and other polymers containing secondary and tertiary amino groups (for example, hydrogen release at thermal oxidation).

The idea of O_2 molecular complex with π-system of heteroaromatic fragment does not exhaust all possibilities of explaining anomalies in thermal behavior of PAI and analogous polymers. Basing on the studies of thermal behavior of some highly heat-resistant nitrogen-containing, organic polymers, the authors [47] suggested transfer of macromolecules possessing developed conjugation system into electronically excited state as the initiation act for the radical-chain oxidation process. Here oxygen plays the role of suppressor – activator of intercombinatorial transitions, i.e. similar to O_2 role in photo processes. Experimental data indirectly prove participation of electronically excited states in the degradation process. Recently, quantum-chemical analysis of thermal behavior of phthalimide cycle as the polyimide model was performed [48]. The energetic possibility of purely thermal transferring the model to the triplet state already at 500°C is shown. From this state radical decays and H-transfers are initiated, which lead to crosslinking of models and destruction of imide cycle with carbon oxides and water release. Excitation levels in conjugated macromolecules are much lower than in low-molecular models. This is confirmed by quantum-chemical calculations of models and polyheteroarylene oligomers [49], and by disposition of the long-wave boundary in UV-spectra. Therefore, macromolecules may be transited into electronically excited state at temperatures lower than indicated by the authors [48]. The absence of O_2 increases the probability of intercombinatorial transition into reactive triple state.

In the present context we may not choose between two mentioned alternatives of pyromellitic fragment activation in PAI or polyimides. Here of importance is another thing: degradation behavior of PAI and other polyheteroarylenes displays a mechanism, conditionally called the "quick response", which leads to local development of degradation with specific product output. Such "quick response" may be performed only in the system of molecular orbitals of polymellitic fragment which, affected by O_2 and temperature, initiates or promotes initiation of radical-chain oxidation, yet developed in fatty chain of PAI. At low temperatures, pyromellitic fragment acts like somewhat a magnifying glass which focuses sunlight rays to a point and ignites the sample. However, as temperature increases to the processing level, degradation is also initiated in this rigid fragment of the macrochain.

Injection of additives is the common method for studying the mechanism of chemical reactions. Phenol antioxidants inhibit oxidation in reactions with peroxy-radicals PO_2 and at 250°C, in concentration of $10^{-3} - 10^{-2}$ mol/kg slows down O_2 absorption in PAI (Figure 22). Kinetics of O_2 absorption by PAI is characterized by the absence of induction period. Therefore, critical concentration of antioxidant may not be determined. The concentration dependence of antioxidant efficiency, determined from the ratio of initial rate constants of O_2 absorption by unstabilized and stabilized PAI is optimal at concentration of 0.008 mol/kg. Though antioxidant effects induced by phenol injection are low (probably, due to high temperature [40] and the absence of optimal composition at the current stage), the fact of slowing down by low additions confirms once again the contribution of radical-chain oxidation to thermal oxidative degradation of PAI.

As shown above, some features of PAI thermal oxidation do not fit traditional schemes. The suggestion about existence of molecular complexes with O_2 or other nontrivial initiation acts leads to analogy of PAI with aliphatic PA or classical polyimides, thermal oxidation of which is hindered by addition of Cu^{2+}/I^- complexes [28] and organophosphorous compounds, respectively. For example, anilidophosphoric acid diphenyl ether (compound BT-5) is the most effective in polyimides [47]. Injection of the mentioned additives to PAI at 250°C slows down O_2 absorption more intensively (Figure 22) than phenol antioxidants. The effectiveness of additives is also observed at higher temperatures (Figure 23), at which phenols are absolutely ineffective.

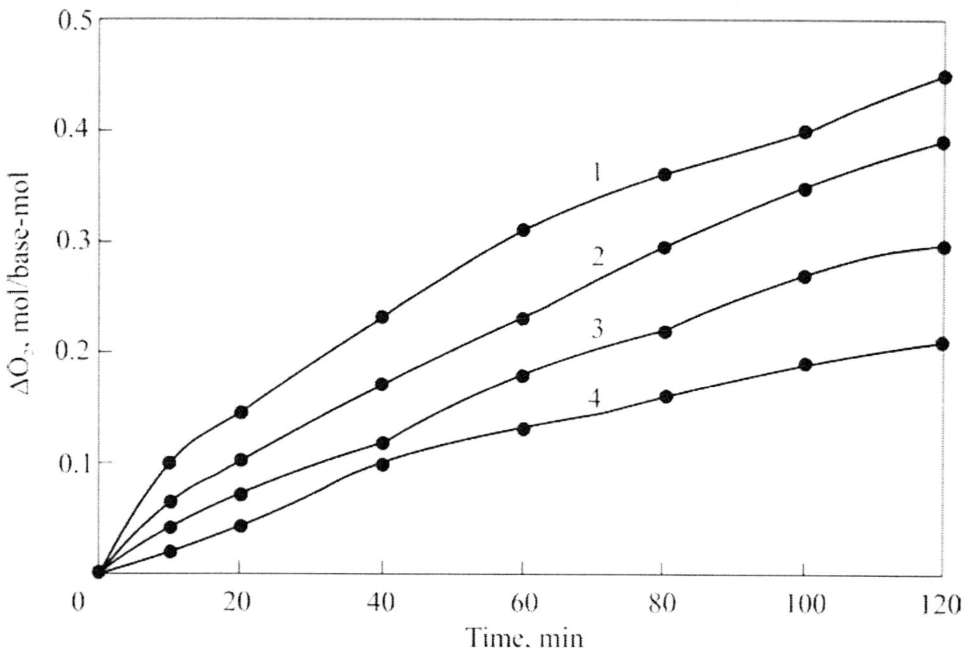

Figure 22. Kinetic curves of O_2 absorption at thermal oxidation of PAI: initial (1), added by 1 wt.% *Irganox 1010* (2), 2 wt.% *BT-5* (3), a mixture of 0.05 wt.% $CuSO_4$ and 2 wt.% *BT-5* (3).

$T = 250°C$, $P_{O_2} = 266.6$ kPa

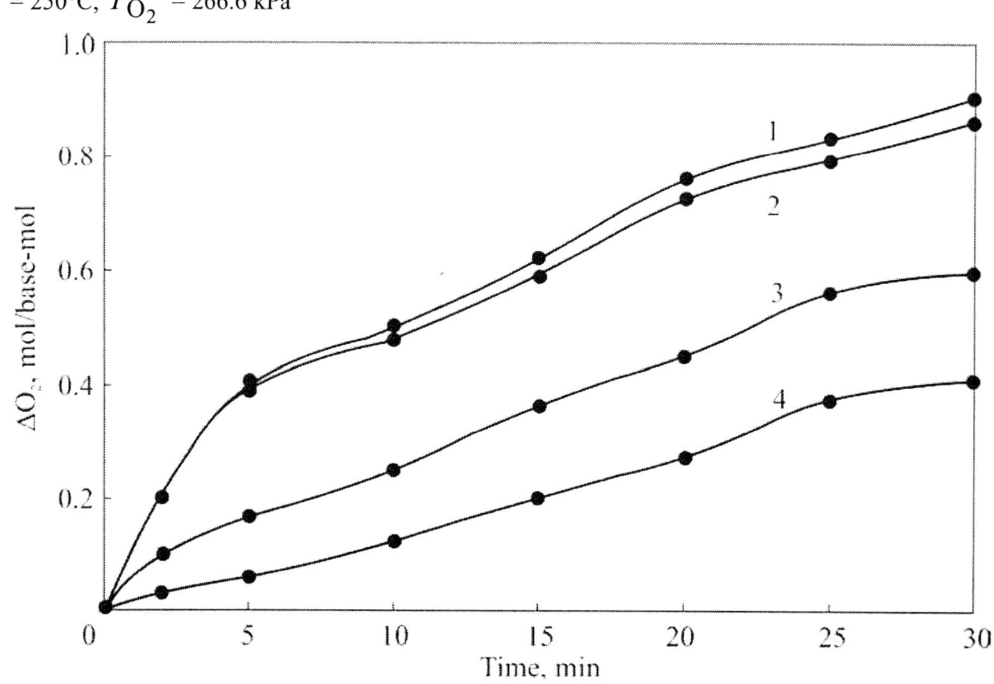

Figure 23. Kinetic curves of O_2 absorption at thermal oxidation of PAI: initial (1), added by 1 wt.% *Irganox 1010* (2), 2 wt.% *BT-5* (3), a mixture of 0.05 wt.% $CuSO_4$ and 2 wt.% *BT-5* (3).

$T = 320°C$, $P_{O_2} = 266.6$ kPa

The dependence of Cu^{2+}/Γ addition efficiency on concentration has a peak (Figure 24), analogous to the case of aliphatic PA. Even extremely low concentrations of Cu^{2+} are highly effective in PAI. On the contrary, injection of this system to PE destabilizes methylene chain at any temperatures (Figure 25). Though methylene chain in PAI is quite long, the test with the additive and comparison with PE definitely indicate importance of a heterofragment in the oxidation process.

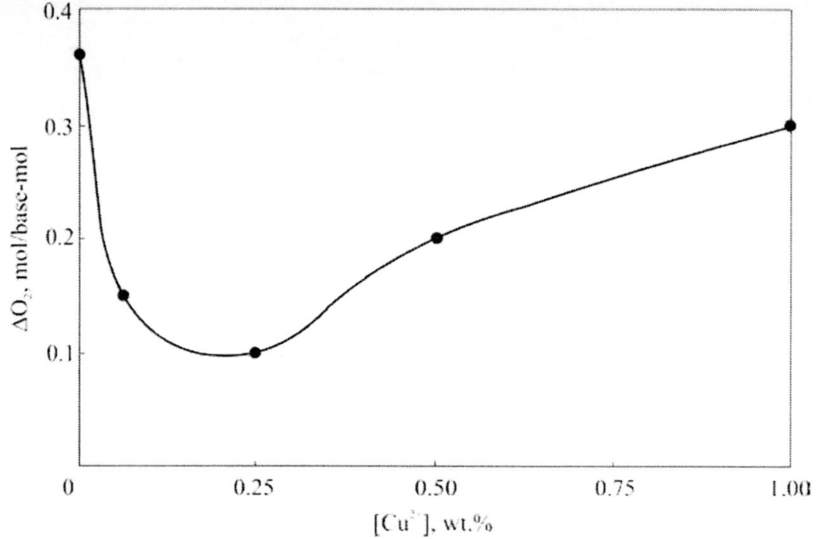

Figure 24. Dependence of PAI thermal stability (by O_2 absorption at 250°C during 1 hour) on Cu^{2+}/Γ concentration

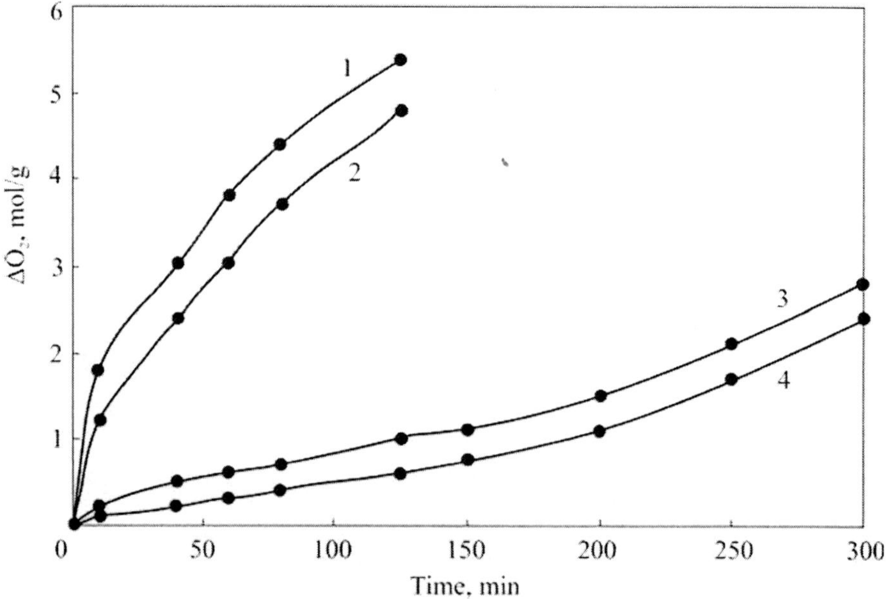

Figure 25. Kinetic curves of O_2 absorption by PE: initial (1, 2) and added by 0.05 wt.% $CuSO_4$ (3, 4) at $T=$ 320 and 200°C, respectively. P_{O_2} = 399.9 kPa

Finally, PAI in its structure has quite long flexible methylene chain, oxygen attackable as any hydrocarbon chain, specially at heating, and a rigid pyromellitic fragment possessing high resistance to oxidation. As compared with other polymers having methylene chains in their structure, PAI are much more stable. This results not from chemical reasons, but from the action of kinetic factors, determined by physical structure of polymeric body, which remains solid up to high temperatures. Physical structure rigidity and limitation of molecular motions in the polymer are determined by pyromellitic fragment. To put it differently, these structures stimulate physical protection mechanisms for an aliphatic chain. On the other hand, from pure chemical positions, if physical limitations are eliminated, pyromellitic fragment increases reactivity of border methylene groups, more active in the reaction with O_2 rather than "internal" groups of methylene chain. Oxidation, hindered in the solid phase, bursts out in the polymer melt affecting the whole aliphatic chain and partly, to a lesser extent, an aromatic fragment.

Some specific features of PAI thermal oxidation are clearly displayed at low temperatures at the background of typical radical-chain hydroperoxide oxidation of hydrocarbon chain and, to the authors' point of view, indicate chemical functionality of pyromellitic fragment in the degradation process. Antioxidant action of classical antioxidants of the phenol type is much lower than that of copper ion and phosphate, seemingly untypical antioxidants. Oxidation leads to pyromellitic fragment emission from macromolecule in the form of pyromellite diimide or to crosslinking with H_2 release, which is fully absent at degradation of methylene chains in hydrocarbon polymers. One may suggest that pyromellitic and phthalamide fragments participate in the degradation process, besides, not only by activation of α-bonds in neighboring flexible chain, but possibly materially. Though direct proofs are absent yet, the authors suppose that, besides classical initiation, for example, of the following type:

$$RH + O_2 = R + HO_2,$$

where PH is methylene chain, thermal oxidation is initiated by an act with participation of an aromatic fragment. It may be formation of a molecular complex from one or several molecules of O_2 with π-system of the fragment or thermal transfer of the fragment to the triplet state with the help of O_2. The following initiation reactions are of importance.

1. Classical Direction

2.   Formation and Decay of Molecular Complex with O_2

3.   Transfer to electronically excited state

Further stages of the process fit traditional radical-chain scheme with the full selection of appropriate degradation products.

REFERENCES

[1]   *US Patent No. 2,760,835*, Fatty-aromatic Polyimides, Publ. June 18, 1955.

[2]   Vinogradova S.V., Churochkina N.A., and Vyigodsky Ya.S., 'Synthesis and properties of some fatty-aromatic polyimides', *Vysokomol. Soed.*, 1971, vol. *A13*(3), pp. 1146 – 1150. (Rus)

[3]   Vohninkel E., *Neue Polumere Werkstoffe fur die Industrielle Andwendung*, Munhen: Carl Hauser Verlag, 1983, S. 257.

[4]   Tuichiev Sh., Kuznetsova A.M., and Mukhametdieva A.M., 'Temperature changes of molecular and permolecular structure of polyalkanimide', *Vysokomol. Soed.*, 1987, vol. *A29*(8), pp. 1756 – 1760. (Rus)

[5]   Vigodsky J.S., Vinogradova S.V., Nagiev S.T., and Korschak V.V., 'A study of copolyimide synthesis, structure and properties', *Acta Polymerica*, vol. *33*(2), pp. 131 - 137.

[6]   Kazaryan L.G., Azriel A.E., Vasil'ev V.A., Annenkova N.G., Pinaeva N.K., and Chernova A.G., "Structure and properties of polyalkanimide', *Vysokomol. Soed.*,1988, vol. *A30*(3), pp. 644 – 647. (Rus)

[7]   Hummel D.O., In: *Analytical Pyrolysis*, Ed. C.S.R. Jenes and C.A. Cramers, Amsterdam: Elsevier Publ., 1977, p. 177.

[8]   Dussel H.J., Rosen H., and Hummel D.O., ‚Feldionen- und Elektronen- stos- Massenspektrometrie von Polymeren und Copolymeren', *Die Macromolekulare Chemie*, 1976, Bd. *177*(8), S. 2343 – 2368.

[9]   Soboleva L.I., Matevosyan Ts.M., Chernova A.G., and Pinaeva N.K., 'Hydrolytic properties of polyalkanimide', In Coll.: *Production and Processing of Plastics and Synthetic Resins*, Moscow, NIITEKhim, 1976, No. 9, pp. 34 – 36. (Rus)

[10]   Chernova A.G., Lebedinskaya M.L., Pinaeva N.K., and Nekrasova L.N., 'Deformation and strength properties of poly(1,12-dodecamethylene pyromellit)imide', In Coll.: *Production and Processing of Plastics and Synthetic Resins*, Moscow, NIITEKhim, 1977, No. 7, pp. 20 – 23. (Rus)

[11]   Chernova A.G., Sedyikh V.M., Lushcheikin G.A., Pinaeva N.K., and Nekrasova L.P., 'Electrical properties of polyalkanimide *AI-1G*', In Coll.: *Production and Processing of Plastics and Synthetic Resins*, Moscow, NIITEKhim, 1979, No. 2, pp. 15 – 17. (Rus)

[12]   *The Library of IR-spectra for Perkin-Elmer Device* (USA), N.Y., 1981.

[13]   Laius L.A. 'The study of PI-film structures by IR-spectroscopy method', Vysokomol. Soed., 1974, vol. *A16*(9), pp. 2101 - 2107. (Rus)

[14]   Nakanisi K., *IR-Spectra and Structure of Organic Compounds*, Moscow, Mir, 1965, 215 p. (Rus)

[15]   Magaril R.Z., *The Mechanism and Kinetics of Homogeneous Thermal Transformations of Hydrocarbons*, Moscow, Khimia, 1970, 224 p. (Rus)

[16]   *Energies of Chemical Bond Break, Ionization Potentials and Affinity to Electron*, Moscow, Nauka, 1974, 351 p. (Rus)

[17]   Blumenfeld A.B., Goglev R.S., Kovarskaya B.M., and Heiman M.B., 'About strength of chemical bonds in polyethers', *Vysokomol. Soed.*, 1972, vol. *A14*(10), pp. 2215 – 2220. (Rus)

[18]   Buchachenko A.L. and Wasserman A.M., *Stable Radicals*, Moscow, Khimia, 1973, 407 p. (Rus)

[19]   Bulgarovskaya I.V., Smelyanskaya E.M., Fedorov Yu.G., and Zvonkova Z.V., 'Crystalline structure of 1:1 pyromellitic N,N-dimethyldiimide – anthracene complex', *Kristallografia*, 1977, vol. *22*(1), pp. 184 - 187. (Rus)

[20]   Mortimer C., *Reaction Heats and Bond Strengths*, Moscow, Mir, 1964, 287 p. (Rus)

[21]   Semenov N.N., *On Some Problems of Chemical Kinetics and Reactivity (Free Radicals and Chain Reactions)*, 2nd Ed., Moscow, AN SSSR, 1958, 686 p. (Rus)

[22]   Thermal Stability of Polymers, Ed. R.T. Conley, vol. *1*, N.Y.: Marcel Dekker, 1970, 347 p.

[23]   Annenkova N.G., Kovarskaya B.M., Gur'yanova V.V., Pshenitsyna V.P., and Molotkova N.N., 'High-temperature oxidation of polyimide', *Vysokomol. Soed.*, 1975, vol. *A17*(1), pp. 134 - 138. (Rus)

[24]   Encyclopedia of Polymers, vol. *3*, Moscow, Sovetskaya Encyclopedia, 1977. (Rus)

[25]   Emanuel N.M., Denisov E.T., and Maizus Z.K., *Chain Reactions of Hydrocarbons Oxidation in the Liquid Phase*, Moscow, Nauka, 1965, 375 p. (Rus)

[26]   Kandratiev V.N. and Nikitin E.E., *Chemical Processes in Gases*, Moscow, Nauka, 1981, 262 p. (Rus)

[27]   Sklyarova A.G., Branzali F., and Wasserman A.M., 'Comparison of C-H-bond reactivity in polyethylene and low-molecular hydrocarbons', *Vysokomol. Soed.*, vol. *B16*(1), 1972, pp. 72 - 78. (Rus)

[28]   Kovarskaya B.M., Blyumenfeld A.B., and Levantovskaya I.I., *Thermal Stability of Heterochain Polymers*, Moscow, Khimia, 1977, 263 p. (Rus)

[29]   Emanuel N.M. and Buchachenko A.L., *Chemical Physics of Polymer Aging and Stabilization*, Moscow, Nauka, 1982, 359 p. (Rus)

[30]   Shlensky O.F., Vainstein E.F., and Matyukhin A.A., 'Dynamic thermal decomposition of linear polymers and its study by thermoanalytical methods', *J. Thermal Anal.*, 1988, vol. *34*(3), pp. 645 - 655.

[31]   Edemskaya V.V., 'The study of high-temperature oxidation of polyethylene and polypropylene', *Candidate Dissertation Thesis*, Moscow, 1973. (Rus)

[32]   Serenkova I.A., Kulagin V.N., Tseitlin G.M., Shlyapnikov Yu.A., and Korshak V.V., 'Kinetics and mechanism of polybenzoxazole thermal oxidation', *Vysokomol. Soed.*, 1974, vol. *B16*(7), pp. 493 - 496. (Rus)

[33]   Neiman M.B., *Aging and Stabilization of Polymers*, Moscow, Nauka, 1964, 332 p. (Rus)

[34]   Denisov E.T., *Oxidation and Degradation of Carbochain Polymers*, Moscow, Khimia, 1990, 286 p. (Rus)

[35]   Do C.H., Pearce E.M., Bulkin B.J., and Reinschulssel H.K., 'FI-IR- Spectroscopic study on the thermal and thermal oxidative degradation of Nylons', *J. Polym. Sci.: Polym. Chem. Ed.*, 1987, vol. *A25*(9), pp. 2409 - 2424.

[36]   Volozhin A.I., Prokopchuk N.R., Yakimtsova A.B., Krut'ko E.T., and Solntsev A.P., 'Thermal stability of poly(4,4'-diphenyloxide)pyromellitimide of bicycle-(2,2',1)-hept-5-ene-2,3-dicarboxylic acid', *Vestsi AN BSSR, Ser. Khim. Nauk*, 1988, No. 5, pp. 109 - 111. (Rus)

[37]   Abramova I.M., Azriel A.E., Vatagina V.A.,Zezina L.A., Zyisk K.Yu., Kazaryan L.G., Pinaeva N.N., Savina M.E., Sade A.G., and Chernova A.G., 'The effect of aging conditions on the structure and properties of polyalkanimide *AI-1G*', In Coll.: *Heat-Resistant Materials*, Moscow, NIITEKhim, 1985, pp. 162 – 167. (Rus)

[38]   Annenkova N.G., 'The Study of Thermal Stability of Aromatic Polyimides', *Candidate Dissertation Thesis*, Moscow, 1973. (Rus)

[39]   Vdovina A.L., 'Thermal Transformations and Stabilization of Polypyromellitimide, Polyphenylquinoxaline and Copolyimidophenyl-quinoxalines', *Candidate Dissertation Thesis*, Moscow, 1988. (Rus)

[40] Shlyapnikov Yu.A., Kiryushkin S.G., and Mar'in A.P., *Antioxidant Stabilization of Polymers*, Moscow, Khimia, 1986, 252 p. (Rus)

[41] Minsker K.E. and Fedoseeva G.T., *Degradation and Stabilization of Polyvinylchloride*, 2nd Ed., Moscow, Khimia, 1979, 272 p. (Rus)

[42] Chien J.C.W. and Wang D.S.T., 'The thermal and thermal oxidative of polyolefines', *Macromolecules*, 1976, vol. *8*, pp. 920 - 929.

[43] Bolland J.L. and Ten Have P., Molecular Complex-O_2 with oxidative *ot* polyolefins, *Trans. Faraday Soc.*, 1947, vol. *43*, pp. 201 - 205.

[44] Andreeva M.B., 'Thermal transformations and stabilization of aliphatic-aromatic polyamides and derived mixtures', *Candidate Dissertation Thesis*, Moscow, 2002. (Rus)

[45] Buchachenko A.L., Complexes of Radicals and Molecular Oxygen with Organic Molecules, Moscow, Nauka, 1970, 489 p. (Rus)

[46] Smirnov L.A., 'Increasing heat and light resistance of caproic fiber with the help of polymer stabilization by metal-containing organic compounds', *Candidate Dissertation Thesis*, Moscow, 1969. (Rus)

[47] Blyumenfeld A.B., Levantovskaya I.I., and Annenkova N.G., 'Problems of thermal stability and stabilization of heterochain polymers', In Coll.: *Progress in Science and Technology, Ser. Navy Chem. Technol.*, vol. *20*, Moscow, VINITI, 1985, pp. 143 – 211. (Rus)

[48] Yakimansky A.V., Milevskaya I.S., Zubkov V.A., and El'yashevich A.M., 'Quantum-chemical analysis of crosslinked system and volatile product formation at thermolysis of polyimides', *Vysokomol. Soed.*, 1989, vol. *A32*(11), pp. 2318 – 2326. (Rus)

[49] Kosobutsky V.A., 'Electron composition and some physical and chemical properties of aromatic polyamides and polyheteroarylenes', *Candidate Dissertation Thesis*, Rostov-na-Donu, 1973. (Rus)

[50] Tager A.A., *Physical Chemistry of Polymers*, Moscow, Gosizdat, 1963, 528 p. (Rus)

In: Physical Organic Chemistry: Theory and Practice
Eds: A. D'Amore and G. E. Zaikov, pp. 211-229

ISBN 1-59454-275-9
© 2005 Nova Science Publishers, Inc.

Chapter 14

EXPERIMENTAL PROOF OF THE CLUSTER MODEL

G. V. Kozlov

Kabardino-Balkarian State University, 173, Chernychevskogo st.,
Kabardino-Balkaria, Nal'chik, 360004, Russia

*G. E. Zaikov**

*N.M.Emanuel Institute of Biochemical Physics, Russian Academy of Sciences,
4, Kosygin st., Moscow, 119991, Russia

INTRODUCTION

As indicated in Chapter 1, existence of the local order in the polymer amorphous state directly or indirectly proves practically all modern experimental techniques. In the current Chapter the authors will briefly consider the experimental observations which prove the cluster model, i.e. type of packing, existence range, etc.

X-RAY DIFFRACTION ANALYSIS

It has been assumed for a long time that techniques applying wide-angle X-ray diffraction give short information about the structure of amorphous polymers, and that interpretation of experimental results obtained is very complicated and ambiguous [1 – 3]. Nevertheless, the study of the totality of polymers with regularly variable structure can give important data on it. In particular, for a group of polymers, comparison of data on the position of their amorphous halo peak with regard to the change of their structures (size of side groups, macromolecule cross-section) has induced a conclusion [1 – 3] about existence of the local order in amorphous polymers. These observations were not studied in detail, which is probably caused by some circumstances and, first of all, by the absence of an appropriate model of amorphous polymer structure considering the presence of the local order in them. Moreover, in accordance with the authors' point of view, in the works [1 – 3] an insufficient attention was paid to analysis of behavior of the amorphous halo intensity, which is also the sensitive structural parameter of the material. The more so, it has been concluded [4] that amorphous halo intensity gives an approximate measure of the local order in the structure.

Recently, a series of works was performed [4 – 9], in which amorphous halos of amorphous and amorphous-crystalline polymers were studied in much more details and, which is most important, quantitatively. Conclusions made in the mentioned works have been already compared with the results, obtained by the authors in Chapter 3 of the present monograph. These comparisons will be continued below, but one important qualitative difference in interpretations of wide-angle X-ray diffraction data, displayed in the works [4 – 9] and discussed below, should be noted. The difference concludes in the absence of any quantitative structural model of the polymer amorphous state in these works. Essentially, [4 – 9] this model was composed on the basis on the data, obtained by analysis of the wide-angle X-ray diffraction. Considered in the previous Chapter were some aspects of X-ray diffraction data applied to structural defects. More general interpretation of this experimental technique applied to the entire cluster model will be given below.

For epoxy polymers, derived from ED-22 resin of amine (EP-1) and anhydride (EP-2) cross-linking type, dependence of the Bragg interval d on the curing agent : oligomer ratio, K_{st}, possesses a maximum near the stoichiometric ratio $K_{st} = 1.0$ (Figure 1). Chemical cross-link network frequency, ν_{nw}, varies similarly (Table 1). Miller and Boyer [2] have suggested the following correlation between d and effective diameter of macromolecule, D (the distance between centers of macromolecules), simulated by a cylinder:

$$D = d^{1.22}. \tag{1}$$

Table 1. Structural Characteristics of Epoxy Polymers [10]

Epoxy polymer	K_{st}	$\nu_{nw} \times 10^{-26}$, m^{-3}	T_g, K	$\nu_{cl} \times 10^{-27}$, m^{-3}
EP-1	0.50	2	326	1.3
	0.75	6	366	1.9
	1.0	17	423	2.7
	1.25	12	405	3.5
	1.50	8	390	2.0
EP-2	0.50	3	333	1.5
	0.75	12	389	2.4
	1.0	19	423	2.9
	1.25	12	408	2.9
	1.50	8	393	2.0

In accordance with the equation (1) the extreme value of $d(K_{st})$ function indicates that the macromolecule effective diameter increases with the cross-linking frequency. This fact is simply understood with the help of simplified planar scheme of the polymer network structure (insertion in Figure 1). If for macromolecule a configuration located between two sectional planes depicted by dashed lines is taken, it will appear similar to the case of linear polymer shaped as a backbone with side groups [11]. It is common knowledge [2] that the latter inhibit dense packing of chains expanding distance between them and, as a consequence, D [10]. Parameter ν_{nw} is increased analogously to the number of side groups, wherefrom, symbate change of D and ν_{nw} follows [10].

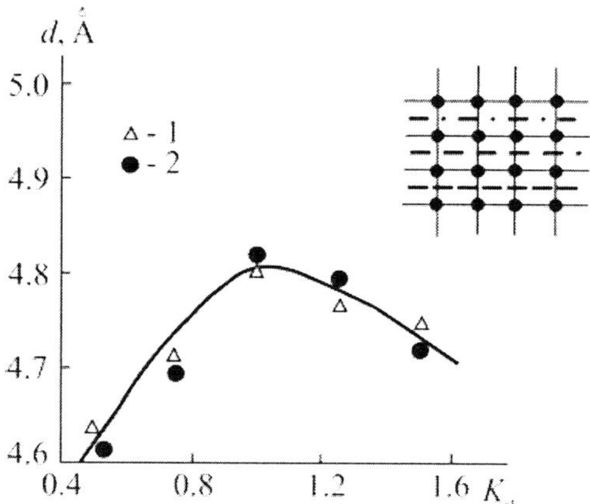

Figure 1. Dependence of the Bragg interval d on curing agent : oligomer ratio K_{st} for EP-1 (1) and EP-2 (2). Insertion presents scheme of the polymer network structure [10]

For linear polymers, it was also suggested to approximate glass transition temperature T_g by the following empirical correlation:

$$T_g = 129\left(\frac{S}{C_\infty}\right)^{1/2}, \text{K},$$

(2)

where S is the macromolecule cross-section, Å^2; C_∞ is the characteristic relation. The equation (2) can also be applied to EP under consideration under the assumption that C_∞ varies with K_{st} and, as a consequence, with v_{nw}. Figure 2 shows results of C_∞ calculations, performed by the above-mentioned method. Clearly C_∞ is decreased monotonously with the cross-linking frequency increase, and compactness of the macromolecular coil increases [13]. According to [2], collectionwise with simultaneous increase of S, this circumstance must promote an increase of the local order parameter.

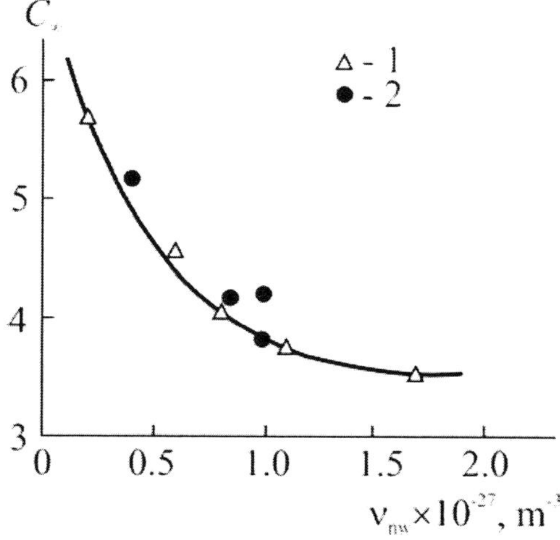

Figure 2. Dependence of the characteristic relation C∞ on the chemical cross-link network frequency vnw for EP-1 (1) and EP-2 (2) [10]

Let us check concordance between this conclusion and ideas of the cluster model [14]. The calculation data on macromolecular entanglement cluster network frequency, v_{cl}, and the relative part of clusters, φ_{cl}, confirm the above statement about the local order increase with v_{nw} (Table 1 and Figure 3). Contrary to assumptions [2], where the change of d is directly associated with the local order change, the authors assume that d is just a factor characterizing D and C_∞, i.e. chain rigidity. Simultaneously, the latter factor determines φ_{cl} variation [10].

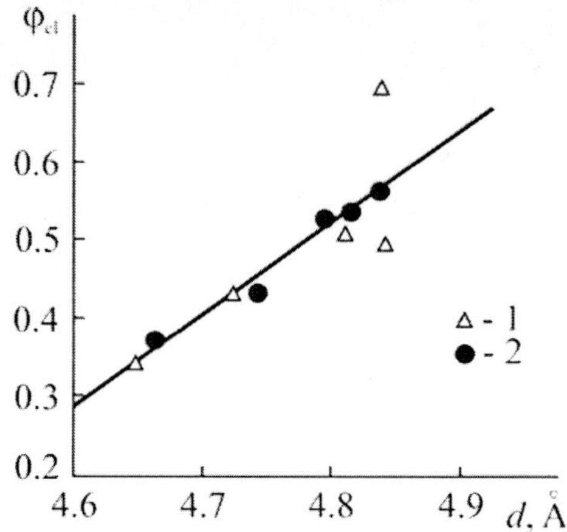

Figure 3. Dependence of the relative fracture of clusters, φ_{cl}, on the Bragg interval, d, for EP-1 (1) and EP-2 (2) [10]

Usually, wide-angle X-ray diffraction halos of amorphous polymers deviate from the ideal shape (asymmetry, unclear maximum, etc.) that allows a supposition about superposition of several simpler dispersions [8, 9]. By studying the shape of amorphous halos for EP-1 and EP-2 epoxy polymers, it has been detected that the best description is obtained under the assumption of two-component structure of the halo. Using ideas of the cluster model [14], one can associate a halo component with the maximum disposed at lower θ with a packless matrix, and the one at higher θ with clusters [15].

The integral intensity of both components and the total halo as well displays an extreme variation with K_{st} possessing a maximum at $K_{st} = 1.0$. Comparison of the dispersion integral intensity from packless matrix ($I_{p.m.}$) with $\varphi_{p.m.}$ indicates linearity of $I_{p.m.}(\varphi_{p.m.})$ dependence and its passing through the origin of coordinates. This gives supplementary grounds for associating the dispersion curve 1 for greater d with the packless matrix [10, 15].

X-ray diffraction analysis of more complicated cross-linked systems was performed [16] on the example of EP-2 epoxy polymer of the anhydride cross-linking type, modified by adamantane carboxylic acids. It is common knowledge that injection of adamantane fragments in EPs significantly affects their characteristics. The action of such carcass structures is considered on the example of EP-2 [17], and interpretation of experimental results within the framework of the cluster model allowed supposition of the existence of two types of clusters in EP-2 modified by adamantane carboxylic acids: stable ones formed by segments and unstable ones formed due to interaction between the adamantane fragments. The conformity degree between these ideas and the real structure of modified EPs was also determined [16]. Two series of EPs modified by adamantane mono- (EP-3) and dicarboxylic (EP-4) acids were studied.

The studies of the halo shape observed for EP-3 and EP-4 indicated their best description in the presence of the halos, whereas for EP-2 with two halos only. Occurrence of the amorphous halo third component for EP with carcass fragments indicates that the structural state of EP-3 and EP-4 is more complicated (inhomogeneous) compared with unmodified EP-2 (Figure 4). Within the framework of the cluster model, occurrence of three halos may be induced by the presence of three structural components: packless matrix and two types of clusters. The latter are formed by collinear segments of EP network and adamantane fragments.

Table 2 presents values of the Bragg interval, d, calculated from position of the maximum of experimentally observed halo and its components, separated after computerized treatment of diffraction patterns (d_1, d_2, d_3) and corresponded amplitude (I^a) and integral (I^i) intensities. A tendency of the Bragg interval growth with modifier concentration C was observed. As a consequence, introduction of carcass fragments induced loosening of EP structure.

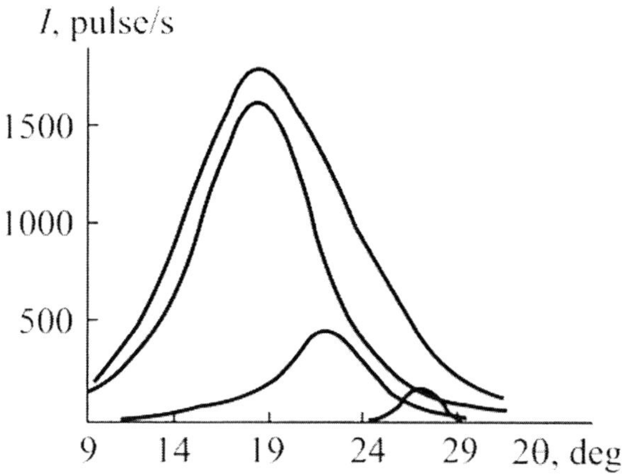

Figure 4. Experimental X-ray diffraction pattern (1) and its factorization for EP-3 [16]

Table 2. The Effect of Modifier Concentration on X-ray Diffraction Characteristics of EP [16]

Epoxy polymer	C, %	d	d_1	d_2	d_3	I_l	I_1^a	I_2^a	I_3^a	I_1^i	I_2^i	I_3^i
				Å				pulse/s ×10⁻³			relative units	
EP-2	0	4.76	4.84	3.86	-	1.65	1.22	0.67	-	7.7	6.8	-
EP-3	1.0	5.0	5.07	3.86	3.03	1.86	1.53	0.96	0.14	10.2	7.9	0.3
	2.5	4.66	4.55	-	3.18	1.91	1.84	-	0.41	15.6	-	2.1
	5.0	5.05	5.04	4.21	3.52	2.12	1.75	0.79	0.57	11.5	3.9	3.3
	10.0	4.93	4.76	-	3.31	1.98	1.99	-	0.42	8.3	-	0.8
	15.0	4.98	5.15	4.27	3.34	1.92	2.12	1.45	0.44	4.3	3.0	0.6
EP-4	1.4	4.78	4.89	4.05	3.29	1.80	1.63	0.47	0.17	14.6	2.6	0.4
	2.9	4.83	4.91	3.96	3.09	1.70	1.69	0.47	0.25	11.6	1.4	0.5
	4.3	4.98	4.83	-	3.16	2.37	2.21	-	0.75	18.1	-	4.4
	9.2	5.19	5.12	3.95	-	1.97	1.88	0.47	-	13.5	3.0	-
	13.2	5.28	5.36	4.54	3.45	2.06	1.43	0.11	0.25	7.6	6.6	1.3

As before, the halo component with the greatest corresponded Bragg interval d_1 should be associated with the packless matrix. In this case, d_2 and d_3 characterize the network of physical entanglements. Because d_2 values for EP-2, EP-3 and EP-4 are quite close (especially at low concentrations of the modifier), this parameter must be corresponded to entanglements (clusters), formed by EP segments. Thus d_3 is corresponded to the network of physical entanglements, formed by interaction of adamantane fragments.

First of all, structural changes, which may be obtained from the analysis of parameters of EP general amorphous halo, should be considered. However, note first that addition of adamantane groups makes the network of chemical cross-links less frequent (Figure 5) that obviously happens due to steric obstacles, formed by them [17]. Intensity and rate of ν_{nw} decrease are independent of the modifier type and the method of its injection, and at concentration $C \approx 15\%$, ν_{nw} apparently reaches the near-border values. Thus $0 \leq C \leq 15\%$ range is of the greatest interest for the study.

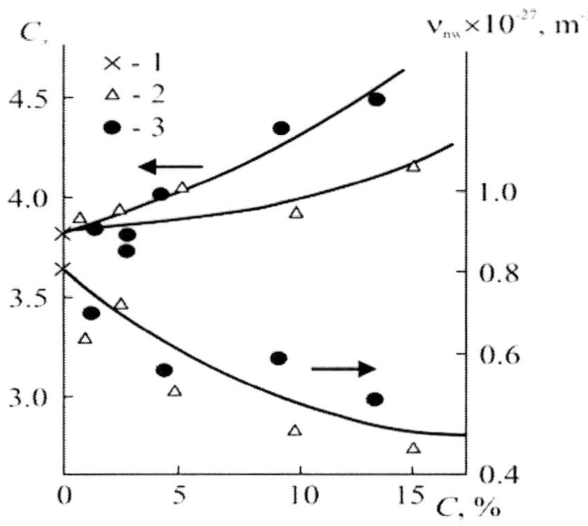

Figure 5. Dependencies of chemical cross-link network frequency ν_{nw} and characteristic relation C_∞ on the modifier concentration C for EP-2 (1), EP-3 (2) and EP-4 (3) [16]

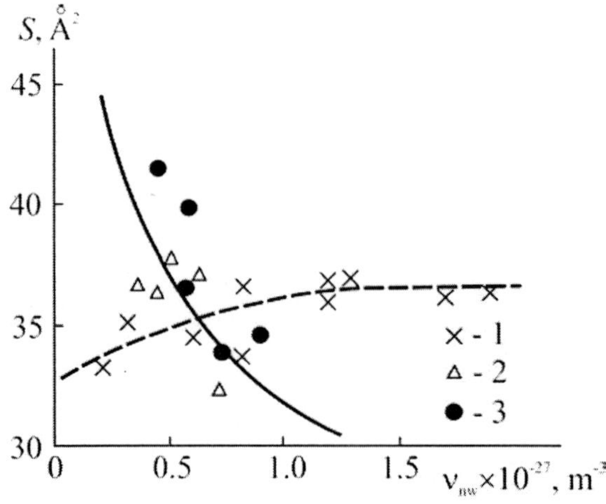

Figure 6. Dependence of macromolecular cross-section S on chemical cross-link network frequency ν_{nw} for EP-2 (1), EP-3 (2) and EP-4 (3) [16]

Figure 6 shows $S(v_{nw})$ dependencies calculated by the equation (1) for EP-3 and EP-4 systems. In contrast with EP-2, they display S decrease with v_{nw} growth. For EP-3 and EP-4, v_{nw} decrease is accompanied by increase of concentration of side groups formed by adamantane fragments (refer to Figure 6), which displays a high influence on S inducing its significant increase.

Dependencies $C_\infty(C)$ for modified EPs, shown in Figure 7, indicate more intensive C_∞ increase for EP-4 rather than for EP-3 due to differences in chemical structures of these systems. In the case of EP-3, adamantane fragment is linked by two bonds to the EP network or partly forms a card polymer structure. For EP-4, card polymer structure with a single bond at the adamantane fragment is the most probable alternative. Thus for macromolecular coil, card elements of EP-4 chain structure are more effective as steric hindrances. They inhibit the macromolecular coil degree similar to EP-2 or EP-3 [16].

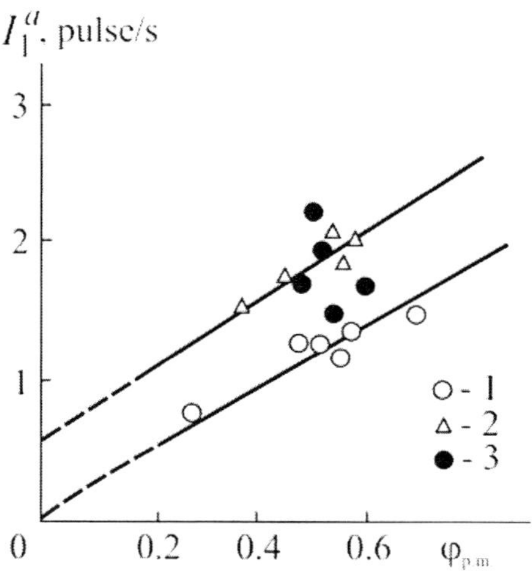

Figure 7. Dependencies of amplitude intensity I_1^a on relative fracture of the packless matrix $\varphi_{p.m.}$ for EP-2 (1), EP-3 (2) and EP-4 (3) [16]

Let us pass to the study of parameters of amorphous halo components (Table 2). Figure 8 shows dependencies of amplitude intensity of the halo first component, I_1^a, on the packless matrix fracture, $\varphi_{p.m.}$. For all EPs, I_1^a increases with $\varphi_{p.m.}$. However, in contrast with $I_1^a(\varphi_{p.m.})$ dependencies for unmodified EP-1 and EP-2, analogous dependence for EP-3 and EP-4 at $\varphi_{p.m.} = 0$ is extrapolated to a non-zero value I_1^a. Obviously the difference is associated with the presence of adamantane fragments in the packless matrix of EP-3 and EP-4, which form the third component of the amorphous halo (Table 2). Clearly for EP-3 and EP-4, extrapolated value I_1^a at $\varphi_{p.m.} = 0$ equals, approximately, 0.6×10^3 pulse/s, and the amplitude intensity of the third component, I_3^a, varies within the range of $(0.14 - 0.75) \times 10^3$ pulse/s (Table 2), i.e. these values are quite close.

Let us now consider the interconnection between parameters of the amorphous halo second component (I_2^a), d_2 and the local order zones. As mentioned above, the presentation of segments composing clusters in the form of linear defects assumes that φ_{cl} increase must induce I_2^a decrease. Actually, such tendency is observed (Figure 8), data for unmodified EP-2 also fitting the general line. The modifier concentration increase induces d_2 growth (Table 2). Cluster formation from segments at greater distance between their axial lines means that the local order zones are formed easier with C increase. That is why a correlation between parameters I_2^a and d_2 should be expected that was observed in the work [16].

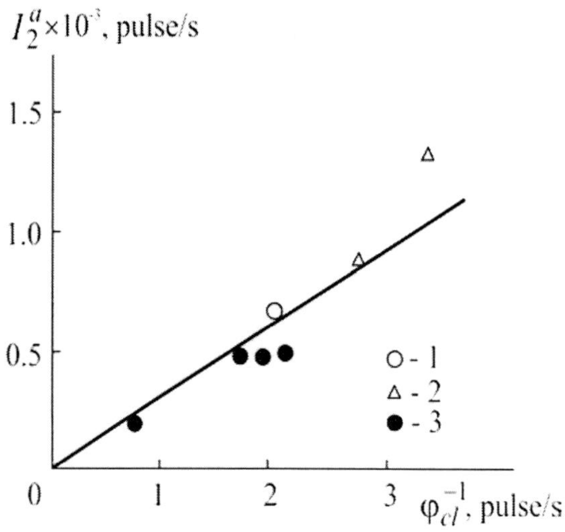

Figure 8. Dependence of the amplitude intensity I_2^a on reverse relative part of clusters φ_{cl} for EP-2 (1), EP-3 (2) and EP-4 (3) [16]

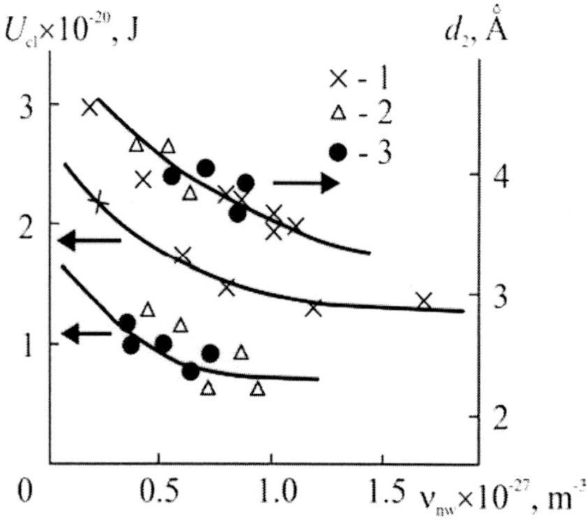

Figure 9. Dependencies of the Bragg interval d_2 and association energy of a couple of segments in the cluster U_{cl} on the chemical cross-link network frequency v_{nw} for EP-2 (1), EP-3 (2) and EP-4 (3) [16]

For modified and unmodified EP, comparison of d and d_2 dependencies on v_{nw} displays a significant difference in their behavior. As the function of v_{nw}, parameter d_2 values fit a single curve for all studied EP [16] (Figure 10). At the same time, in the case of unmodified EP, d is weakly increased with v_{nw} and intensively decreased for modified EP (refer to Figure 6). Probably, though changing the chain structure, adamantane fragments do not directly present in the local order zones, formed by segments of epoxy polymeric network.

Dependencies $U_{cl}(v_{nw})$ for studied EP [16] were found symbate (Figure 9); however, absolute values of U_{cl} for modified epoxy polymers are much lower. This means that formation of the local order zones in EP-3 and EP-4 proceeds much easier. Apparently, strong interactions created by adamantane fragments promote for collinear disposition of EP segments that simplifies formation of the fluctuation cluster network [7].

Dependencies of the entanglement cluster network frequency, v_{cl}, on U_{cl} fit a single curve for EP-2, EP-3 and EP-4, whereas dependence $I_2^a (U_{cl})$ is linear and passes through the origin of coordinates (Figure 10). As a consequence, intensity of the amorphous halo second component characterizes the energetic state of clusters: the higher U_{cl} is, the greater I_2^a is. Increasing U_{cl} reduces probability of the local order formation (Figure 10) that explains I_2^a decrease with φ_{cl} growth (refer to Figures 8).

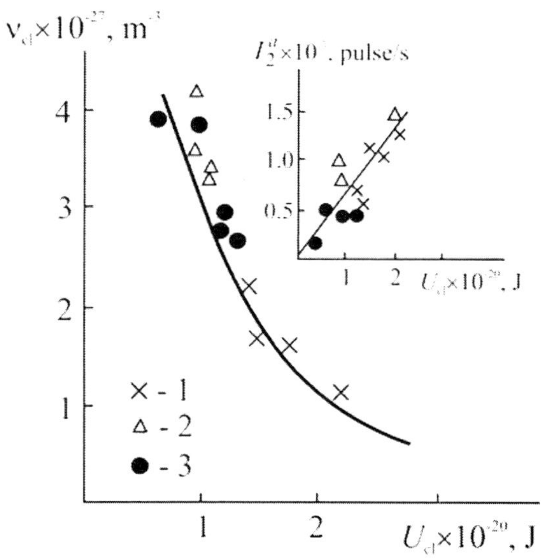

Figure 10. Dependence of entanglement cluster network frequency, v_{cl}, on the association energy of a couple of segments in a cluster, U_{cl}, for EP-2 (1), EP-3 (2) and EP-4 (3). Insertion shows dependence of amplitude intensity I_2^a on U_{cl} [16]

Figure 11 shows dependencies of amplitude intensity and Bragg interval of the amorphous halo third component corresponded to the entanglement network of adamantane fragments on the modifier concentration C. The parameters mentioned vary extremely. Dependence $d_3(C)$ indicates that in the area of small C modifier concentration increase and corresponded v_{nw} decrease (refer to Figure 5) significantly simplify formation of a network from the adamantane fragments: it can be formed at more and much more higher d_3.

However, at $C > 5\%$ parameter d_3 increase is stopped and a slight decrease testifying about the reverse effect is observed [16].

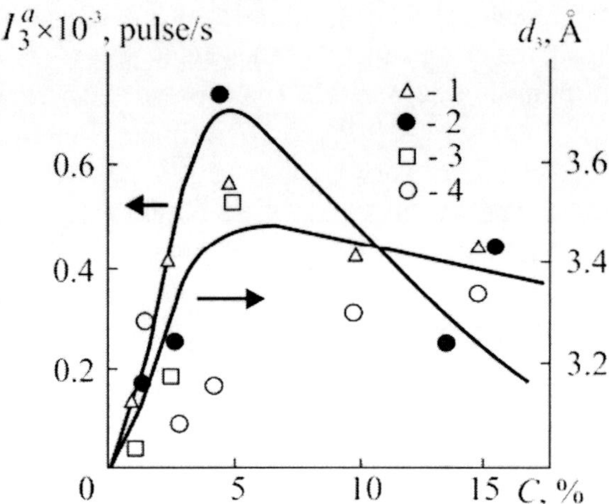

Figure 11. Dependence of amplitude intensity I_3^a (1, 2) and the Bragg interval d_3 (3, 4) on modifier concentration C for EP-3 (1, 3) and EP-4 (2, 4) [16]

Calculation of average I_2^a and I_3^a values for EP-3 and EP-4 displays two-fold exceeding of the former over the latter (Table 2). Figure 10 shows that higher difference will be observed in U_{cl} of the mentioned types of networks. A decrease of U_{cl} for a network formed due to strong interactions of adamantane fragments means the possibility of easier decomposition of clusters composing it. The highest stability of the latter is reached at optimal modifier concentration equal $C \approx 5\%$. As $C < 5\%$, the lack of adamantane fragments impedes formation of stable clusters, and at $C > 5\%$ their excess creates steric hindrances to the mentioned process.

Lower values of U_{cl} for clusters from adamantane fragments are caused by their shorter length, compared with EP clusters (the length of the latter equals l_{st}). This means that the length l_a of adamantane fragments in the cluster is K times smaller than l_{st}, where K is the ratio of appropriate energies U_{cl}. It can be simply calculated from the data in Figure 10. Then taking the mean value of K and calculating the relative part of clusters of adamantane fragments, φ_A, by the known technique [18], the dependence $I_3^a\left(\varphi_A^{-1}\right)$ for EP-3 and EP-4 is obtained. The results are shown in Figure 12, where the dashed line shows the run of $I_3^a\left(\varphi_A^{-1}\right)$ dependence, plotted by the data from Figure 8.

Note also that analogous dependence of "backlashes" [19], presented in coordinated of the Figure 12, will be shaped as a straight line parallel to the abscissa axis, because frequency of backlashes is constant [14, 18]. As a consequence, by energetic parameters, the network of macromolecular entanglements composed from adamantane fragments is the intermediate stage between EP cluster network and the "backlash" one. Two points corresponded to EP-3 ($C = 5\%$) and EP-4 ($C = 4.3\%$) fit the line 1 in Figure 12. As mentioned above, these epoxy polymers possess the most stable clusters of adamantane fragments, characteristics of which are close to EP clusters. The rest points fit the straight line 2 by slope more close to the abscissa axis, i.e. clusters in these epoxy polymers, composed by adamantane fragments, are rather incomplete local order zones approaching the properties of macromolecular backlashes [19].

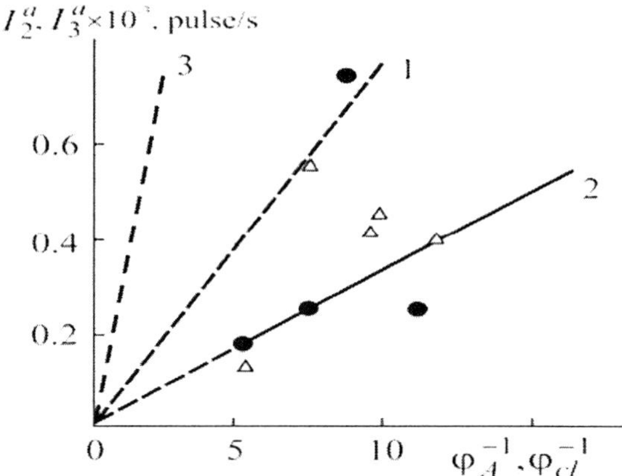

Figure 12. Dependence of amplitude intensities I_3^a (1, 2) and I_2^a (3) on reverse relative fractures of clusters from adamantane fragments φ_A (1, 2) and segments φ_{cl} (3). Refer to Figure 4.12 for the legend [16]

Thus results discussed in this Section indicate that application of the wide-angle X-ray diffraction data involving ideas of the cluster model allow quite thorough analysis of structural organization of complex polymeric systems, for example, epoxy polymers modified by carcass fragments [10, 11, 15, 16].

IR-SPECTROSCOPY

IR-spectroscopy is the well-known and successful method for studying the structure of polymeric materials [20, 21]. The detailed analysis of IR-spectroscopy data within the framework of a model presented in [20 – 24] for HDPE + Z composites. However, this analysis is of the qualitative type, whereas application of the cluster model allows obtaining of quantitative relations of structural parameters and characteristics of IR-spectra of these composites [25]. Further on, such relations can be used for direct assessment of the structure of crystallized polymer non-crystalline zones by IR-spectroscopy methods.

It should be noted that Wedgewood and Seferis [21] have also suggested that the change in relations of extinction coefficients $e_{1,303}/e_{1,352}$ and $e_{1,303}/e_{1,368}$ can be direct consequence of intensification of differences between structures of amorphous and interphase zones of crystallized polymers. Thus the decrease of mentioned relations, B, is stipulated by increase of macromolecular unit concentration in stretched *trans*-conformations and, as a consequence, by increase of the regular order in the interphase zones. Quantitatively, the mentioned degree can be expressed via ν_{cl} [25]. Therefore, there are grounds for expecting a correlation between ν_{cl} and B^{-1}. Figure 13 shows the dependence $\nu_{cl}(B^{-1})$, composed by IR-spectroscopy data, that proves this conclusion and allows the local order assessment for non-crystalline zones of HDPE + Z composites.

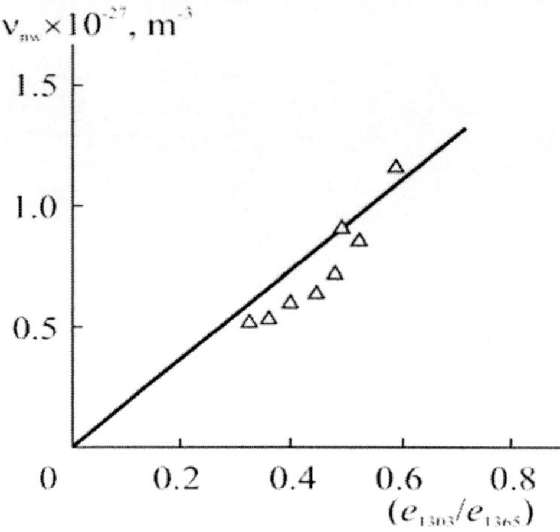

Figure 13. Dependence of entanglement cluster network frequency, ν_{cl}, on relation of extinction coefficients $e_{1,303}/e_{1,368} = B$ for HDPE + Z composites [25]

If the interpretation suggested is correct, it should be expected that decrease of $e_{1,303}/e_{1,352}$ and $e_{1,303}/e_{1,368}$ relations is associated with increase of the relative part of interphase zones, υ_{ip} [26]. Figure 14 shows dependence of υ_{ip} on reverse $e_{1,303}/e_{1,352}$ relation. Quite good linear correlation between these parameters proves the interpretation suggested [25].

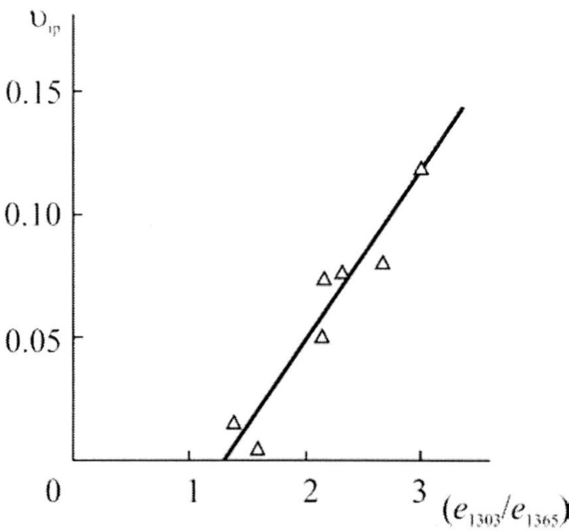

Figure 14. Dependence of interphase zones relative part, υ_{ip}, on the reverse relation of extinction coefficients, $e_{1,303}/e_{1,352}$, for HDPE + Z composites [25]

Figure 14 shows $\nu_{cl}(\upsilon_{ip})$ correlation. It should be noted that for HDPE + Z composite ν_{cl} increase (the ordering degree) with υ_{ip} is observed. As this dependence is extrapolated to $\upsilon_{ip} = 0$, $\nu_{cl} = 0$ is obtained. Thus for HDPE + Z composites with $\upsilon_{ip} \neq 0$, all clusters are concentrated in the interphase zones. However, the initial linear HDPE possesses $\nu_{cl} \neq 0$ at υ_{ip}

= 0. This means that the absence of interphase zones is not equal to the absence of the local order in non-crystalline zones of polyethylenes; therefore, in this case clusters are concentrated in the amorphous phase [25].

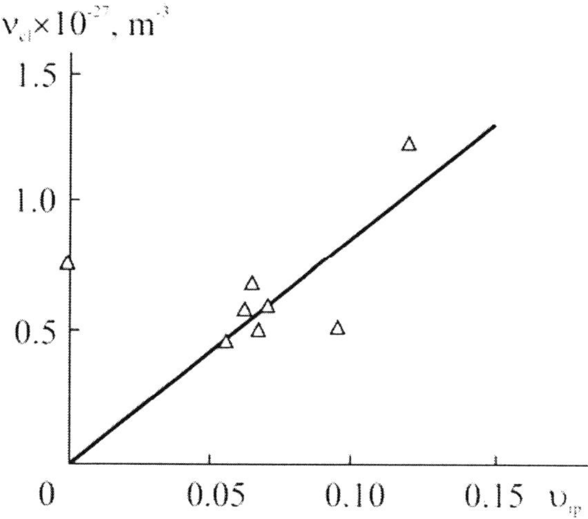

Figure 15. Correlation between entanglement cluster network frequency, v_{cl}, and the relative part of interphase zones, υ_{ip}, for HDPE + Z composites [25]

Three studied IR-bands (1,030, 1,352 and 1,368 cm^{-1}) are induced by oscillations of CH$_2$-groups in a specific sequence which includes several groups, some of which are present in the *gauche*-conformation [20, 21, 27]. For CH$_2$-groups in *trans*- and *gauche*-conformations located in non-crystalline zones, the band at 1,078 cm^{-1} is related to the stretching mode in C-C bond [27]. Changes of macromolecular unit conformations induced by introduction of Z modifier into HDPE can be assessed using the relative extinction coefficients, e^{rel}, which corresponds to the specific mass of non-crystalline component. The value of e^{rel} can be calculated as follows [27]:

$$e^{rel} = \frac{e}{\upsilon_{nc}},\tag{3}$$

where υ_{nc} is the relative part of non-crystalline component, determined using crystallinity degrees, estimated by DCS method [24].

Dependencies of e^{rel} for 1,078, 1,303 and 1,352 cm^{-1} IR-absorption bands were also calculated [23]. For the bands at 1,303 and 1,352 cm^{-1} at Z concentration of 0.05 wt.%, a minimum of e^{rel} is observed, whereas for the band at 1,078 cm^{-1} e^{rel} is independent of Z concentration within the limits of experimental error (Figure 16). For three above-mentioned IR-bands, these dependencies of e^{rel} characterize extreme decrease of the number of CH$_2$-groups in *gauche*-conformation with simultaneous increase of their number in *trans*-conformations. Since all IR-bands used in the analysis are corresponded to non-crystalline zones, one may conclude that for 1,303 and 1,352 cm^{-1} bands at Z concentration of 0.05 wt.%, e^{rel} decrease indicates the increase of tie chain frequency in stretched conformations due to the presence of loops [27]. In the framework of the cluster model, frequency of tie chains, V_{tr}, can be calculated by the following equation [28]:

$$V_{tr} = \frac{2v_{cl}}{F}.$$

(4)

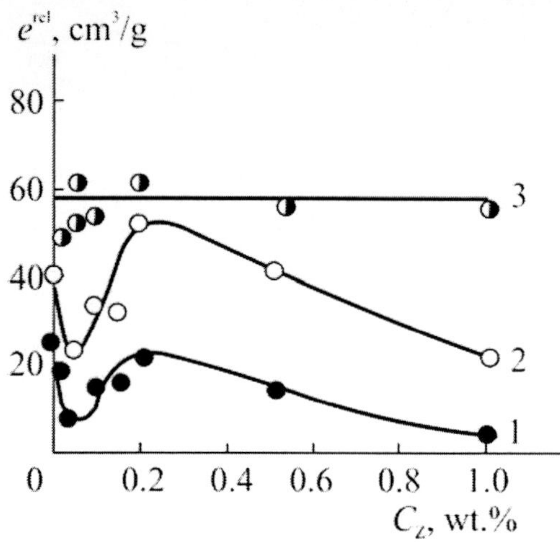

Figure 16. Dependencies of relative extinction coefficient, e^{rel}, on modifier Z concentration, C_Z, in HDPE + Z composites for IR-bands at: 1,303 cm^{-1} (1), 1,352 cm^{-1} (2) and 1,078 cm^{-1} (3) [23]

Figure 17 shows dependence of V_{tr} on reverse e^{rel} of the band at 1,352 cm^{-1} for HDPE + Z composites. Good correlation proving the above assumption was again obtained.

Thus, combined application of the cluster model of the polymer amorphous state structure with IR-spectroscopy data allows obtaining of quantitative information about structure of non-crystalline zones in HDPE + Z composites. Figures 13, 14 and 17 present graphs, which can be used for calibration for this purpose [25].

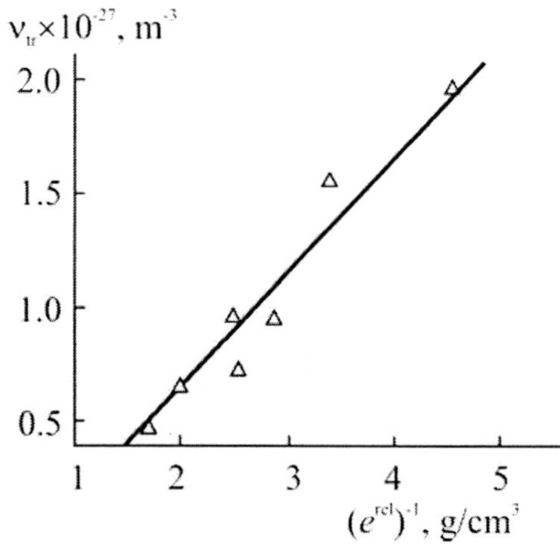

Figure 17. Dependence of tie chain frequency, V_{tr}, on reverse relative extinction coefficient, e^{rel}, of IR-band at 1,352 cm^{-1} for HDPE + Z composites [25]

LOW-FREQUENCY RAMAN SCATTERING

One of the properties of interest, general for all glassy-like systems and absent for the appropriate crystals, is the presence of excessive low-energy density of oscillating states in the acoustic zone of the spectrum. For low energies the spectrum area with energies in the range of $2 - 10$ MeV (≤ 100 cm^{-1}) are accepted [29]. For crystals in the mentioned frequency range, the Debye model of oscillatory actuations is usually applied well. As for amorphous bodies, the situation is different. Besides acoustic oscillations, additional quasi-local oscillatory actuations are present in the low-energy area of the spectrum of these materials. These actuations induce anomalous low-temperature properties of amorphous bodies. Being quite different from the properties of crystals, these low-temperature properties of amorphous bodies, nevertheless, are universal, which is expressed in independence of particular chemical composition and type of the short range order of glasses.

At the present time, data on density of oscillatory states in the low-frequency zone are displayed for a series of metallic glasses, amorphous and amorphous-crystalline polymers, inorganic glasses, and amorphous silicon and germanium [30 – 35].

To the point of view of Malinovsky, Novikov and Sokolov [30], there are grounds for associating excessive low-energy density of oscillatory states in glasses with the presence of typical length, which is the mean (intermediate) order radius of the nanometer size, in them. In the model presented by these authors, low-energy oscillatory actuations responsible for excessive density of oscillatory states are assumed localized on nano-inhomogeneities of the structure.

These observations should be discussed in more detail and compared with characteristics of the cluster model. A model based on the results of neutron and low-frequency Raman scattering was presented [36], in which discrete structure of glasses was suggested. The "boson" peak in the Raman scattering is associated with the excessive density of oscillatory states. As assumed, these two interrelated features represent the result of oscillations, localized in structural units (blobs), which form the glass structure. The formula determining the blob size, $2a$, was deduced as follows [36]:

$$\omega_0 = S_f\left(\frac{\upsilon}{2a}\right), \tag{5}$$

where ω_0 is the "boson" peak frequency; S_f is the blob shape factor (for spherical particles S_f is close to 0.8; for linear objects $S_f = 0.5$); υ is the sonic speed in the blob.

For inorganic As_2S_3 and SiO_2 glasses, the blob size $2a$ was estimated as 21 Å and 22 Å, respectively. These values are in good concordance with the size of structural units approximately equal 20 Å, estimated from width of the first sharp diffraction peak for As_2S_3 [36].

An interesting interpretation of inelastic scattering of neutrons was suggested by Novikov [37]. He has assumed that it is stipulated by the presence of oscillations localized on disclination defects, which are added to usual Debye oscillations. He has shown that the number of "defect" atoms which form disclinations equals $10 - 15\%$ of the total number of atoms. Of special attention is correspondence of this interpretation and the concept of structural defect in the cluster model.

Some ambiguity of the model presented is also accepted [36]. The main and the most frequently put forward objection is the following: the suggested cluster structure of glasses was not yet directly and clearly displayed, for example, by data of electron microscopy of small-angle neutron and X-ray diffraction analysis. Nevertheless, it is possible that the discrete structure observed in oscillation dynamics does not display well-resolved fluctuation density. The reason for existence of blobs is in weaker bonds between blobs than between atoms in them. Moreover, blobs display not a simple shape, but consist of more or less branched chains [36].

The structure of amorphous polymers (PETP and PMA) and tension effect on it were studied with the help of low-frequency Raman scattering [38]. Structural characteristics of these and some other polymers were calculated by equation (5), i.e. it is assumed that similar to [36] the structure of these polymers consists of blobs. As observed, the blob size, $2a$, does not significantly differ from the distance between macromolecular entanglements (backlashes) in appropriate polymeric melts, determined by the plateau modulus [39, 40]. The distance between entanglements, ζ, is given by the following relation [40]:

$$\zeta = N_l^{1/2} C_\infty^{1/2} l_c ,\qquad\qquad (6)$$

where N_l is the average number of bonds l_c long between entanglements.

Quite close conformity between the blob size $2a$ and the distance ζ was obtained [36]. On this basis a conclusion was made that macromolecular "backlash"-specific memory is preserved at the melt – solid polymer transition. In this connection, two remarks should be made. Firstly for unspecified polymer, distances between "backlashes" and clusters are approximately equal. This means correctness of approximate equality of $2a$ and ζ for the entanglement cluster network also. Secondly, $2a$ value for a series of polymers, calculated in the work [38], varies within the range of 40 – 70 Å which conforms to the cluster structure dimensions [41].

Stretching of a polymer causes increase of the longitudinal blob size with stretch [38]. The same effect, namely, R_{cl} increase, is observed in the framework of the cluster model [42].

It is common knowledge that the presence of the first sharp diffraction peak of X-ray and neutron structural factor and the boson peak are frequently assigned to structural correlations of an intermediate scale. However, dimensions of structural units obtained from these experimental data are different, though linear correlation between them (R_c and $2a$, respectively) is observed [43]. It is assumed that $4a$ equals $(3 – 5)R_c$. Note also that structural parameters R_c and $2a$ are identified by different techniques. The value R_c is determined by the width of the first sharp diffraction peak in accordance with the Scherrer formula. It is common knowledge that for crystalline materials this formula gives the size of regular zones (microcrystallites). The value $2a$ is calculated by the boson peak position (equation (5)) and characterizes the structure periodicity [38]. Making comparison with the cluster model, it should be indicated that for the latter R_c is analogous to radius r_{cp} of closely packed local order zone (cluster), which represents the amorphous analogue of microcrystallite with stretched chains [14]. Therefore, $2a$ value corresponds to the distance between clusters, R_{cl}. Figure 19 shows the interrelation between R_{cl} and r_{cp} for five amorphous and five amorphous-crystalline polymers. Clearly for both groups of polymers linear approximations possessing different slopes are obtained. According to the data present [43], for inorganic materials this slope varies from 0.40 to 0.66. The mentioned limits correlate well with the slopes determined from data in Figure 19: 0.31 and 0.70, respectively. Different slopes of linear approximations in Figure 18 are obviously stipulated by different mechanisms of the cluster formation for the mentioned polymer classes.

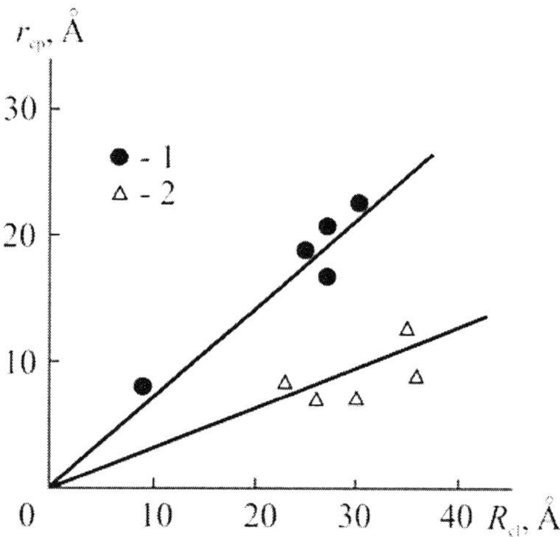

Figure 18. Correlation between the radius r_{cp} of closely packed zone and the distance between clusters R_{cl} for amorphous (1) and amorphous-crystalline (2) polymers [41]

Finally, structural mechanism of elastic deformation for PC was studied by the low-frequency Raman spectroscopy method [45]. It has induced a conclusion that that invariability of the boson peak position at low test temperatures assumes invariability of zones with higher cohesion during elastic deformation. As a consequence, this deformation is realized in the zones with lower cohesion, where molecular reconstruction proceeds. The zones with higher cohesion are oriented parallel to the deformation axis. Since in the framework of the cluster model the zones with higher cohesion are identified with clusters and the ones with lower cohesion with packless matrix, the model suggested [45] completely correlates with description of the structural mechanism of elastic deformation in the framework of the cluster model [46].

For PC, at temperatures close to T_g elastic deformation is also realized by partial orientation of macromolecular rovings in the zones with lower cohesion and general deformation of the zones with higher cohesion [45]. This does also correlate with interpretation of the cluster model, because at temperatures mentioned (above T_g') the packless matrix is thermally devitrified [44, 46].

Thus the data applying ideas of the discrete structure of amorphous materials [30, 34, 36 - 38, 43] completely correlate with the cluster model statements both qualitatively and quantitatively.

REFERENCES

[1]   Miller R.L., Boyer R.F., and Heijboer J., *J. Polymer Sci.: Polymer Phys. Ed.*, 1984, vol. *22*(12), pp. 2021 - 2041.

[2]   Miller R.L. and Boyer R.F., *J. Polymer Sci.: Polymer Phys. Ed.*, 1984, vol. *22*(12), pp. 2043 - 2050.

[3]   Miller R.L., *Proc. 17th Int. Symp. "Order Amorphous State"*, Midland, Mich., August 18-21, 1985, New York – London, 1987, pp. 33 - 49.

[4]   Song H.-H. and Roe R.J., *Macromolecules*, 1987, vol. *20*(11), pp. 2723 - 2732.

[5]   Lin C., Busse L.E., Nagel S.R., and Faber J., *Phys. Rev. B.*, 1984, vol. *29*(9), pp. 5060 - 5062.

[6]   Murthy N.S., Minor H., Bedna'rczyk C., and Krimm S., *Macromolecules*, 1993, vol. *26*(7), pp. 1712 - 1721.

[7]   Kumar S. and Adams W.W., *Polymer*, 1987, vol. *28*(8), pp. 1497 - 1504.

[8]   Murthy N.S. and Minor H., *Polymer*, 1990, vol. *31*(6), pp. 996 - 1002.

[9]   Murthy N.S., Correale S.T., and Minor H., *Macromolecules*, 1991, vol. *24*(5), pp. 1185 - 1189.

[10]   Kozlov G.V., Beloshenko V.A., Kuznetsov E.N., and Lipatov Yu.S., *Ukrainski Khimicheski Zhurnal*, 1997, vol. *63*(7), pp. 52 – 61. (Rus)

[11]   Kozlov G.V., Beloshenko V.A., Kuznetsov E.N., and Lipatov Yu.S., *Doklady NAN Ukrainy*, 1994, No. 12, pp. 126 – 128. (Rus)

[12]   Shogenov V.N., Beloshenko V.A., Kozlov G.V., and Varyukhin V.N., *High Pressure Physics and Technology*, 1999, vol. *9*(3), pp. 30 – 36. (Rus)

[13]   Aharoni S.M., *Macromolecules*, 1983, vol. *16*(9), pp. 1722 - 1728.

[14]   Kozlov G.V. and Novikov V.U., *Uspekhi Fizicheskikh Nauk*, 2001, vol. *171*(7), pp. 717 – 764. (Rus)

[15]   Kozlov G.V., Kuznetsov E.N., Beloshenko V.A., and Lipatov Yu.S., *Doklady NAN Ukrainy*, 1995, No. 11, pp. 102 – 104. (Rus)

[16]   Beloshenko V.A., Kozlov G.V., Kuznetsov E.N., Stroganov I.V., and Stroganov V.F., *High Pressure Physics and Technology*, 1997, vol. *7*(3), pp. 62 – 71. (Rus)

[17]   Beloshenko V.A., Kozlov G.V., Stroganov I.V., and Stroganov V.F., *High Pressure Physics and Technology*, 1994, vol. *4*(3-4), pp. 113 – 117. (Rus)

[18]   Sanditov D.S., Kozlov G.V., Belousov V.N., and Lipatov Yu.S., *Ukrainian Polymer J.*, 1992, vol. *1*(3-4), pp. 241 - 258.

[19]   Wu S., *J. Polymer Sci.: Part B: Polymer Phys.*, 1989, vol. *27*(4), pp. 723 - 741.

[20]   Okada T. and Mandelkern L., *J. Polymer Sci., A-2*, 1967, vol. *5*(2), pp. 239 - 242.

[21]   Wedgewood A.R. and Seferis J.C., *Pure and Appl. Chem.*, 1983, vol. *55*(5), pp. 873 - 892.

[22]   Mashukov N.I., Gladyishev G.P., Kozlov G.V., and Mikitaev A.K., *Thes. Rep. VI All-Union Coord. Meeting of Polymer Spectroscopy*, 1989, Minsk, p. 81. (Rus)

[23]   Mashukov N.I., Serdyuk V.D., Kozlov G.V., Ovcharenko E.N., Gladyishev G.P., and Vodakhov A.B., *Stabilization and Modification of Polyethylene by Oxygen Acceptors* (Preprint), 1990, Moscow, IchF AS USSR, 64 p. (Rus)

[24]   Mashukov N.I., Gladyishev G.P., and Kozlov G.V., *Vysokomol. Soedin.*, 1991, vol. *A33*(12), pp. 2538 – 2546. (Rus)

[25] Malamatov A.Kh., Mashukov N.I., and Kozlov G.V., *Izv. KBNTs RAN*, 1999, No. 3, pp. 65 – 68. (Rus)

[26] Mashukov N.I., Serdyuk V.D., Kozlov G.V., and Khatsukova M.A., *Interphase Zone Structure and Blow Viscosity of High Density Polyethylenes Modified by High-Dispersion Fe/FeO Mixture*, Manuscript deposited to VINITI RAS, Moscow, June 20, 1994, No. 1538-V94. (Rus)

[27] Glenz W. and Peterlin A., *J. Macromol. Sci. Phys.*, 1970, vol. *B4*(3), pp. 473 - 490.

[28] Flory P.J., *Polymer J.*, 1985, vol. *17*(1), pp. 1 - 12.

[29] Kozlov G.V. and Sanditov D.S., *Anharmonic Effects and Physicomechanical Properties of Polymers*, 1994, Novosibirsk, Nauka, 261 p. (Rus)

[30] Malinovsky V.K., Novikov V.N., and Sokolov A.P., *Uspekhi Fizicheskikh Nauk*, 1993, vol. *163*(5), pp. 119 – 124. (Rus)

[31] Bagryansky V.A., Malinovsky V.K., Novikov V.N., Pushchaeva L.M., and Sokolov A.P., *Fizika Tverdogo Tela*, 1988, vol. *30*(8), pp. 2360 – 2366. (Rus)

[32] Zemlyanov M.G., Malinovsky V.K., Novikov V.N., Parshin P.P., and Sokolov A.P., *Zhurnal Eksperimental'noi I Teoreticheskoi Fiziki*, 1992, vol. *101*(1), pp. 284 – 293. (Rus)

[33] Klinger M.I., *Fizika I Khimia Stekla*, 1989, vol. *15*(3), pp. 377 – 396. (Rus)

[34] Karpov V.G. and Parshin D.A., *Pis'ma v ZhETF*, 1983, vol. *38*(11), pp. 536 – 539. (Rus)

[35] Dyadyina G.A., Karpov V.G., Soloviev V.N., and Khrisanov V.A., *Fizika Tverdogo Tela*, 1989, vol. *31*(4), pp. 148 – 154. (Rus)

[36] Duval E., Boukenter A., and Achibat T., *J. Phys.: Condens. Matter*, 1990, vol. *2*(11), pp. 10227 - 10234.

[37] Novikov V.N. In Coll.: *Advanced Solid State Chemistry*, Amsterdam, Elsevier, 1989, p. 349.

[38] Achibat T., Boukenter A., Duval E., Lorentz G., and Etienne S., *J. Chem. Phys.*, 1991, vol. *95*(4), pp. 2949 - 2954.

[39] Graessley W.W. and Edwards S.F., *Polymer*, 1981, vol. *22*(10), pp. 1329 - 1334.

[40] Kavassalis T.A. and Noolandi J., *Macromolecules*, 1988, vol. *21*(9), pp. 2869 - 2879.

[41] Kozlov G.V., Belousov V.N., and Mikitaev A.K., *High Pressure Physics and Technology*, 1998, vol. *8*(1), pp. 101 – 107. (Rus)

[42] Aloev V.Z., Kozlov G.V., and Beloshenko V.A., *Izvestiya VUZov, Severo-Kavkazsky Region, Estestvennye Nauki*, 2001, No. 1(113), pp. 53 – 56. (Rus)

[43] Sokolov A.P., Kisliuk A., Soltwisch M., and Quitmann D., *Phys. Rev. Lett.*, 1992, vol. *69*(10), pp. 1540 - 1543.

[44] Belousov V.N., Kozlov G.V., Mikitaev A.K., and Lipatov Yu.S., *Doklady AN SSSR*, 1990, vol. *313*(3), pp. 630 – 633. (Rus)

[45] Mermet A., Duval E., Etienne S., and G'Sell C., *Polymer*, 1996, vol. *37*(4), pp. 615 - 623.

[46] Kozlov G.V., Beloshenko V.A., Gazaev M.A., and Novikov V.U., *Mekhanika Kompozitnyikh Materialov*, 1996, vol. *32*(2), pp. 270 – 278. (Rus)

In: Physical Organic Chemistry: Theory and Practice
Eds: A. D'Amore and G. E. Zaikov, pp. 231-236

ISBN 1-59454-275-9
© 2005 Nova Science Publishers, Inc.

Chapter 15

ADVANCES IN THE FIELD
OF HIGH-PURITY REAGENTS

A. M. Yaroshenko[*]
State Research Institute of Chemical Reagents and Highli Pure Chemical Substances,
Moscow, Russia
G. E. Zaikov
State Research Institute of Chemical Reagents and Highli Pure Chemical Substances
N.N. Emanuel Institute of Biochemical Physics,RAS, Moscow, Russia

INTRODUCTION

The scientific and engineering activities of the Institute of Chemical Reagents and Special-Purity Substances (IREA) have paralleled the development of novel fields of science and technology for eighty five years [1,2].

The scientific and technological potentials of the IREA grew exponentially until the 1990s. That period involved several stages connected with the progress of metallurgy, nuclear power engineering, semiconductor technology, microelectronics, space technology, optical-glass preparation processes, optical fiber communication, and laser systems.

The rapid development of the IREA was stimulated by the many challenges presented by semiconductor electronics. This important stage culminated by the 1990s in the development of processes for the large-scale production of a wide range of special-purity reagents. In addition, a large number of high-sensitivity analytical procedures were devised. Subsequently, the research activities of the institute declined because of the fall in the demand for high-purity reagents.

At present, research efforts, supported in part by the Government, are concentrated on a number of promising areas.

For example, in the field of fine purification of liquids, attention is focused on the purification of acids and other substances with a higher purity level (on the order of 10^{-8} to 10^{-10} wt % of homogeneous impurities), containing 10-50 cm^3 of microparticles 0.1-0.3 Jim in size. In particular,

---

[*] e-mail: bessarabov@irea.org.ru

new approaches have been devised for the preparation of high-purity mineral (except for HNO_3) and acetic acids surpassing in quality the existing grades. Such acids are used to devise environmentally friendly processes with the use of equipment made from advanced structural fluoropolymer materials [3]. This technology is based on the hydroionic convection effect, which involves a number of phenomena: hydroionic repulsion at the fluoropolymer/acid interface, which gives rise to acid separation into nonaque-ous and aqueous (enriched in ionic impurities) phases; the formation of hydroionic clusters; thermal diffusion of the clusters to the evaporation surface; and ionic entrainment of hydroclusters. The hydroionic convection process is run with the use of special fluoropolymer equipment with a large evaporation surface, at certain temperature and concentration gradients (on the surfaces of the structural material and a flowing thin liquid film) [4].

A novel process was devised for the preparation of high-quality orthophosphoric acid; high-purity meta-, pyro-, and polyphosphoric acids; and transparent salts, including ortho-, pyro-, and polyphosphates, which are of great interest for biotechnological applications.

In those studies, the state of harmful microimpurities and their behavior in fine-purification processes were assessed, and the process conditions were optimized for the recovery of stoichiometric compounds with the preset degree of protonation from solutions. Many of those processes were checked on a production scale.

Significant advances have recently been made in nanotechnology and the preparation of ultrafine, high-purity metallic and oxide powders for advanced lumi-nophors, capacitor and piezoelectric ceramics, and crystal growth of oxides. Considerable effort has been focused on the preparation of organic compounds of germanium, aluminum, silicon, and other elements. Special requirements are imposed on the preparation of substances with a limited content of the main substance [5].

The institute resumed the research concerned with the technology of high-purity rare-earth oxides, fluorides, and phosphates for electronic engineering and optical-glass preparation [6].

New advances have been made in the preparation of special-purity barium metasilicate [7] and calcium, strontium, and barium formates [8]. To this end, multi-component water-salt systems have been investigated systematically at different temperatures. Studies were also directed to develop environmentally friendly processes [9].

In recent years, extensive studies have been devoted to the effect of various atmospheric factors on the characteristics of engineering and structural materials for foundations, roadbeds, and bridges. As a result, chlorine-free materials have been designed which minimize the destructive effects of deicers (e.g., sodium and potassium chlorides) and enhance the strength of carbonate structural materials

One of the most important areas of research is the development of the alkoxide route to high-purity compounds for optics and electronics [10].

In the field of organic chemistry, efforts are focused primarily on the synthesis of new macrocycles and processes for the fractionation of mixtures of di-, tri-, and tetraethyleneglycols and their dichlorides with the aim of extending the range of tailor-made organic reagents. For example, pentaethyleneglycol, a costly reagent, could be prepared via the condensation of triethylene-glycol dichloride and ethyleneglycol in aqueous sodium hydroxide solutions during heating. An efficient process for the purification of pentaethyleneglycol and its dichloride was devised. Conditions were found for the synthesis of benzo-18-crown-6 (reaction between pentaethyleneglycol dichloride and pyrocate-chin) in yields no lower than 50%. The process for the preparation of benzo-18-crown-6 was used to organize small-scale production of various benzo-18-crown-6 derivatives.

A process was developed for the synthesis and purification of 4'-bromine-, 4'-nitro-, and 4'-carboxybenzo-18-crown-6 and functionally substituted derivatives of benzo-18-crown-6. This promising catalyst and selective reagent has not yet been produced in Russia on a commercial scale.

A commercially viable process was devised for the synthesis of the dibenzo-21-crown-7 ether, which is a selective reagent for the extraction of ^{137}Cs from nitrate solutions [11].

Efforts aimed at creating a combined preparation containing microelements in chelate form for agricultural applications culminated in the production of a new, high-efficiency composition, Khelaton-I, in which a phosphor-containing ligand is used as a chelating agent [12].

A continuous process was developed for the first time for the synthesis of a fodder addition: an iron(III) complex of diethylenetriaminopentaacetic acid, dipro-tonated without preliminary isolation of the acid [13].

A unique, high-quality additive was developed—an oxylphosphor-containing chelating agent for the preparation of enhanced-stability hydrogen peroxide [14].

The institute implemented the production of (ami-dinothio)acetic acid—a sulfur-containing organic ligand, which was used to produce a monoethanolamine salt of thioglycol acid [15].

One of the most important areas of research was the development of a new family of organic luminophors with an anomalously large Stokes shift and a process for the preparation of such luminophors. The process was patented in many countries. Such luminophors are successfully used in luminescent flaw detection, criminology, and the protection of bonds and documents [16].

New organic colorless luminophors were created which luminesce throughout the visible range. They were used to produce felt-tip ink which makes it possible to protect documents just after filling.

A daylight-invisible luminescent ink triad for jet printers was developed which allows one to print full-color latent images.

A method was proposed for producing a hidden label in ethyl alcohol for technical applications, which is of considerable practical importance [17].

A greenhouse film containing a red organic lumino-phor has recently been patented.

The institute continues to pay a great deal of attention to the synthesis of chelating agents and related compounds for technological applications. For example, an inhibitor of mineral salt deposition, IOMS-I, has been developed for stabilizing water treatment in water-supply systems of industrial and power engineering plants [18].

In recent years, great advances have been made in the technology of organic solvents for microelectronic and medical applications. In particular, the limit on the water content, specified for each particular solvent, was brought to the minimum possible level: 0.1 wt % in eth-anol, 0.01 wt % in isopropanol, 0.03 wt % in chloroform, etc. The limit on the content of suspended microparticles in high-purity solvents was introduced: 10-50 particles 0.1-0.3 (im in size per milliliter. Our research is conducted in collaboration with electronics and other manufacturers.

Of all the chemical institutes, only the IREA retained a group concerned with the cybernetics of chemical processes and information technology. This group focuses on a new area of research—theoretical foundations of systems analysis and computer simulation in the technology of special-purity substances.

Recently, two new directions have attracted their attention: the development and implementation of advanced information (CALS) technologies (IO-10303 STEP) in the chemical industry [19] and the systems analysis of the dynamics of the innovative potential of chemical and petrochemical institutions between 1990 and 2002 [20]. Both directions have been sponsored by the RF Ministry of Industry and Science and the INTAS (grant no. 971-30770).

The institute always paid much attention to chemical equipment. In recent years, intensive research effort has been concentrated on the design of flexible automatic systems and modular apparatuses from advanced structural materials.

Extensive studies have made it possible to develop the physicochemical principles of the application of low- and high-melting fluoropolymers in environmentally friendly, high-efficiency processes in aggressive liquid media such as halogen, sulfuric, and acetic acids and also in the technology of high-purity water. A quantitative method was proposed for assessing the physicochemical stability in aggressive liquid-fluoropolymer systems with the use of physicochemical relations based on entropy changes in the system [21, 22].

The activities of the IREA in the fields of standardization, quality control, and product certification are closely related to environmental, economical, and other important issues and take advantage of the latest achievements in analytical chemistry.

Particular attention is paid to the methodology of combined processes based on the use of high-efficiency group concentration approaches in conjunction with ICP atomic emission spectroscopy and the use of fluo-ropolymer autoclaves in filtered-air zones.

A novel method was proposed and tested for analyzing small samples (0.05-0.1 ml). The central feature of this method is that the sample is introduced into an excitation source at a reduced rate, which improves the analytical accuracy of the method by about one order of magnitude [23].

To analyze high-purity liquids (organic solvents, water, acids, and etchants), use was made of the distillation concentration of 50- to 100-ml samples in special apparatuses with a concentration coefficient of up to 10^3. This allowed the detection limits for many elements to be reduced to a level of 10^{-8}-10^{-9} wt %.

For analysis of high-purity solid substances (oxides, salts, and composites), a new procedure was devised for stripping samples with the use of bench scale autoclaves of original design [23].

In spite of economic problems, the IREA has retained its status as a state institution and has maintained the high level of its scientific and technological potential.

REFERENCES

[1]   Ryabenko, E.A., On the Occasion of the 75th Anniversary of the Institute of Chemical Reagents and Special-Purity Substances, *Vysokochist. Veshchestva*, 1993, no. 1, pp. 7-18.

[2]   Ryabenko, E.A., 80th Anniversary of the Institute of Chemical Reagents and Special-Purity Substances, *Khim. Prom-st.* (Moscow), 1996, no. 10, pp. 611-619.

[3]   Ryabenko, E.A., Blyum, G.Z., and Yaroshenko, A.M., *Physicochemical Durability of Structural Fluoropolymers in High-Purity Liquids*, Neorg. Mater., 1999, vol.35, no. 11, pp. 1382-1385 [Inorg. Mater. (Engl.Transl.), vol. 35, no. 11, pp. 1183-1185].

[4]   Blum, G.Z., Ryabenko, E.A., Yaroshenko, A.M., et al., Use of Fluoropolymers, Unique Properties for Design ofNew, Highly Effective, Ecologically Friendly Technologies for Fine Purification of Halogen Hydrides and TheirAqueous Solutions, *J. Tribal. Assoc.*, 2002, vol. 8, no. 1, pp. 14-21.

[5]   Efremov, A.A., Grinberg, E.E., and Fedorov, V.A., *High-Purity Organoelement Compounds for Microelectronics,* Vysokochist. Veshchestva, 1988, no. 3, pp. 5-12.

[6]   Shklover, L.P., Vasil'eva, L.V., Chicherina, G.P., et al, USSR Inventor's Certificate no. 1564117, Byull. Izo-bret., 1990, no. 18.

[7]   Fakeev, A.A., Akhtamova, A.Shch., Kon'kova, O.V., et al., *Phase Equilibria in the $Sr(HCOO)_2$-$Mn(HCOO)_2$-H_2O System at 25°C,* Zh. Neorg. Khim., 1997, vol. 42, no. 5, pp. 814-817.

[8]   Fakeev, A.A., Orlova, I.V., Iskhakova, A.G., and Zhadanov, B.V., *Phase Equilibria in the $Ni(HCOO)_2$-$Ca(HCOO)_2$-H_2O System at 25°C,* Zh. Prikl. Khim. (S.-Peterburg), 2000, vol. 73, no. 10, pp. 1581-1584.

[9]   Fakeev, A.A., Orlova, I.V., Iskhakova, A.G., and Zhadanov, B.V., *Phase Equilibria in the $Co(HCOO)_2$-$Ca(HCOO)_2$-H_2O System at 25°C,* Zh. Prikl. Khim.(S.-Peterburg), 1997, vol. 70, no. 10, pp. 1602-1605.

[10]   Belov, E.P., Gerlivanov, V.G., Ryabenko, E.A., et al., RF Patent 2129984, 1999.

[11]   Polosin, V.M. and Ershova, T.N., RF Patent 2178788,2001.

[12]   Shkol'nikova, L.M., Sotman, S.S., Tsirul'nikova, N.V., et al., *Synthesis and Structure of Anhydrous 4-Phospho-nomethyl-2-Oxo-l-Piperazinyl Acetic Acid,* Kristal-lografiya, 1990, vol. 35, no. 5, pp. 1154-1159.

[13]   Rudakova, G.Ya., Tsirul'nikova, N.V., Dyatlova, N.M., et al., USSR Inventor's Certificate no. 1365691, Byull.hobret., 1988, no. 15.

[14]   Tsirul'nikova, N.V., Zhadanov, B.V., Matkovskaya, T.A., et al, *Synthesis and Physicochemical Properties of 1-Carboxymethyl-2-Oxopiperazine-4-Methyl PhosphonicAcid, Koord.* Khim., 1994, vol. 20, no. 7, pp. 557-560.

[15]   Tsirul'nikova, N.V., Danielyan, D.G., and Temkina,V.Ya., *Synthesis of Sulfur-Containing Chelating Agents Basedon Iminothiazolidone, in Khimiya,* kompleksonov i ikhprimenenie (Chemistry and Application of ChelatingAgents), Moscow: Inst. of Chemical Reagents and Special-Purity Substances, 1985, pp. 32-35.

[16]   Bolotin, B.M., Savenkova, E.B., Apanovich, N.A., et al, RF Patent 2139925, 1999.

[17]   Bolotin, B.M., Savenkova, E.B., Apanovich, N.A., et al, RF Patent 2151789, 2000.

[18]   Tsirul'nikova, N.V., Karandeeva, I.V., Zhadanov, B.V., et al, *Structure of Methylphosphorylated Ethylenedi-amine-W, N'-Diacetic Acid,* Org. Khim., 1990, vol. 60,no. 11, pp. 2636-2637.

[19]   Bessarabov, A.M. and Afanas'ev, A.N., *CALS Technology for Designing Promising Chemical Processes,* Khim. Tekhnol, 2002, no. 3, pp. 26-30.

[20]   Bessarabov, A.M., Ryabenko, E.A., Safonova, T.A., andEfimova, V.P., *Innovative Potential of Chemical Scientific Institutions in Russia,* Khim. Rynok, 2002, no. 3, pp. 42-^6.

[21]   Blum, G.Z., Ryabenko, E.A., Yaroshenko, A.M., et al, *Fluoropolymers in Technologies of Hydrogen Halides, Polym.-Plastics Technol Eng.,* 2000, no. 39 (5), pp. 835-843.

[22]   Blum, G.Z., Ryabenko, E.A., Yaroshenko, A.M., et al, *Ecological Aspects of Manufacture and Application of Highly Pure Liquid Substances,* Nova Science, 2002,vol. 2, pp. 53-58.

[23]   Fakeev, A.A., Kurmanguzhina, A.A., and Sukha-novskaya, A.I., *Purification of Calcium Nitrate Solutionsvia Impurity Coprecipitation on Hydrous Alumina*, Zh.Prikl Khim. (S.-Peterburg), 2002, vol. 75, no. 5, pp. 724-729.

In: Physical Organic Chemistry: Theory and Practice
Eds: A. D'Amore and G. E. Zaikov, pp. 237-245

ISBN 1-59454-275-9
© 2005 Nova Science Publishers, Inc.

Chapter 16

MODULAR TECHNOLOGY OF HYDROFLUORIC ACID

G. Z. Blyum,[*] *A. M. Yaroshenko*
Research Institute of Chemical Reagents and Special-Purity Substances, Moscow, Russia
G. E. Zaikov
N.N. Emanuel Institute of Biochemical Physics
Russian Academy of Sciences, Moscow, Russia

ABSTRACT

It is shown that the modular technology of high-purity hydrofluoric acid can be optimized along two lines: replacement of the module for the chemical treatment of the acid by a more efficient one and the introduction of a hydroionic convection module for reducing the concentration of ionic impurities.

Key words: hydrofluoric, purity hydrofluoric acid, chemical technology, azeotropic solution

INTRODUCTION

The modular technology of high-purity hydrofluoric acid is best suited for both automating the production process and improving the quality of the high-purity product.

The existing processes for the commercial production of OSCh 27-5 and OSCh 20-6 hydrofluoric acid, derived from the approaches devised at the Research Institute of Chemical Reagents and Special-Purity Substances in cooperation with the PO Galogen, meet the demands of various application areas. The technology of special-purity hydrofluoric acid, implemented with the use of a self-contained modular plant (Proton type) made of Teflon, makes it possible to improve the quality of the product to the level corresponding, in terms of the main regulated impurities and particles, to the specifications used in Merck, Canto Chemical, and other leading chemical companies [1].

---

[*] e-mail: bessarabov@irea.org.ru

The self-contained modular plant incorporates the following modules and units: module for the chemical treatment of hydrofluoric acid with a potassium permanganate solution (to combine microimpurities of volatile arsenic, phosphorus, sulfur, and other compounds into nonvolatile complexes); unit for separating concentrated hydrofluoric acid into gas (hydrogen fluoride) and liquid (azeotropic solution) flows, containing a module for hydrogen fluoride filtration through a wetted self-cleaning fluoroplastic-fiber membrane; module for rectification of the azeotropic solution; module for the absorption of the purified hydrogen fluoride by the purified azeotropic solution (to give the product with a preset hydrogen fluoride concentration); module for acid microfiltration in order to remove suspended microparticles; module for collecting and mixing the final product; and unit for packing the high-purity product in a transport module (with supply of filtered air).

However, advances in modern technology impose more and more stringent requirements on the quality of high-purity hydrofluoric acid. In connection with this, new approaches to this problem and new optimization methods are needed. Any modern technology must incorporate at least three basic components: theory and practice of physicochemical processes; a scientific approach to design issues, ensuring efficient means of implementing the process; and the optimization of the process with the use of computer methods and statistical data processing.

Accordingly, the existing technology is being updated along the following lines:

(1) design of computer-based systems for optimizing the manufacturing modules and a computer-based automatic control system;

(2) development of a new module for the chemical treatment of hydrofluoric acid with the aim of oxidizing volatile microimpurities to nonvolatile complexes in order to ensure a continuous, closed-loop process;

(3) development of an additional module for the fine purification of the azeotropic solution with the aim of removing ionic impurities;

(4) creation of a unit for packing the product in a transport module.

Consider these points in detail.

(1) Two steps of investigation along this line are planned. To enhance the efficiency of each module in the production process, we use new programs designed to optimize the process. The next step is to create an overall computer-based automatic control system that would coordinate the operation of all the constituent modules.

(2) This direction involves the development of a new module for the chemical treatment of hydrofluoric acid using a novel approach: photochemical reaction between dissolved oxygen and azeotropic clusters of hydrofluoric acid in solution according to the scheme

$$3H_2O.2HF + O_2 + \Sigma h\nu \xrightarrow{-t^\circ,C} 2H_2O.H^+ + 2HF + O_3$$

The energy for this reaction is provided by radiation from a mercury lamp (180-360 nm).

Recall that, in the existing process [2-4], hydrofluoric acid is treated with a potassium permanganate solution. The reaction by-products-volatile compounds of intermediate-valence manganese oxyfluorides---contaminate the hydrofluoric acid being purified. Moreover, the stillage residue contains a large amount of manganese salts. In contrast, the new process for the oxidation of arsenic impurities is based on the above photochemical reaction.

The proposed process involves a homogeneous reaction between the oxygen dissolved in hydrofluoric acid and the water molecules of azeotropic clusters at about -10C. The ozone forming in the HF layer oxidizes arsenic microimpurities by the reaction

$$AsF_3 + O_3 + H_2O \rightarrow AsOF_3.H_2O + O_2,$$

converting arsenic into nonvolatile complexes. Concurrently, other volatile impurities, such as phosphorus and sulfur, are also oxidized.

An important aspect of any technology is optimization of the production process. To this end, a number of problems must be resolved.

To implement the new process, a chemically stable, UV-transparent material is needed. According to our results, among the possible materials are the fluoroplastics F-4, F-4MB, and F-50, which surpass quartz in UV transmission (150--380 nm). Irradiation of HF solutions is however complicated by the strong scattering and absorption of UV radiation by the solution. Therefore, the layer of hydrofluoric acid saturated with oxygen at the optimal temperature must be sufficiently thin. This can be achieved using a purpose-designed apparatus. Two possible designs are represented schematically in Figures 1 and 2. In addition, it is necessary to adjust the total power of the UV source and to optimize the temperature of the overall process, which includes ozone generation and the oxidation of arsenic microimpurities.

Figure 3 shows the temperature dependence of the optimal yield of ozone for the oxidation of arsenic microimpurities. The optimal UV irradiation time depends on the height of the apparatus, the number of recirculation cycles, and the velocity of the liquid flow in the reaction zone.

To determine the time-dependent ozone concentration $C_{O3} = f(\tau)$ @C@O@3(@[tau], we used a differential equation of the form

$$\frac{d\,C_{O3}}{d\,\tau} = b - ac$$

where a is the rate constant for the ozone concentration, and b is the ultimate ozone concentration, which depends on the oxygen solubility:

$$b = f(C_{O2})$$

The solution has the form

$$C_{O3} = b(1 - e^{-a\tau});$$

Figure 4 illustrates a graphical procedure for kinetic optimization of photochemical ozone generation.

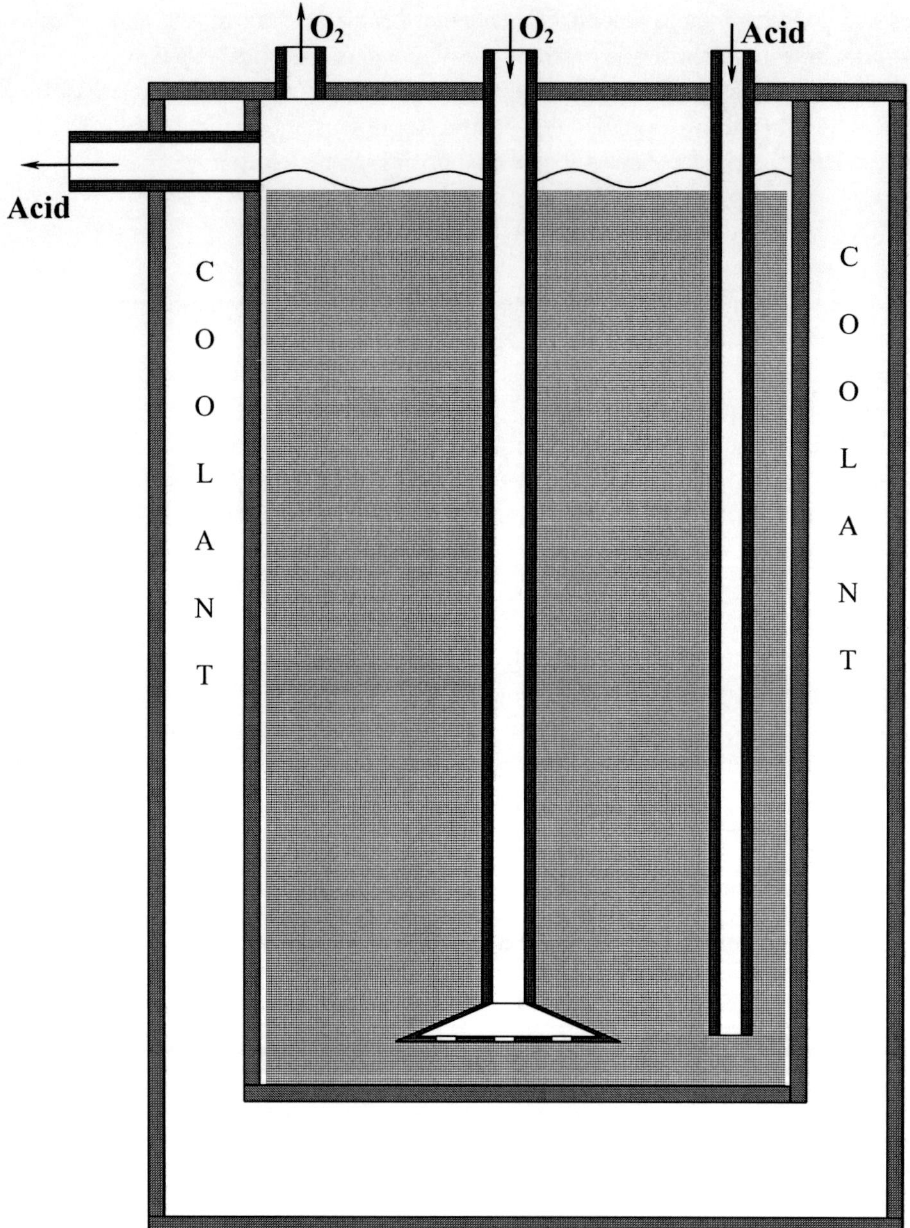

Figure 1. Apparatus for saturating hydrofluoric acid with oxygen gas at the optimal temperature. Key: 1. Acid; 2. Coolant

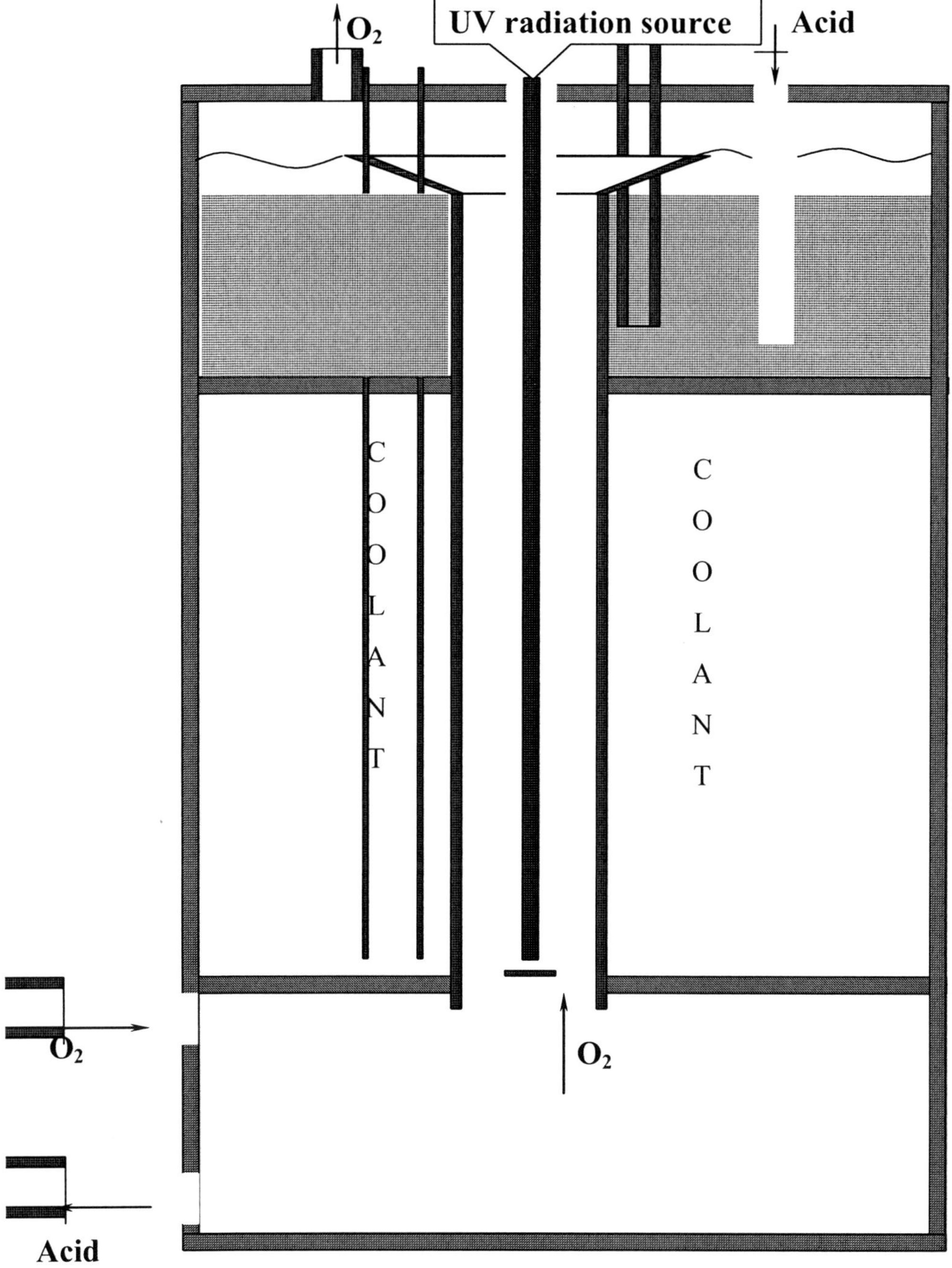

Figure 2. Apparatus for producing ozone via photochemical reaction and subsequent oxidation of arsenic microimpurities. Key: 1. Acid; 2. Coolant; 3. UV radiation source

**Temperature dependence of ozone concentration
in a solution of hydrofluoric acid**

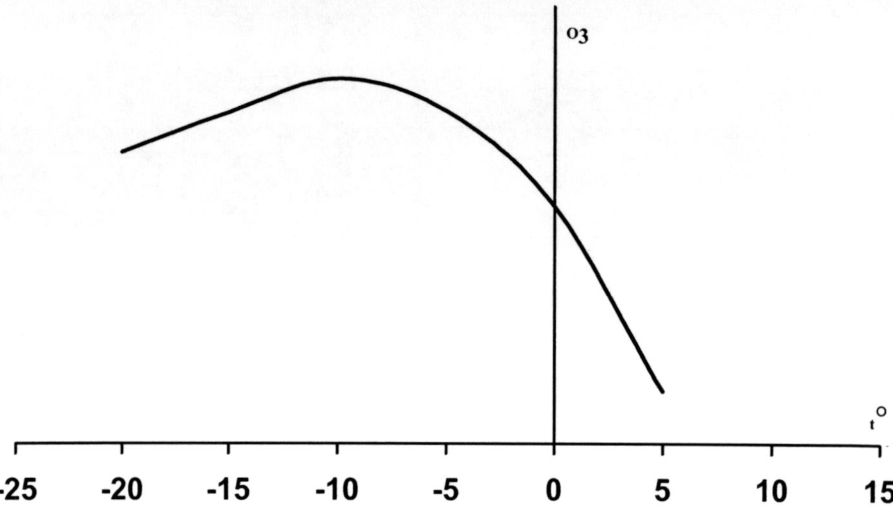

Figure 3. Temperature dependence of ozone concentration in a solution of hydrofluoric acid.

Graphical optimization of photochemical ozone generation

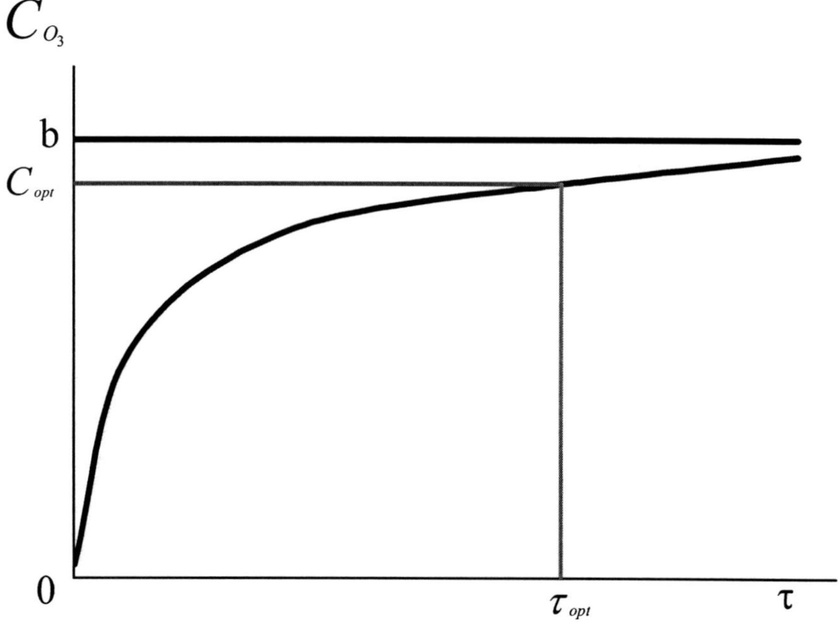

Figure 4. Graphical optimization of photochemical ozone generation. Key: 1. optimal

Temperature optimization of the hydroionic convection process

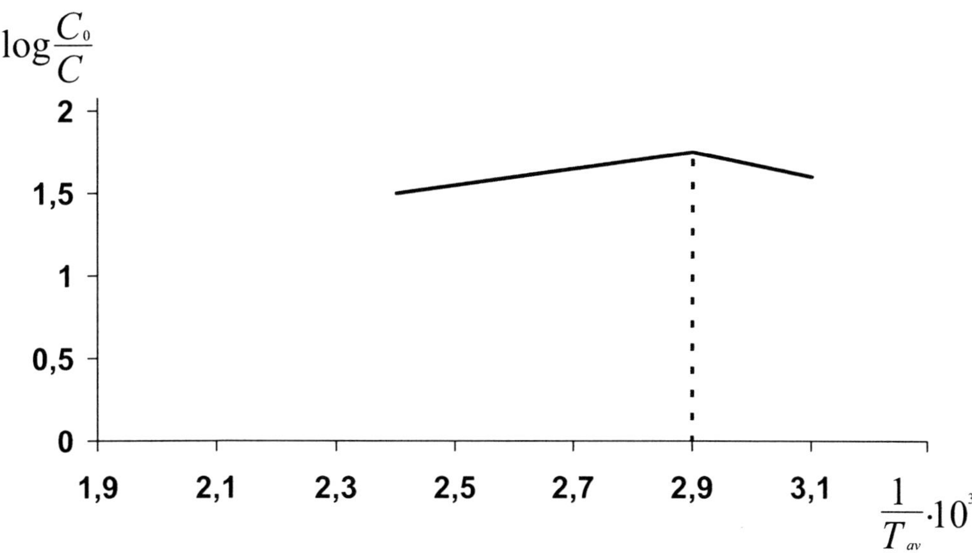

Figure 5. Temperature optimization of the hydroionic convection process. Key: 1. log; 2. av

(3) This direction involves the development of an additional module for the fine purification of the azeotropic solution with the aim of removing ionic impurities. Additional purification of the azeotropic composition of hydrofluoric acid with the aim of removing ionic impurities to a level below 0.1--0.01 ppm is needed because, even during slow boiling or vaporization without boiling, HF vapor is contaminated with hydroionic clusters. In the simplest case, this process is called ionic entrainment [5, 6]. In the solution, microimpurity ions form hydrates (clusters), which are rather stable (bonding energy of about 4190 J/mol), whereas their bonds with other dipoles in the solution are weaker by an order of magnitude (about 419 J/mol).

During vaporization of a liquid layer, hydroclusters of ions are the first to vaporize. The question that arises in this context is how to ensure the delivery of hydroionic clusters to the surface layer of the liquid film flowing down over the inner wall of, e.g., a cylindrical apparatus. To this end, we take advantage of two effects, in addition to ionic entrainment: hydroionic repulsion (resulting from the poor wettability of the fluoropolymer surface by the liquid) and thermodiffusional transport. Thus, the total internal work in the system is

$$\Delta E = \Delta E_1 + \Delta E_2 + \Delta E_3 ,$$

where ΔE_1, ΔE_2, and ΔE_3 are the energies of cluster formation, hydroionic repulsion, and thermodiffusional transport, respectively. The driving force of the last process is the temperature gradient between the inner (adjacent to the apparatus) and outer surfaces of the flowing liquid film (δ).

The vaporization of hydroionic clusters from the surface of an acid solution requires heat supply (Q). This process, called hydroionic convection, involves changes in ion concentrations in the liquid being purified (azeotropic mixture) and can be represented by the exponential

$$C = C_0 \cdot e^{-\frac{\Delta E - Q}{RT}} ,$$

where C_0 is the initial ion concentration in the mixture, and C is the current concentration.

The above equation can be written in the form

$$\ln \frac{C_0}{C} = \frac{\Delta E - Q}{RT}$$

Recall [6] that

$$\Delta E_1 = K_1 \cdot \rho \cdot \xi ,$$

where $\cdot \rho$ [rho] is the charge on the ion, $\cdot \xi$ is the degree of hydration of the ion, and K_1 is the constant of hydrocluster formation;

$$\Delta E_2 = K_2 \cdot \frac{1 + \cos \theta_a}{1 + \cos \theta_w}$$

where θ[theta]a[roman] and θ[theta]w[roman] are the fluoropolymer-acid and fluoropolymer-water contact angles, respectively, and K_2 is the hydroionic repulsion constant; and

$$\Delta E_3 = K_3 \, \Delta t / \delta ,$$

where $\Delta t / \delta$ is the temperature gradient across the liquid film, and K_3 is the thermodiffusional transport constant.

Thus,

$$\ln \frac{C_0}{C} = \frac{1}{RT} \left(K_1 \cdot \rho \cdot \xi + K_2 \cdot \frac{1 + \cos \theta_a}{1 + \cos \theta_w} + K_3 \, \Delta t / \delta \right)$$

Constants K_1, K_2 and K_3 can be assessed from experimental data.

The value of $\ln(C_0/C)$ for the overall process is from 1.5 to 2. This implies that the total decrease in the concentration of ionic impurities attains two orders of magnitude. The total concentration of ionic microimpurities in the purified azeotropic mixture is 10@--2 to 10@--4 ppm.

As can be seen from the above equation, the process in this module can be optimized by adjusting the average temperature t and temperature gradient $\Delta t / \delta$ (Figure 5), and also by selecting an appropriate fluoropolymer (with the required contact angle θ[theta] a[roman]). For hydrofluoric acid, these parameters are $t = 70 \, ^0C$ and $\Delta t / \delta = 30$—$50 \, ^0C$/mm. At a liquid flow velocity of 20-30 cm/s, the height of a vertical apparatus must be about 100 cm. Thus, the residence time of the liquid in the evaporation zone is 3-5 s. At a mean thickness of the liquid layer below 1 mm and a 30-mm diameter of a cylindrical apparatus, the production rate of one element of the module is 2--3 cm3/s, or 30--50 tons per year. The fabrication of a multielement apparatus will make it possible to increase the production rate to the required level.

(4) We made important modifications to the unit for packing the high-purity product in a transport module in the clean zone.

Thus, our results offer significant advances in producing high-purity hydrofluoric acid surpassing the world's best analogs in quality.

REFERENCES

[1] Ryabenko, E.A., Blyum, G.Z., Yaroshenko, A.M., et al., *Principles of Devising Flexible Technologies of Special-Purity Substances on the Basis of Modular Apparatuses,* Vysokochist. Veshchestva, 1990, no. 6, pp. 12-28.

[2] Blyum, G.Z., Vinogradov, G.G., Vasil'eva, Yu.V., et al., *USSR Inventor's Certificate no. 59 172,* 1977.

[3] Ryabenko, E.A., Blyum, G.Z., Vinogradov, G.G., et al., *USSR Inventor's Certificate no. 666 678,* 1979.

[4] Ryabenko, E.A., Blyum, G.Z., and Myagkov, B.I., *USSR Inventor's Certificate no. 731 673,* 1980.

[5] Ryabenko, E.A., Blyum, G.Z., Yaroshenko, A.M., et al., *Environmental Issues in the Production of High-Purity Hydrohalic Acids,* Khim. Prom-st. (Moscow), 1996, no. 10, pp. 671-675.

[6] Blyum, G.Z., Ryabenko, E.A., Yaroshenko, A.M., et al., *The Use of Specific Properties of Fluoroplastics in the Production of Innovative, High-Efficiency, Environmentally Friendly Processes for Fine Purification of Hydrogen Halides and Their Aqueous Solutions, Plast. Massy,* 1999, no. 10, pp. 121-124.

In: Physical Organic Chemistry: Theory and Practice
Eds: A. D'Amore and G. E. Zaikov, pp. 247-260

ISBN 1-59454-275-9
© 2005 Nova Science Publishers, Inc.

Chapter 17

BIOACTIVE ORGANIC–INORGANIC HYBRID MATERIALS SYNTHESIZED BY SOL-GEL METHOD

M. Catauro[1], A. D'Amore[2] and A. Marotta[1]

[1]Department of Materials and Production Engineering, University of Naples Federico II, Piazzale Tecchio, 80125 Naples, Italy

[2]Department of Aerospace and Mechanical Engineering, Second University of Naple-SUNs, Via Roma 21, 81031 Aversa (CE), Italy

ABSTRACT

Novel organic/inorganic, polycaprolactone (PCL)/MO_2 with M = Si or Ti, hybrid materials were synthesized by sol-gel method from multicomponent solutions containing tetramethylortosilicate (TMOS) or titanium butoxide (TBT), polycaprolactone (PCL), water, and methilethylketone (MEK). The structure of interpenetrating network is realized by hydrogen bonds between M-OH (M = Si or Ti) group (H donator) in the sol-gel intermediate species and carboxylic group (H-acceptor) in the repeating units of the polymer. The presence of hydrogen bonds between organic/inorganic components of the hybrid materials was proved by FTIR analysis. The morphology of the hybrid materials was studied by scanning electron microscope (SEM). The structure of a molecular level dispersion has been disclosed by atomic force microscope (AFM), pore size distribution and surface measurements. The bioactivity of the synthesized hybrid materials has been showed by the formation of a layer of hydroxyapatite on the surface of the samples soaked in a fluid simulating the composition of the human blood plasma.

1. INTRODUCTION

The study of organic-inorganic nanocomposites networks has recently become an expanding field of investigation [1, 2]. At first glance, these materials are considered as biphasic materials, where the organic and inorganic phase is mixed at the nm to sub-μm scales. Nevertheless, it is obvious that the properties of these materials are not just the sum of the individual contributions from both phases; the role of the inner interfaces could be predominant. The nature of the interface has been used recently to divide these materials into two distinct classes [3]. In class I, organic and inorganic compounds are embedded and only weak bonds (hydrogen, van der Waals or ionic bonds) give the cohesion to the whole structure. They result from much research work emerging from sol-gel and polymer chemists and these materials will present a large diversity in their structures and final properties. In class II materials, the phases are linked together through strong chemical bond (covalent or ionic-covalent bonds).

The sol-gel process has proved to be versatile and has been widely used in the preparation of organic/inorganic hybrid materials, [4-6] non-linear optical materials [7], and mesomoporous materials [8]. The sol-gel chemistry is based on the hydrolysis and polycondensation of metal alkoxides $M(OR)_x$, where M = Si, Sn, Ti, Al, Mo, V, W, Ce, Zr and so forth. There is considerable interest in organic-inorganic hybrid composite materials prepared via the sol-gel process. A variety of organic polymers have been introduced into inorganic networks to afford the hybrid or composite materials with or without covalent bonds between the polymer and inorganic, components, respectively.

The sol-gel reactions are affected by the value of many parameters such as structure and concentration of the reactants, solvents, and catalysts as well as reaction temperature and rate of removal of by-products and solvents [9, 10]. In particular the presence of organic components modifies the morphology and physical properties of the sol-gel products. For example, the base-catalyzed sol-gel reaction usually results in translucent or opaque products with visible organic-inorganic phase separation. Under acid catalysis and carefully controlled reaction conditions, transparent and monolithic hybrid/composite materials can be obtained.

A key issue that remains unresolved in these organic-modified materials is the degree of mixing of the organic-inorganic components, i.e., the phase homogeneity. The high optical transparency to visible light indicates that the organic-inorganic phase separation, if any, is on a scale of $\leq 400nm$. Many conventional methods for analyzing composite materials have not proved to be effective. For example, the changes in and disappearance of well-defined glass transition of the polymer component as measured by differential scanning calorimeter (DSC) or dynamic mechanical analysis (DMS) suggests the diminution of phase separation but offer little quantitative information [11, 12]. Transmission electron microscopy (TEM) often fails to provide useful morphological data because of weak contrast [11]. Recently, there have been several reports with encouraging examples of applying atomic force microscopy (AFM) in the analysis of sol-gel materials [11, 13].

It is known that SiO_2 and TiO_2 glasses are bioactive, i. e. they are able to bond to living bone [14]. As reported in the literature [14, 15], the essential condition for glasses and glasses-ceramics to bond to living bone is the formation of a bone-like apatite layer on the surfaces. "In vitro" studies are performed [14, 15] by soaking the glasses in a simulated body fluid (SBF) to study hydroxyapatite formation on the surface.

The aim of the present paper is the sol-gel synthesis, characterization and bioactivity of polycaprolactone/SiO_2 and polycaprolactone/TiO_2 hybrid materials. The organic component was chosen taking into account the biodegradable and biocompatible nature of the polycaprolactone (PCL).

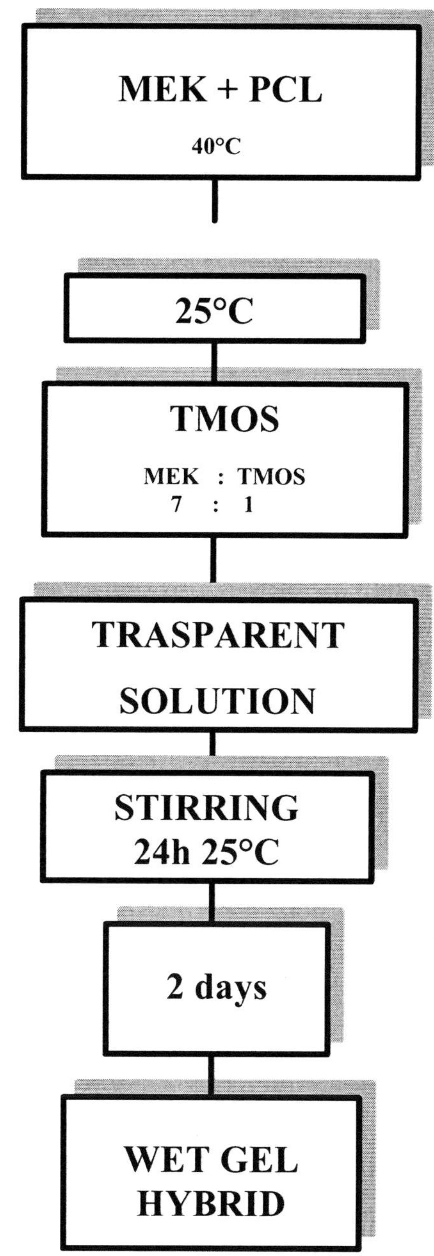

Figure 1. Flow chart of PCL/SiO$_2$ gel synthesis.

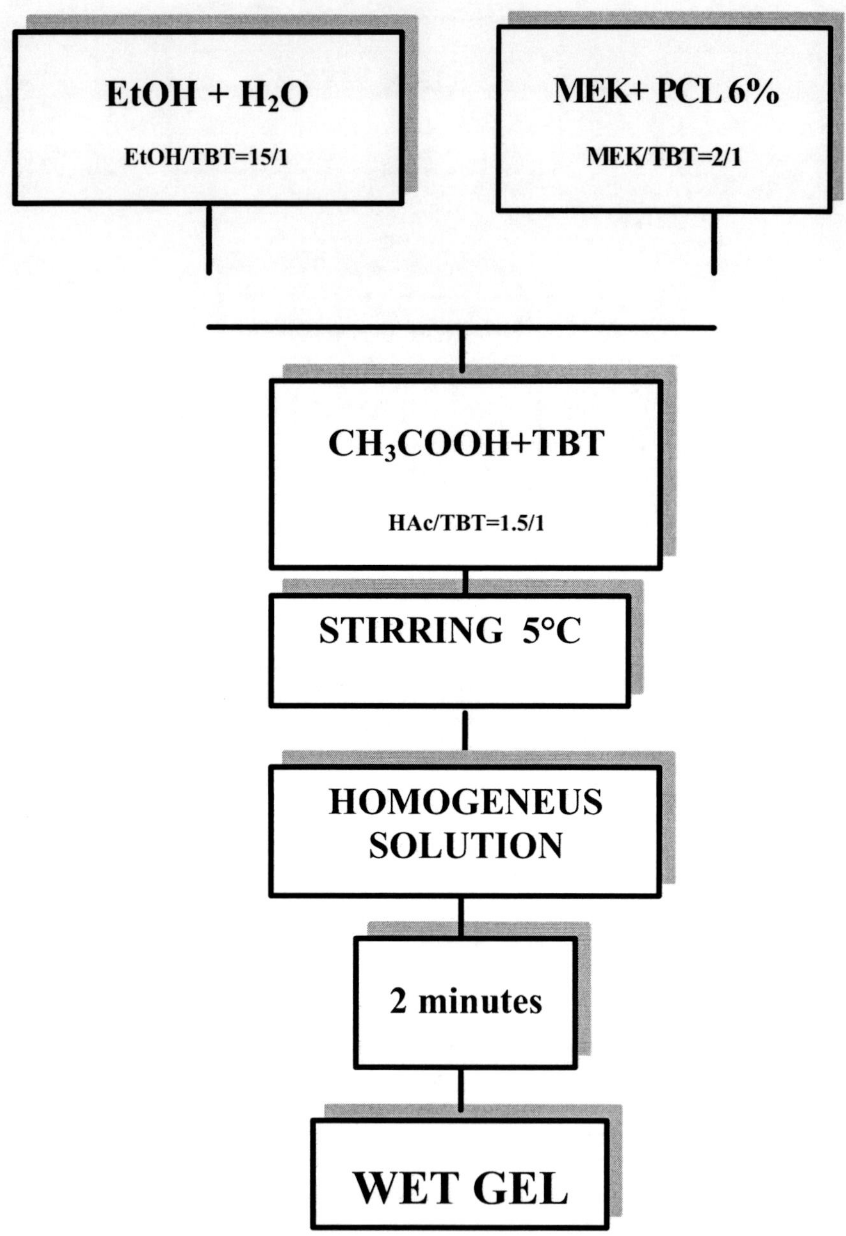

Figure 2. Flow chart of PCL/TiO₂ gel synthesis.

2. EXPERIMENTAL

Organic/inorganic, polycaprolactone (PCL)/MO$_2$ with M = Si or Ti, hybrid materials were prepared by means of sol-gel process from analytical reagent grade of polycaprolactone (PCL) as organic component and tetramethylortosilicate (TMOS) or titanium butoxide (TBT) as inorganic component The amount of the organic component was 6 wt% in both PCL/SiO$_2$

and PCL/TiO$_2$ hybrid materials. Figures 1 and 2 show the flow charts of hybrid synthesis by the sol-gel method. Molar ratio of the starting materials is indicated in the figures.

The presence of hydrogen bonds between organic/inorganic components of the hybrid materials was ascertained by FTIR analysis. Fourier transform infrared (FTIR) transmittance spectra were recorded in the 400-4000 cm^{-1} region using a Mattson 5020 system, equipped with a DTGS KBr (Deuterated Tryglycine Sulphate with potassium bromide windows) detector, with resolution of 2 cm^{-1} (20 scans). KBr palletized disks containing 2 mg of sample and 200 mg KBr were made. The FTIR spectra have been elaborated by Mattson software (FTIR Macros).

The nature of MO$_2$ gel, polycaprolactone (PCL) and PCL/MO$_2$ hybrid materials were ascertained by X-ray diffraction (XRD) analysis using a Philips diffractometer. Powder samples were scanned from $2\Theta = 5°$ to $60°$ using CuK$_\alpha$ radiation.

The microstructure of the synthesized gels has been studied by a scanning electron microscopy (SEM) Cambridge model S-240 on samples previously coated with a tin Au film and by a Digital Instruments Multimode atomic force microscopy (AFM) in contact mode in air.

In order to study their bioactivity, samples of the studied hybrid materials were soaked in a simulated body fluid (SBF) with ion concentrations, as reported elsewhere [15], nearly equal to those of the human blood plasma. During soaking the temperature was kept fixed at 37°C. Taking into account that [16, 17] the radio of the exposed surface to the volume solution influences the reaction, a constant ratio was respected of 50 mm^2 m^{-1} of solution, as in ref. 15.

The ability to form an apatite layer was ascertained by submitting to FTIR analysis PCL /SiO$_2$ reacted samples and to SEM and EDS analysis PCL/TiO$_2$ reacted samples.

An electron microscope (Cambridge Stereo scan 240) equipped with an energy dispersive analytical system (EDS) LINK AN 10000 was used in order to verify the morphology of the coated sample and to make a qualitative elemental analysis.

3. RESULTS AND DISCUSSION

Gelation is the result of hydrolysis and condensation reactions according to the following reactions

$$M(OR)_4 + nH_2O \Rightarrow M(OR)_{4-n} (OH)_n + nROH \tag{1}$$

$$-MOH + RO\text{-}M\text{-} \Rightarrow -M\text{-}O\text{-}M\text{-}+ ROH \tag{2}$$

$$-M\text{-}OH + OH\text{-}M \Rightarrow -M\text{-}O\text{-}M\text{-} + H_2O \tag{3}$$

where M = Si R = CH$_3$ for PCL/SiO$_2$ and M = Ti R = C$_4$H$_9$ for PCL/TiO$_2$

The reaction mechanism is not known in very detail; however, it is generally accepted that they proceed through a second order nucleofic substitution [18].

The reaction 4 shows the formation of hydrogen bond between the carbossilic group of organic polymer and the hydroxyl group of inorganic matrix.

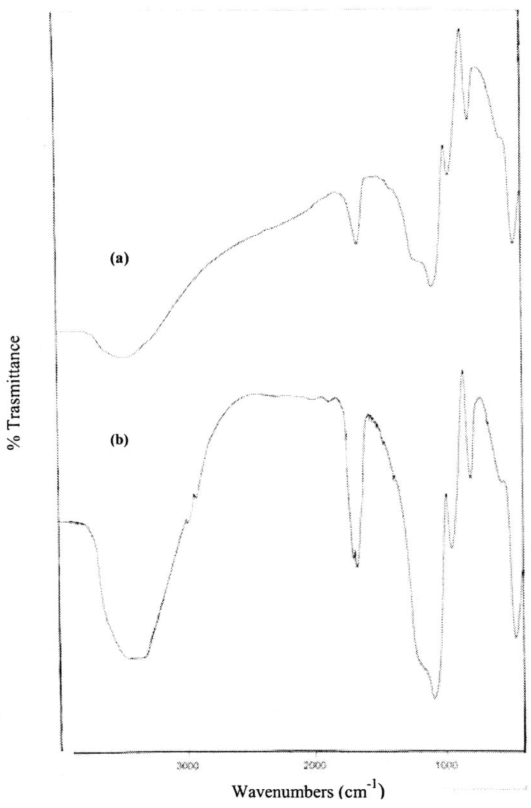

$$\tag{4}$$

The existence of hydrogen bonds was proved by FTIR measurements.

Fig. 3 shows the infrared spectrum of (a) the SiO_2 gel and (b) the PCL/SiO_2 gel. In Fig. 3 (a) the bands at 3400 and 1600 cm^{-1} are attributed to water. The bands at 1080 and 470 cm^{-1} are due to the stretching and bending modes of SiO_2 tetrahedra. In the Fig. 3(b) the bands at 2950 and 2853 cm^{-1} are attributed to the symmetric stretching of $-CH_2-$ of policaprolattone. c) The broad band at 3200 cm^{-1} is the characteristic O-H group of hydrogen bands.

Figure 3. FTIR spectra of a) SiO_2 gel and b) PCL/SiO_2 gel.

Fig. 4 shows the infrared spectrum of (a) the TiO_2 gel and (b) the PCL/TiO_2 gel. In Fig. 4 (a) the absorption peak may be described like the vibration of TiO_6 octahedra with different kinds of ligand. The band around 800 cm^{-1} is suggested to be vibrations of polyhedra TiO_n with coordination number less than six. The doublet around 1500 cm^{-1} corresponds to the bidental acetate ligand linked to the titanium. In the Fig. 4(b) the following additional bands can be detected: a) The band at 1715 cm^{-1} is due to the characteristic carboxylic group shifted to low wave numbers. b) The bands at 2950 and 2853 cm^{-1} are attributed to the symmetric stretching of $-CH_2-$ of policaprolattone. c) The broad band at 3200 cm^{-1} is the characteristic O-H group of hydrogen bands.

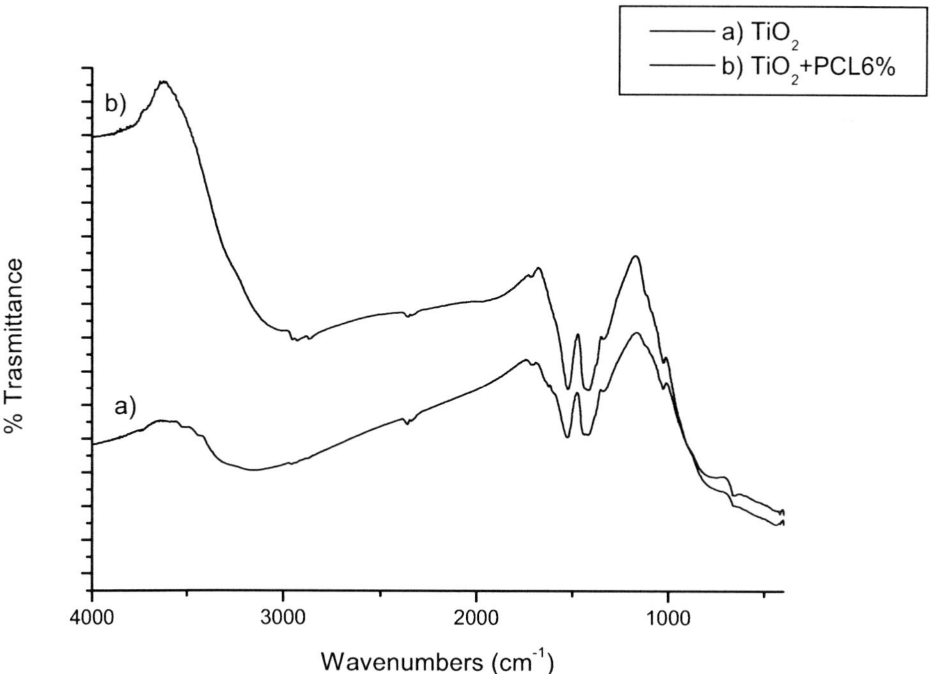

Figure 4. FTIR spectra of a) TiO_2 gel and b) PCL/TiO_2 gel.

The nature and microstructure of the PCL/SiO_2 and PCL/TiO_2 hybrid material have been studied by X-ray diffraction (XRD) and scanning electron microscopy (SEM).

The diffractograms in Fig. 5 show that TiO_2 gel exhibits broad humps characteristic of amorphous materials, fig. 5a, while sharp peaks can be detected on the diffractogram of polycaprolactone typical of a crystalline material, Fig. 5b. On the other hand the XRD spectrum of PCL/TiO_2 hybrid material gel Fig. 5c, exhibits broad humps characteristic of amorphous materials is amorphous as well that of TiO_2 gel.

SEM micrographs of a TiO_2 gel sample and of PCL/TiO_2 gel sample are shown in figures 6 and 7. No appreciable difference between the morphology of the two amorphous materials can be observed. This result confirms the XRD results i. e. the crystalline nature of the organic component was lost during the synthesis. A similar behavior was detected on XRD patterns and SEM micrographs of a SiO_2 gel sample and of PCL/SiO_2 gel sample.

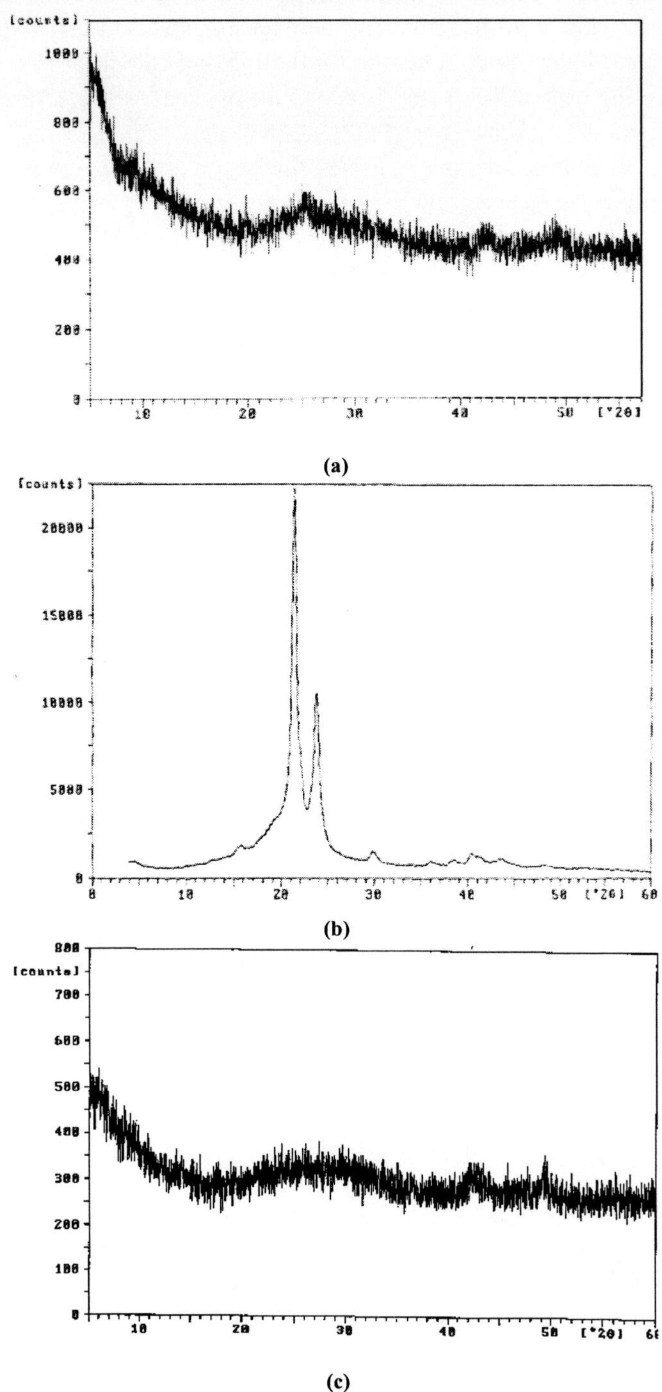

Figure 5. XRD spectra of a) TiO_2 gel, b) PCL and c) PCL/TiO_2 gel.

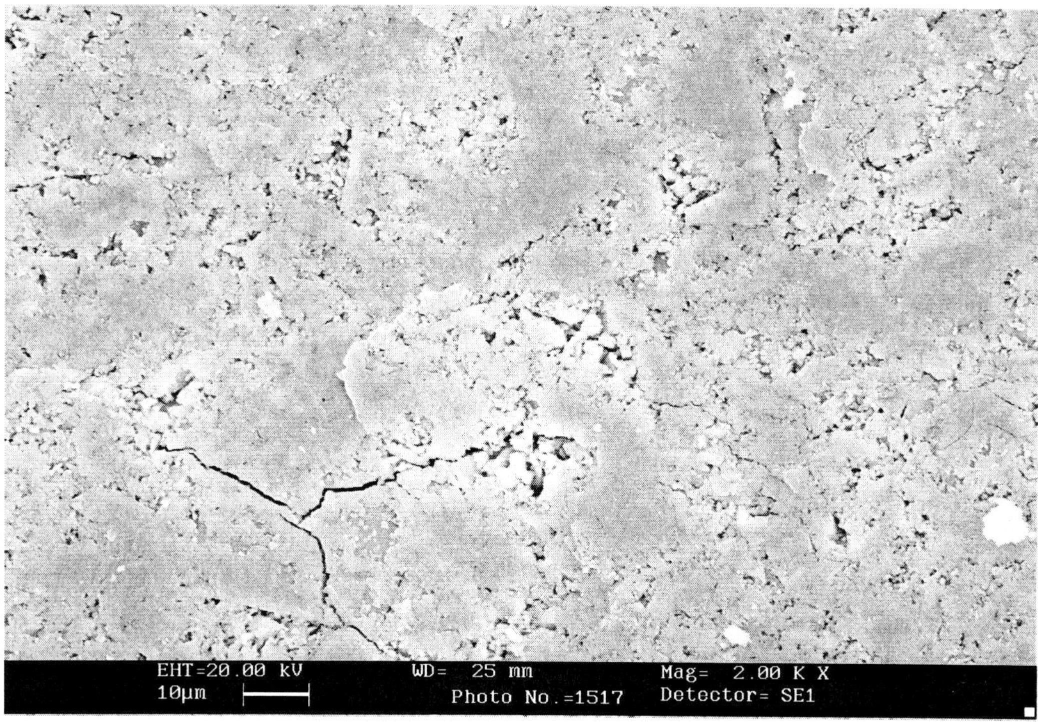

Figure 6. SEM micrograph of TiO$_2$ gel.

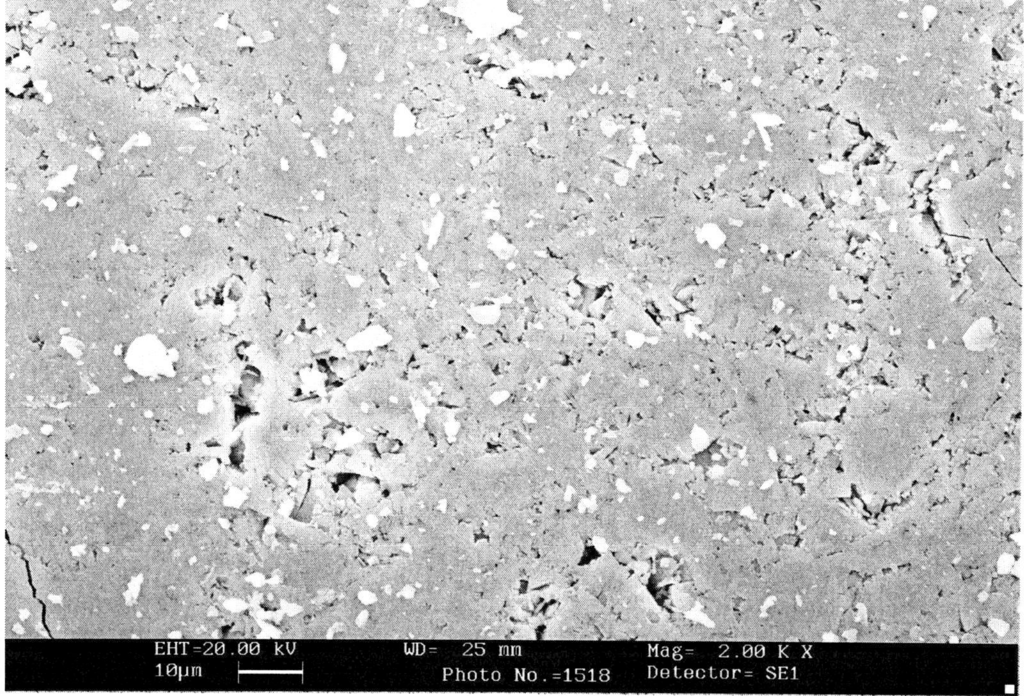

Figure 7. SEM micrograph of PCL/TiO$_2$ gel.

The degree of mixing of the organic-inorganic components, i. e. the phase homogeneity has been ascertained applying the atomic force microscopy (AFM) in the analysis of the hybrid materials. The AFM contact mode image can be measured in the height mode or in the force mode. Force images (z range in nN) have the advantage that they appear sharper and richer in contrast and that the contours of the nanostructure elements are clearer. In contrast, height images (z range in nm) show a more exact reproduction of the height itself [19]. In this work the height mode has been adopted to evaluate the homogeneity degree of the hybrid materials.

The AFM topographic image of PCL/SiO_2 and PCL/TiO_2 gel samples are shown in Fig. 8a and 8b respectively. As can be observed the average domain size is less than 130 and 40 nm. This result confirms that the PCL/SiO_2 and PCL/TiO_2 gels synthesized can be considered organic/inorganic hybrid materials as suggested by literature data [20]

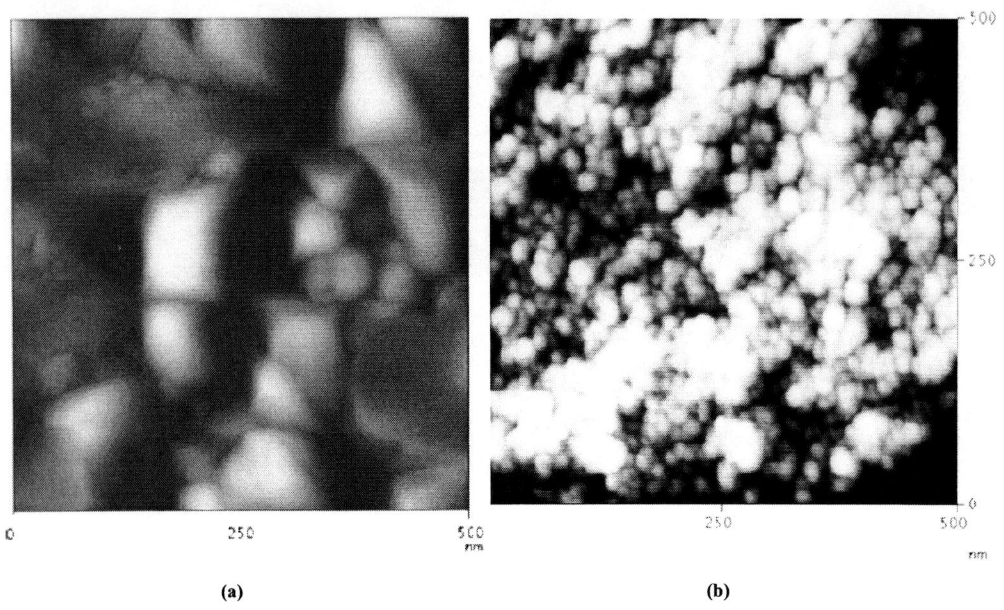

Figure 8. AFM images showing the microstructure of a) PCL/SiO_2 gel and b) PCL/TiO_2 gel.

The hybrid materials were finally soaked in SBF, as indicated in ref. 21, for "in vitro" bioactivity tests.

The bioactivity of PCL/SiO_2 gel was ascertained by FTIR measurements on sample soaked in SBF for 7, 14 and 21 days as shown in Fig. 9. Evidence of formation of an hydroxyapatite layer is given by the appearance of the 1116 and 1035 cm^{-1} bands, usually assigned to P-O stretching [22], and of the 580 cm-1 band usually assigned to the P-O bending mode [22]. The splitting, after already 7 days soaking, of the 580 cm^{-1} band into two others at 610 and at 570 cm^{-1} can be attributed to formation of crystalline hydroxyapatite [21]. Finally the band at 800 cm^{-1} can be assigned to the Si-O-Si band vibration between two adjacent tetrahedra characteristic of silica gel [22]. This supports the hypothesis that a surface layer of silica gel forms as supposed in the mechanism proposed in the literature for hydroxyapatite deposition [23, 24].

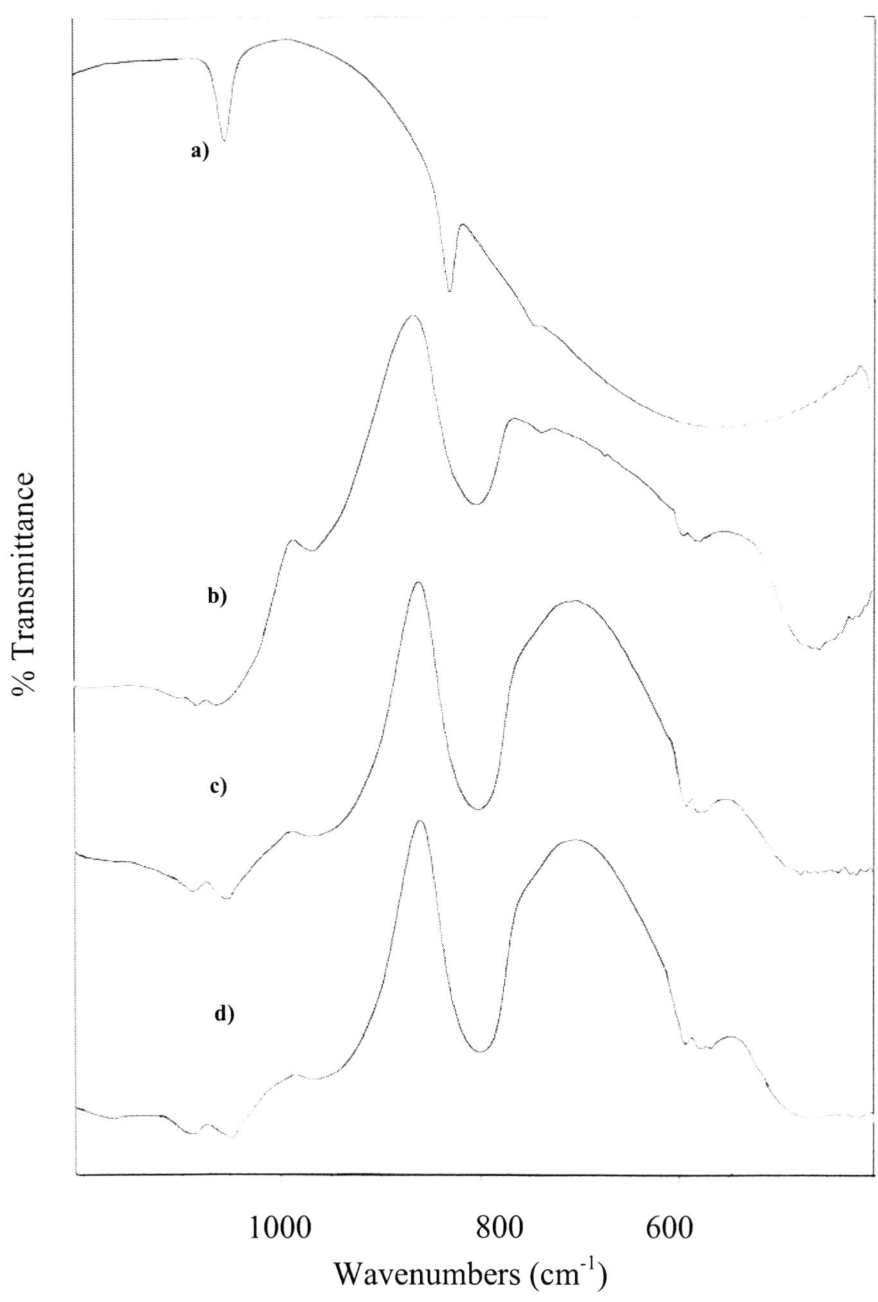

Figure 9. FTIR spectra of PCL/SiO$_2$ gel samples after different times of exposed to SBF: a) not exposed; b) 7 days; c) 14 days; d) 21 days.

Figure 10(a) shows the SEM micrographs of a PCL/TiO$_2$ gel sample soaked in SBF for 21 days. The characteristic apatite globular crystals are clearly visible. As it can be seen, the

EDS reported in Fig. 10(b) confirms that the surface layer observed in the SEM micrographs are composed of calcium phosphate.

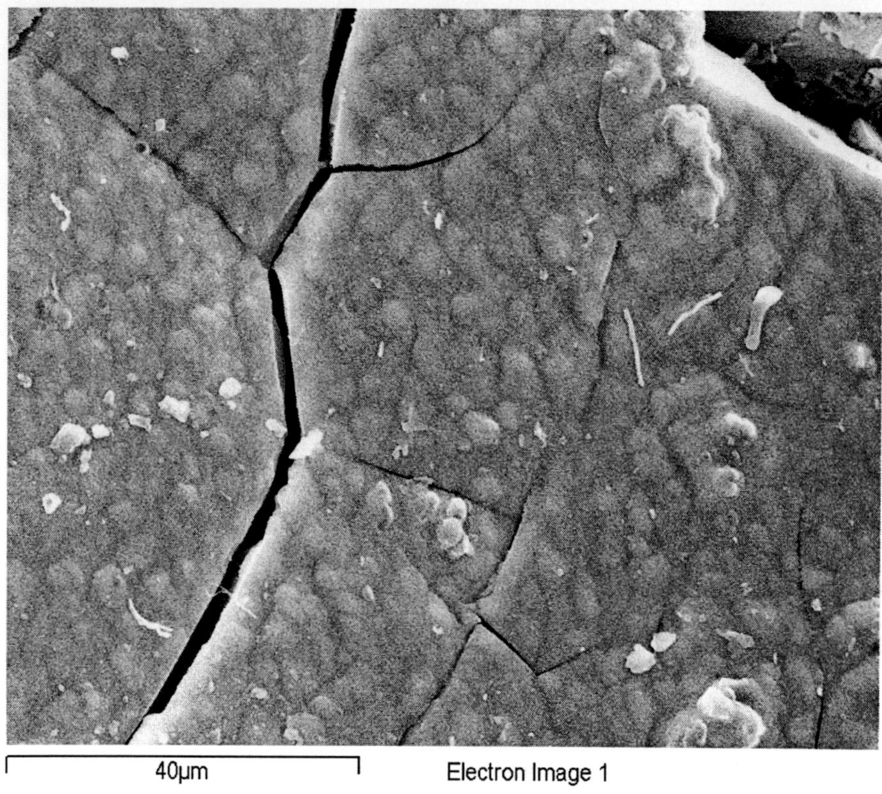

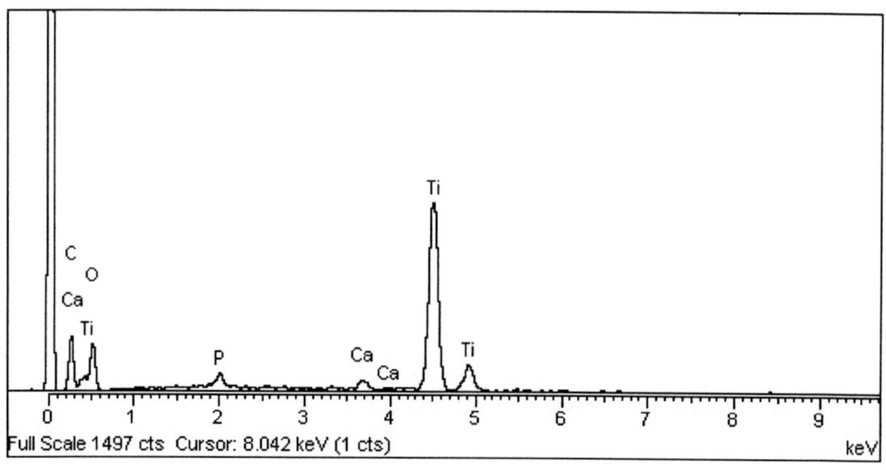

Figure 10. a) SEM micrographs and b) EDS of PCL/TiO$_2$ gel sample soaked in SBF for 21 days.

4. CONCLUSIONS

The polycaprolactone (PCL)/MO$_2$ with M = Si or Ti materials, prepared via sol-gel process, were found to be organic/inorganic bioactive hybrid materials.

The polymer (PCL) was incorporate into network by hydrogen bonds between the carboxylic groups of organic polymer and the hydroxyl groups of inorganic matrix. The formation of hydrogen bonds was ascertained by FTIR measurements.

Moreover the AFM and SEM analysis confirm that PCL/MO$_2$ can be considered homogenous organic/inorganic hybrid materials because the average domains are less than 400 nm in size.

Finally the formation of a layer of hydroxyapatite on the surface when samples of the PCL/MO$_2$ were soaked in SBF (fluid simulating body) showed by SEM micrograph and related EDS or detected on FTIR spectra indicates that the PCL/MO$_2$ can be considered bioactive materials.

Further investigations are required for evaluating the mechanical properties of the synthesized hybrid materials.

REFERENCES

[1]   C. Sanchez and F. Ribot, in *Proceedings of the First European Workshop on Hybrid Organic-inorganic Materials* (Synthesis, Properties, Applications), Bierville-France, 1993.

[2]   *Hybrid Organic-inorganic Materials* edited by L.L Klein and C. Sanchez Special Issue of J. Sol-Gel Sci. Tech. **5** (1995)

[3]   C. Sanchez and F.Ribot, New *.J .Chem.* **18** (1994) 1007.

[4]   P. Judeinstein and C. Sanchez, *J. Mater. Chem.* **6**, 511 (1996)

[5]   L.L Hench and J. K. West, *Chem. Rev.* **90**, 33 (1990).

[6]   B. M. Novak, *Adv. Mater.* **5**, 422 (1993).

[7]   G.H. Hsiue, J. K. Kuo, R. J. Jeng, J.I. Chen, X.L. Jiang, S. Marturunkakul, J.Kumar and S. K. Tripathy, *Chem. Mater.* **6**, 884 (1994)

[8]   Y. Wei, J. Xu, H. Dong, J. H. Dong , K. Y. Qiu and S. A. Janson-Varnum, *Chem. Mater.* **11**, 2023 (1999).

[9]   C.J. Brinker and G.W. Scherer, in *Sol-Gel Science, the Physics and Chemistry of Sol-gel Processing* (Academic Press: San Diego 1989).

[10]   L. C. Klein in *Sol – Gel Technology* (Noyes Publications: Park Ridge N. J. 1988).

[11]   I.A. David and G.W. Scherer, *Chem. Mater.* **7**, 195 (1995).

[12]   J. J. Fitzgerald, C. J. T. Landry and J. M. Pochan, *Macromolecules* **25**, 3715. (1992)

[13]   T. Saegusa, *Macromol. Symp.* **98**, 719 (1995).

[14]   T. Kokubo, Boll. Soc. Esp. Ceram.Vid., *Proceedings of the XVI Int. Cong.on Glass*, Madrid, 1992, **1(31-C1),** 119 (1999).

[15]   C. Ohtsuki, T. Kokubo and T.Yamamuro, *J. Non - Cyst. Solids* **143**, 84 (1992).

[16]   L.L. Hench and D.E. Clark, *J. Non - Cryst. Solids* **28,** 83 (1978).

[17]   A. Paul. *Chemistry of Glasses*, 2nd edition (Chapman and Hall, London, New York 1990) p.184.

[18] C. Sanchez, J. Livage, M. Henry and F. Babonneau, *J. Non - Cryst. Solids* **100**, 650 (1988).

[19] E. Radlein and G.H. Frischot, *J. Non - Crist. Solids*, **222,** 69 (1997).

[20] Y. Wei, J. Danliang, D. Brennan, D. N. Rivera, Q. Zhuang, N. J. Di Nardo and K. Qiu, *Chem. Mater.* **10**, 769 (1998)

[21] D.S. Wang and C.G.Pantano, *J. Non - Cryst. Solids* **147&148,** 115 (1992).

[22] . L.L. Hench, *J. Amer. Soc.* **74**, 1487 (1991).

[23] Y. Kim and A. E. Clark, *J. Non - Cryst. Solids* **113**, 195 (1989).

INDEX

A

absorption, 52, 70, 71, 85, 128, 130, 131, 132, 142, 144, 161, 165, 168, 173, 184, 185, 186, 187, 188, 189, 192, 194, 200, 202, 203, 204, 223, 238, 239, 253

absorption bands, 71, 85, 144, 168, 173, 187, 223

acetate, 13, 112, 113, 115, 253

acetic acid, 36, 128, 232, 233, 234

acetone, 36

acidity, 20, 39, 40, 59, 66, 67, 68

acrylic acid, 112, 117

activation energy(ies), 31, 102, 147, 148, 156, 157, 170, 173, 174, 175, 176, 185

active centers, 11, 62, 68

active radicals, 149

adamantane, 214, 215, 216, 217, 219, 220, 221

adamantane fragments, 214, 215, 216, 217, 219, 220, 221

additives, 21, 38, 71, 72, 94, 193, 202

adhesion, 2, 23, 115

adsorption, 73, 127

aging, 153, 154, 155, 156, 157, 158, 191, 193, 194, 198, 199, 208

alcohol(s), 14, 36, 42, 82, 84, 88, 91, 156, 157, 186, 188

aldehydes, 186, 188, 189

aliphatic compounds, 178

aliphatic polymers, 183

alkane, 190

alkylation, 21, 23, 39, 40, 46, 52, 53, 54, 56

aluminum (Al), 17, 20, 26, 53, 67, 76, 77, 78, 79, 144, 153, 158, 232, 248

aluminum oxide, 144

amines, 121, 126

amorphous phases, 31

amorphous polymer(s), 211, 214, 226

amorphous-crystalline polymers, 212, 225, 227

antioxidant(s), 52, 126, 201, 202, 205

argon, 171, 180, 200

aromatic compounds, 39, 40

aromatic hydrocarbons, 70

aromatic polyimide(s), 176, 177, 178, 186, 192, 206

aromatic rings, 148, 183

Arrhenius dependence, 185

Arrhenius equation, 100

ash, 36

asymmetry, 214

atomic force microscopy, 248, 251, 256

autocatalysis, 100

autooxidation, 195

autooxidation of polyolefins, 195

B

bacteria, 85, 88

bactericides, 81

bending, 34, 187, 189, 252, 256

benzene, 15, 32, 40, 88, 174

binary blends, 1, 2, 6

birefringence, 2

bisphenol, 1, 2, 4, 84, 91

blends, 1, 2, 3, 4, 5, 6, 7, 111, 120

block-copolymer, 17, 20, 23, 25, 38, 70

blocks, 11, 12, 14, 23, 71, 197

blood, 247, 251

blood plasma, 247, 251

branching, 169, 194

butadiene, 13, 15, 19, 20, 21, 24

C

calcium, 232, 258

calibration, 73, 74, 75, 224

calorimetry, 142

capillary viscometry, 112

carbon, 36, 68, 97, 127, 128, 129, 130, 131, 133, 134, 135, 136, 151, 170, 172, 174, 176, 177, 180, 184, 185, 186, 188, 202

carbon atoms, 180

carbon tetrachloride, 36

carbonization, 107, 144

carbonyl groups, 187

carboxylic acids, 214

carboxylic groups, 169, 189, 259

catalysts, 18, 20, 40, 53, 54, 62, 66, 248

cation, 12, 13, 19, 52, 53, 70

C-C, 173, 174, 175, 178, 199, 223

cellulose, 112

CH_2, 31, 37, 51, 70, 71, 172, 175, 186, 189, 196, 198, 199, 223, 252, 253

CH_2-groups, 31, 186, 196, 198, 199, 223

CH_3COOH, 37

chain branching, 181

chain propagation, 9, 12, 19, 20, 21, 39, 68

chain rigidity, 214

chain transfer, 173, 175, 197

characteristic viscosity, 72

charring, 93, 94, 98, 107

chemical bonds, 207

chemical cross-links, 216

chemical interaction, 119, 146

chemical properties, 37, 132, 209

chemical reactions, 38, 202

chemical stability, 23

chemical structures, 172, 194, 217

chlorinated hydrocarbons, 36

chlorination, 38, 56, 144

chlorine, 20, 144, 146, 147, 149, 164, 232

chloroform, 1, 4, 6, 56, 73, 84, 200, 233

chromatograms, 74, 75

chromatographic analysis, 82

chromatography, 46, 73, 74

cluster model, 211, 212, 214, 215, 221, 224, 225, 226, 227

cluster network, 214, 219, 220, 222, 223, 226

CO_2, 38, 172, 176, 177, 184, 185, 186, 188, 239

coke, 147, 148, 150, 173

coke formation, 150

combustibility, 149

combustion, 107, 141, 142, 148, 149, 150, 151, 158

combustion decelerators, 149

compatibility, 1, 2, 4, 6, 7, 126

composite(s), 12, 25, 94, 113, 120, 128, 129, 130, 131, 132, 133, 134, 135, 139, 141, 142, 143, 145, 146, 147, 148, 149, 151, 153, 168, 169, 221, 222, 223, 224, 234, 248

compound I, 82, 85, 88

condensation, 23, 45, 46, 72, 81, 82, 85, 88, 91, 135, 169, 232, 251

conducting polymer composites, 127, 130

conductivity, 33, 128, 129, 133, 142, 150

constant load, 133, 134

constant rate, 70

conversion, 10, 54, 55

copolymerization, 9, 12, 13, 14, 15, 17, 18, 19, 20, 21, 23, 25

copolymerization reaction, 9

copolymers, 1, 4, 7, 9, 11, 12, 14, 18, 19, 20, 21, 22, 23, 25, 68, 71

copper, 33, 143, 205

corrosion, 127

CPC, 1, 2, 3, 4, 5, 6, 7, 127, 128, 131

cross links, 124

cross-linked polymer(s), 88, 148

crosslinking, 94, 202, 205

cross-linking, 37, 38, 144, 148, 153, 212, 213, 214

cross-linking of macromolecules, 148

cross-linking reaction, 153

crystallinity, 4, 35, 154, 164, 169, 223

crystallization, 154, 181

curing, 212, 213

cyclohexanone, 33

cyclopentadiene, 21, 24

D

damage, 112, 188

decomposition, 35, 37, 60, 68, 69, 70, 71, 107, 172, 176, 220

defects, 225

deformation, 71, 90, 113, 114, 115, 117, 133, 134, 136, 187, 227

degradation, 37, 38, 39, 40, 56, 57, 59, 60, 61, 62, 63, 64, 65, 66, 68, 69, 71, 93, 94, 98, 99, 107, 110, 128, 147, 148, 149, 154, 156, 168, 169, 170, 171, 172, 173, 176, 177, 178, 179, 180, 181, 182, 185, 186, 188, 190, 191, 193, 194, 195, 196, 199, 202, 205, 206

degradation process, 155, 156, 202, 205

degree of crystallinity, 154

dehydrochlorination, 148, 153, 154

density, 20, 31, 112, 129, 150, 154, 225, 226

depolyalkylation, 39, 40, 52

depolymerization, 37, 39, 40, 52, 57, 59, 60, 61, 62, 63, 64, 66, 68, 175, 181

desorption, 19

destruction, 121, 122, 124, 138, 192, 199, 202

destruction processes, 121, 122

diamines, 167

dichloroethane, 15, 56
dichlorsilanes, 81, 91
dielectric permeability, 132
differential scanning, 1, 4, 6, 248
differential scanning calorimeter (DSC), 1, 4, 6, 169, 248
diffusion, 33, 93, 94, 98, 99, 100, 106, 107, 155, 183, 184, 232
diffusion process, 107
dimerization, 53
dimethylformamide, 14
dimethylsulfoxide, 14
dispersity, 131, 132
dissolved oxygen, 238
distillation, 82, 234
DTA curve(s), 70, 180, 182, 183

E

efficiency, 52, 53, 59, 142, 143, 148, 149, 201, 202, 204, 233, 234, 238
elastic deformation, 227
elasticity, 31, 135
elastomers, 31, 128, 132
electrical resistance, 127, 128, 130, 133, 138, 191
electron microscopy, 97, 191, 226, 248
electrophilic catalysts, 39, 40, 59, 60, 66
elongation, 33, 60, 154, 155, 156, 191
energetic parameters, 220
energy, 2, 20, 31, 33, 37, 68, 102, 121, 130, 131, 132, 142, 143, 144, 145, 148, 150, 151, 155, 174, 175, 176, 178, 218, 219, 225, 238, 243, 251
engineering, 25, 133, 168, 184, 231, 232, 233
entanglement network, 219
entanglements, 216, 226
enthalpy, 151
entropy, 36, 234
environment, 156, 177, 197
EP, 212, 213, 214, 215, 216, 217, 218, 219, 220
EP-1, 212, 213, 214, 217
EP-2, 212, 213, 214, 215, 216, 217, 218, 219
EP-3, 214, 215, 216, 217, 218, 219, 220
epoxy polymer(s), 212, 214, 219, 220, 221
ESR, 37, 71, 121, 124, 126, 173
ESR spectra, 121
ester, 14, 17, 18, 19, 33, 84, 177, 178
estimating, 142, 153, 160
ethanol, 81, 82, 172
etherification, 43, 44, 81, 82, 91
ethyl alcohol, 82, 84, 233
ethylene, 16, 24, 25, 71, 84, 112, 170, 172, 175, 176
evaporation, 4, 88, 232, 244
extrusion, 167, 169, 182

F

F-4MB, 239
fast Fourier transform infrared (FTIR), 187, 247, 251, 252, 253, 256, 257, 259
fibers, 161, 167
filler(s), 111, 112, 113, 114, 115, 116, 117, 118, 119, 127, 128, 129, 130, 131, 132, 133, 134, 135, 136, 138, 142, 143, 144, 145, 146, 147, 148, 154, 155, 158
filler particles, 130, 131, 136, 138
films, 4, 23, 88, 167
filtration, 82, 238
fire hazard, 36
fire retardants, 107
flame, 36, 107, 147, 148, 149, 150, 151
flammability, 94, 107
fluoroalkylmethacrylates, 164
fluoroplastics, 239
fluoropolymers, 234
food industry, 36
formaldehyde, 184, 185, 186
France, vii, 259
free energy, 32
free radicals, 20, 37, 57, 61, 122, 124
free rotation, 32
free volume, 36, 128
functionalization, 72

G

gel formation, 170, 193
gel permeation chromatography (GPC), 4, 5
gel-fraction, 122, 124, 169, 170, 182, 194
glass transition, 2, 4, 34, 164, 213, 248
glycol, 24, 25, 232
grafted copolymers, 21, 23, 57
graphite, 128, 133
gravimetric analysis, 82, 88

H

halogen(s), 21, 22, 144, 147, 148, 149, 234
hardener, 128
hardness, 18, 135, 167
heat capacity, 33, 150
heat conductivity, 142, 150, 151
heat release, 145, 181, 182
heat shield, 141, 142, 144, 145, 146, 157, 158
heat transfer, 151
heptane, 16, 19, 33, 40, 70, 178
heterocycle, 172, 173, 177, 181, 185, 190

heterogeneity, 12, 17
hexane, 14, 18, 41
high density polyethylene (HDPE), 178, 221, 222, 223, 224
high-molecular compounds, 44, 46
high-temperature oxidation, 184, 194, 195, 208
homogeneity, 131, 248, 256
homogeneous catalyst, 53
homopolymerization, 10, 13, 19
homopolymers, 2, 14, 19
hydrocarbons, 36, 37, 38, 52, 59, 66, 71, 72, 170, 172, 173, 175, 176, 178, 182, 199, 208
hydrofluoric acid, 237, 238, 239, 240, 242, 243, 244, 245
hydrogen bond(s), 154, 182, 247, 251, 252, 259
hydrogen chloride, 147, 153
hydrogen fluoride, 238
hydrogen peroxide, 37, 233
hydrolysis, 42, 81, 85, 248, 251
hydroperoxide(s), 193, 194, 195, 197, 201, 205
hydrosilylation, 39
hydroxyl, 135, 251, 259
hydroxyl groups, 135, 259

I

incompatibility, 2, 3
induction period, 154, 202
inhibition, 70, 178, 201
inhibitor, 124, 125, 148, 149, 233
inhibitor molecules, 124
inhomogeneity, 154
initial state, 198
inorganic fillers, 112, 113, 117
Instron, 112
insulation, 167
interface, 160, 163, 232, 248
intermolecular interaction, 155, 156
ionic impurities, 232, 237, 238, 243, 244
ionic polymerization, 12
ions, 9, 12, 13, 18, 40, 52, 61, 172, 176, 189, 190, 191, 243
irradiation, 121, 124, 143, 150, 151, 161, 162, 163, 164, 239
IR-spectra, 71, 169, 172, 173, 187, 188, 189, 192, 207, 221
IR-spectroscopy, 4, 168, 185, 188, 189, 207, 221, 224
isobutylene, 13, 14, 17, 18, 19, 20, 21, 22, 23, 24, 25, 31, 34, 37, 38, 39, 40, 52, 57, 59, 60, 68, 71, 72, 74
isomerization, 42, 43, 45, 46, 71, 176, 197, 198
isoprene, 13, 21, 23, 24, 68, 72

isothermal heating, 109

K

kinetic curve(s), 11, 177, 184
kinetic model, 93, 100, 103, 106
kinetic parameters, 64, 66, 103
kinetics, 12, 93, 170, 171, 178, 181, 182, 183, 184, 185, 186, 193, 194

L

Lewis acids, 21, 22, 45
light scattering, 72
light transmission, 159
linear defects, 218
linear dependence, 14, 180, 184
linear polymers, 208, 213
liquid chlorine, 56
local order, 211, 213, 214, 218, 219, 221, 222, 223, 226
logarithmic coordinates, 184
low density polyethylene, 69
low temperatures, 32, 66, 129, 173, 177, 183, 196, 198, 202, 205
low-molecular substances, 46, 147
lubricants, 71

M

macromolecular coil, 213, 217
macromolecular entanglements, 220, 226
macromolecular systems, 121
macromolecules, 12, 25, 31, 32, 37, 38, 39, 52, 56, 57, 59, 61, 62, 63, 64, 66, 68, 69, 71, 73, 121, 129, 130, 135, 136, 148, 154, 169, 170, 173, 178, 201, 212
macroradicals, 37, 175
mass loss, 98, 102, 107, 147, 149, 170, 180, 181, 186
material surface, 149, 150, 151
mechanical and thermal properties, 127
mechanical properties, 35, 107, 112, 115, 119, 135, 139, 259
mechanical stress(es), 154
melt(s), 6, 34, 94, 164, 167, 169, 170, 175, 185, 186, 194, 195, 226
melting, 4, 112, 117, 122, 124, 126, 128, 169, 179, 180, 181, 183, 200, 234
melting temperature, 126, 181
metallurgy, 231
metals, 20, 23, 143, 201
methyl groups, 71

methylene chloride, 15, 16
methylthienylsiloxanes, 81, 82
Mg, 66, 67
Microcalorimetric measurements, 4, 6, 169
microdefects, 155
microimpurity, 243
microstructures, 138
MM, 34, 38, 53, 54
model system, 169
modeling, 12, 142, 174
moisture, 112, 161
molar ratios, 1, 4
molecular mass, 4, 12, 21, 31, 33, 34, 35, 38, 40, 54, 57, 59, 63, 64, 66, 68, 69, 72, 73, 74, 75, 76, 191, 195
molecular mobility, 175, 181, 194
molecular oxygen, 189
molecular structure, 1, 154
molecular weight, 2, 4, 190
MOM, 82, 95
monomer, 3, 9, 10, 11, 12, 13, 18, 19, 20, 21, 23, 37, 40, 57, 59, 60, 62, 68
monomeric systems, 20
monomers, 1, 9, 10, 11, 12, 13, 14, 18, 19, 20, 21, 22, 23, 25, 71, 164
monomolecular, 175
morphology, 94, 95, 97, 129, 182, 191, 247, 248, 251, 253

N

nanocomposites, 93, 94, 95, 97, 107, 247
nanolayers, 98
Netherlands, 165
nitrile, 171, 172, 176, 186
nitrobenzene, 45
nitrogen, 36, 98, 117, 119, 149, 178, 186, 191, 201
NMR, 4, 72, 171, 181, 186, 191, 192, 199
non-stationary, 12, 70
nucleation, 100
nucleophilicity, 13

O

octane, 36, 55, 72
oil(s), 21, 36, 52, 55, 72, 81, 170, 171
olefin(s), 21, 23, 25, 38, 52, 53, 55, 71
oligomers, 23, 72, 74, 173, 190, 191, 192, 195, 198, 202
optical fiber, 159, 160, 231
optical microscopy, 1, 4, 6
optical properties, 2, 143, 144, 164

optimization, 238, 239, 242, 243
optimization method, 238
organic compounds, 144, 177, 199, 209, 232
organic polymers, 201, 248
organic solvents, 85, 122, 233, 234
orientation, 130, 227
oxidation, 55, 57, 122, 126, 144, 148, 150, 153, 154, 155, 178, 180, 181, 182, 183, 184, 185, 186, 188, 189, 190, 191, 192, 193, 194, 195, 196, 197, 198, 199, 200, 201, 202, 204, 205, 208, 239, 241
oxidation products, 186, 190
oxidation rate, 181, 185, 192
oxidative reaction, 178
oxides, 144, 145, 146, 148, 149, 151, 153, 156, 170, 172, 176, 177, 184, 185, 186, 202, 232, 234
oxygen, 36, 38, 57, 94, 98, 99, 142, 147, 148, 149, 154, 177, 178, 182, 183, 185, 186, 188, 191, 192, 193, 200, 202, 205, 239, 240
ozonation, 38, 57
ozone, 37, 38, 57, 70, 73, 239, 241, 242
ozonolysis, 72

P

permeability, 33, 36, 132, 191
peroxide(s), 57, 124, 133, 134, 135, 153
peroxide radical, 124
PET, 181, 183
phase transitions, 180, 184
phenol(s), 40, 41, 42, 43, 44, 46, 84, 121, 124, 126, 202, 205
phenyl esters, 41
phosphorylation, 38, 56
physical and mechanical properties, 111
physical properties, 32, 35, 127, 141, 248
physicochemical properties, 70
plasticity, 113
plasticizer, 146, 148, 154
PM, 128, 177, 178, 192, 193, 198
PMMA, 161
polyamides, 178, 201, 209
polycarbonate(s), 1, 2, 4, 5, 6, 7, 167
polycondensation, 1, 4, 23, 81, 82, 112, 167, 168, 169, 248
polycondensation methods, 23
polydienes, 23
polyester, 184
polyethylene(s), 32, 68, 144, 146, 147, 148, 149, 150, 151, 153, 154, 157, 158, 173, 208, 223
polyheteroarylenes, 200, 202, 209
polyimides (PI), 177, 179, 202, 207, 209
polyisobutylene, 20, 21, 23, 24, 25, 31, 32, 33, 34, 35, 36, 37, 38, 39, 40, 43, 44, 45, 46, 52, 53, 54,

55, 56, 57, 59, 60, 61, 62, 63, 64, 65, 66, 68, 69, 70, 71, 72, 73, 74, 75

polymer amorphous state, 211, 212, 224

polymer amorphous state structure, 224

polymer blends, 112

polymer chains, 70, 128, 155

polymer combustibility, 149

polymer combustion, 158

polymer destruction, 122

polymer materials, 112, 133

polymer matrix, 107, 112, 115, 116, 119, 130, 131, 132

polymer melt(s), 164, 172, 175, 183, 205

polymer molecule, 94, 99

polymer nanocomposites, 94

polymer oxidation, 155

polymer properties, 69

polymer structure, 95, 154, 211, 217

polymer synthesis, 71

polymeric chains, 38, 39, 153

polymeric composites, 154, 156

polymeric materials, 2, 153, 191, 221

polymeric melt, 226

polymeric products, 19, 20, 21, 23, 59, 62, 66, 68, 72, 74, 189

polymerization, 2, 12, 13, 14, 17, 19, 20, 21, 23, 25, 31, 39, 59, 60, 62, 64, 65, 68, 69, 70, 71, 72, 82, 127, 128, 129, 133, 134, 165, 175

polyolefins, 63, 66, 68, 70, 182, 183, 186, 195, 197, 201, 209

polyphosphates, 232

polypropylene (PP), 32, 93, 107, 121, 208

polystyrene, 1, 2

precipitation, 198

prediction, 2

propagation, 11, 12, 13, 18, 21, 173, 181

propylene, 24

PTFE, 164

purification, 231, 232, 233, 238, 243

PVC, 196

PVC dehydrochlorination, 196

pyrolysis, 107, 151, 170, 172, 173, 174, 175, 176, 177, 185, 190

pyromellitic dianhydride, 168, 169

Q

quality, 18, 69, 106, 173, 183, 232, 233, 234, 237, 238, 245

quantum-chemical calculation(s), 179, 202

R

radiation, 15, 20, 37, 38, 69, 94, 121, 123, 126, 143, 144, 150, 151, 153, 154, 161, 163, 164, 238, 251

radical copolymerization, 9, 14

radical mechanism, 37, 56, 59, 60, 71

radical polymerization, 19, 21

radicals, 12, 37, 61, 71, 81, 98, 122, 124, 125, 126, 153, 154, 178, 193, 194, 198, 201, 202

rate constant(s), 13, 57, 61, 62, 63, 66, 68, 69, 147, 184, 185, 202, 239

reactants, 100, 248

reaction order, 195

reaction rate, 12, 154

reaction temperature, 62, 248

reaction zone, 40, 239

reagents, 43, 231, 232

rectification, 238

reduced combustibility, 142, 144, 145, 146, 153

refractive index, 159, 160, 161

refractive indices, 161

regeneration, 54

regression, 103, 106

reinforcement, 191

relaxation processes, 32

relaxation times, 36

reproduction, 256

resistance, 33, 94, 107, 127, 130, 133, 134, 135, 138, 164, 191, 200, 205, 209

room temperature (RT), 4, 37, 100, 133, 134, 147, 153, 167, 200, 244

rubber(s), 20, 128, 129, 130, 131, 132, 133, 134, 136, 138, 139, 144, 149

rubbery state, 147

S

scanning electron microscopy, 6, 251, 253

second virial coefficient, 33

SEM micrographs, 253, 257, 258

sensitivity, 161, 173, 188, 231

silica, 256

silicon, 44, 81, 82, 85, 127, 128, 144, 145, 146, 148, 155, 156, 225, 232

smoke, 36, 107

sodium hydroxide, 232

solid phase, 183, 192, 195, 205

solid state, 100

solubility, 2, 3, 126, 239

solvent(s), 4, 12, 13, 18, 21, 32, 37, 69, 70, 72, 73, 88, 169, 233, 248

spectroscopy, 121, 173, 191, 192, 221, 227, 234

spherulites, 155
spherulitic structures, 154
stabilization, 20, 40, 93, 94, 107, 110, 121, 184, 185, 201, 209
stabilizers, 52, 121, 122, 124
stable radicals, 20, 178
starch, 112
strength, 2, 33, 34, 112, 113, 114, 115, 116, 117, 119, 146, 154, 155, 156, 157, 173, 174, 175, 177, 178, 198, 199, 207, 232
stretching, 31, 130, 133, 134, 135, 136, 138, 223, 252, 253, 256
structural changes, 216
structural defects, 212
structure formation, 181
styrene, 13, 15, 16, 17, 19, 23, 24, 25, 57
substitution reaction, 59
substrates, 199
supermolecular structure, 154
surface energy, 94
surface layer, 151, 243, 256, 258
synthesis, 10, 19, 20, 22, 23, 25, 71, 75, 81, 84, 88, 93, 121, 165, 168, 169, 173, 198, 207, 232, 233, 248, 249, 250, 251, 254
synthetic polymers, 117

T

temperature dependence, 128, 184, 239
tensile strength, 93, 156
ternary blends, 1
tetrachloroethane, 169
TG, 97, 98, 101, 104, 105
thermal aging, 195
thermal analysis, 1, 122, 141, 142, 144, 182
thermal decomposition, 71, 208
thermal degradation, 40, 60, 93, 94, 95, 99, 100, 103, 107, 144, 147, 177
thermal oxidation, 178, 181, 182, 183, 184, 185, 186, 187, 188, 189, 190, 191, 192, 193, 194, 195, 196, 198, 199, 200, 201, 202, 203, 205, 208
thermal oxidative degradation, 178, 199, 202, 208
thermal properties, 146
thermal resistance, 81, 91, 93, 94
thermal stability, 2, 70, 88, 178, 186, 201, 204, 209
thermal treatment, 4, 5
thermodynamic parameters, 59
thermodynamic properties, 32
thermograms, 201

thermogravimetric analysis (TGA), 88, 94, 95, 99, 107, 171, 178, 179, 181
thermogravimetry, 94
thermolysis, 209
thermooxidative, 148
thermooxidative degradation, 148
thermostability, 70
thin films, 144
toluene, 32, 36, 40, 54, 56, 82
transformation degrees, 10
transformation process(es), 155
transparency, 161, 163, 164, 248
transport, 180, 238, 243, 244, 245
trapezium, 18

U

urethane, 25
UV, 37, 52, 56, 70, 200, 202, 239, 241
UV radiation, 56, 239, 241

V

vinyl monomers, 23
vinylchloride, 20, 24, 196
vinylidene chloride, 20
viscosity, 4, 34, 38, 69, 73, 76, 88, 117, 119, 169, 170, 182, 191, 195
volatilization, 94
vulcanizates, 148, 149
vulcanization, 21, 128, 133, 134, 135, 147, 148, 154, 155, 156

W

water, 19, 36, 37, 42, 82, 84, 85, 91, 112, 128, 143, 147, 155, 184, 185, 202, 232, 233, 234, 239, 244, 247, 252
WAXS, 94, 95
wettability, 243
WT, 61

X

X-ray analysis, 191
X-ray diffraction, 168, 169, 181, 211, 212, 214, 215, 221, 226, 251, 253
X-ray diffraction data, 169, 212, 221